The Promise and Performance of
ENVIRONMENTAL CONFLICT RESOLUTION

EDITED BY
ROSEMARY O'LEARY AND LISA B. BINGHAM

RESOURCES FOR THE FUTURE
WASHINGTON, DC

An RFF Press book
Published by Resources for the Future
1616 P Street, NW
Washington, DC 20036–1400
USA
www.rffpress.org

Library of Congress Cataloging-in-Publication Data

The promise and performance of environmental conflict resolution / Rosemary O'Leary and Lisa B. Bingham, editors.
    p.   cm.
    Includes bibliographical references and index.
    ISBN 1–891853–65–1 (cloth : alk. paper) — ISBN 1–891853–64–3 (pbk. : alk. paper)
    1. Environmental law—United States. 2. Dispute resolution (Law)—United States. 3. United States Institute for Environmental Conflict Resolution. I. O'Leary, Rosemary, 1955–   II. Bingham, Lisa (Lisa B.), 1955–
KF3775 .P758   2003
344.73'046—dc21   2003001369

f e d c b a

The paper in this book meets the guidelines for permanence and durability of the Committee on Production Guidelines for Book Longevity of the Council on Library Resources.

This book was designed and typeset in Minion by Betsy Kulamer. It was copyedited by Paula Bérard. The cover was designed by Rosenbohm Graphic Design.

ISBN 1–891853–65–1 (cloth)   ISBN 1–891853–64–3 (paper)

About Resources for the Future *and* RFF Press

Resources for the Future (RFF) improves environmental and natural resource policymaking worldwide through independent social science research of the highest caliber.

Founded in 1952, RFF pioneered the application of economics as a tool to develop more effective policy about the use and conservation of natural resources. Its scholars continue to employ social science methods to analyze critical issues concerning pollution control, energy policy, land and water use, hazardous waste, climate change, biodiversity, and the environmental challenges of developing countries.

RFF Press supports the mission of RFF by publishing book-length works that present a broad range of approaches to the study of natural resources and the environment. Its authors and editors include RFF staff, researchers from the larger academic and policy communities, and journalists. Audiences for RFF publications include all of the participants in the policymaking process—scholars, the media, advocacy groups, nongovernment organizations, professionals in business and government, and the general public.

Contents

PART III
Midstream Environmental Conflict Resolution

PART IV
Downstream Environmental Conflict Resolution at the State and Federal Levels

PART V
Downstream Environmental Conflict Resolution and Outcome Measures

PART VI
Conclusion

Foreword

Pioneering efforts to mediate environmental and natural resources conflicts began almost 30 years ago, with a settlement in the first case in 1974. Initial successes in the 1970s and 1980s led to a major period of expansion, which began in the late 1980s and continues today. In recent years, the use of mediation has expanded into other public policy issues as well.

One can infer from the growing use of mediation by government institutions and other parties that the approach produces value for the users. However, surprisingly little in the way of systematic evidence exists about what that value really is and how it corresponds to what people are seeking, let alone what actions correlate or contribute to these actual outcomes. My 1986 book, *Resolving Environmental Disputes: A Decade of Experience*, was the first effort to collect aggregate, descriptive information about mediated environmental conflicts, examining information from almost 200 cases mediated in the 1970s and early 1980s. With a few exceptions, little else in the way of aggregate data about mediated environmental and public policy issues (or other consensus-building efforts) has been collected on a systematic, comprehensive basis in the past 20 years, particularly for research purposes.

The expansion and maturation of the field make the need for such information greater than ever. Thousands, not hundreds, of cases now have been mediated, affecting large numbers of people and many important public decisions. A wide variety of related processes for building consensus on public issues have emerged. Federal, state, and local governments have invested millions of dollars in conflict resolution programs. Legislation and regulatory guidance exist to shape how mediation is used in public decisionmaking, which broadly affects power relationships, processes, and outcomes for entire categories of disputes. Establishing such conflict resolution procedures is constructive, but more needs to be learned and discussed before anyone can be sure that the right policies and procedures are in place.

The body of literature about mediation of environmental and other public policy issues also has grown considerably over the past decades. Numerous case studies about mediated public disputes have been written, as have thoughtful analyses describing the lessons that authors have learned from these cases and their experiences. Both kinds of research and writing remain important in the field, and excellent examples can be found in this book. Encouragingly, explicit attention has been given more recently to the design of evaluation measures themselves. However, achieving the full value of empirical research is hampered by the current lack of a coherent, shared framework for what we are measuring. The approaching 30th anniversary of the first mediated environmental dispute presents an important milestone and a challenge to practitioners and researchers alike to do more. We should not let three decades of practice pass by without a significant, collective focus on the claims for public policy mediation and on improving our measures of those outcomes.

Many—perhaps too many—claims have been made for why mediation and other conflict resolution processes are valuable social investments. This book is a call to think more carefully about the claims themselves and to do so as part of a dialogue within the field and with those who fund and use mediation and other consensus-building services. What outcomes does the mediation community believe can be achieved? For what outcomes does society ask us to be accountable? How do they compare to the outcomes the parties to disputes are seeking? What measures can be crafted for those outcomes? This discussion needs to answer two categories of questions: what outcomes ought to be expected from specific case interventions? and what outcomes ought to be expected from the general societal investment in these new tools?

The Promise and Performance of Environmental Conflict Resolution explores where the thinking in the field has been and raises questions about where it needs to go. In particular, the book emphasizes why it is important to talk about outcomes and suggests some next steps in doing so. It assumes that the field needs to establish a clearer and more widely accepted framework for talking about goals or outcomes that should be sought and can be achieved through consensus-building processes such as mediation and how to define relevant, valid measures of those outcomes. Differences over what mediation should seek to achieve will always be important because they will spur new thinking and reflection. Rosemary O'Leary, Lisa Bingham, and their contributors, however, lay out a rationale for being more explicit, more systematic, and perhaps, more selective in our focus about outcome research than we have been.

Why Is It Important To Talk about Outcome Measures?

Active discussion of desired outcomes and their measures is important for many reasons. First, researchers, practitioners, and parties all benefit from greater self-awareness of their individual assumptions, motivations, and biases. Self-awareness of one's assumptions is critical in making responsible choices. The classic example in the conflict resolution field is that mediators hold themselves to an ethic of neutrality, often even referring to themselves as neutrals. However, as important as this is, it can be a shallow norm if it is not examined carefully. Neutrality means more than not taking sides on the issues themselves. Think of the decisions that are ostensibly "only" about process, such as who is a party and who is not, what issues are and are not "on the table" for discussion, and more. These process choices critically affect what outcomes occur.

Similarly, mediators' assumptions about what constitutes a successful outcome affect their choices implicitly or explicitly. As a result, mediators can make different, and potentially conflicting, decisions depending on their values about the process and what outcomes it is intended to achieve. For example, if the intended benefit of mediation (either to public decisionmaking in general or to the parties in the case) is to increase efficiency (a cheaper and faster process), could the conscious or unconscious expectation of speed reduce the time and attention given to efforts that are needed to achieve joint gains or improved quality in the agreements reached? The trade-offs are real. If public institutions invest in a mediation program with the expectation of increasing overall settlement rates, could that reduce the choice to spend the time and effort necessary to be more inclusive and create a more "just" process in individual cases with highly polarized minority voices? Parties need this same level of self-awareness about the outcomes they are seeking because they are the ultimate consumers. We hope that an educated consumer will articulate what he or she wants strongly enough for the rest of us to listen.

Second, parties deserve an honest assessment of what is realistic for them to expect from their often substantial investment in mediated processes. It should go without saying that any profession has the ethical responsibility of clearly articulating what benefits its clients or consumers should expect. Parties make a significant investment of time and money in mediation. Most controversial environmental and public policy decisions take many months, and occasionally years, to mediate. Even when a government agency pays for the mediator's services, the parties invest a significant amount of time and money to participate. Parties also take other risks when they agree to participate in mediation. They have other process choices with different possible outcomes, both in the degree to which their interests are achieved and across other measures such as transaction costs and access to the decisionmaking process. They can wait and comment during public hearings later, push for legislative action, urge a sympathetic executive-branch agency to make a decision in their favor, litigate, launch a media campaign, or simply stand aside. We currently are not giving parties as clear a framework for making those choices as we might. Ultimately, these are choices that depend significantly on what outcomes are most important to the parties. Such priority setting is more difficult if the choices—and the trade-offs among choices— are not clearly understood. It is also more difficult because conversations about success commonly lack a distinction between the process of mediation and its outcome. Parties may care about process variables, but process professionals (such as mediators) may focus too much on them. Mediators must always ask themselves (and the parties) what ends "good" mediation practice should serve.

Third, with the expansion of mediation and other consensus-building tools in public decisionmaking, government agencies, foundations, and others are being asked to make large investments both in conflict resolution programs and in the mediation of specific disputes. They ought to expect clarity of thought about accountability for the results achieved. Federal, state, and local governments are making large investments of taxpayers' dollars in establishing programs and offices to foster and manage the use of mediation for resolving controversial environmental and public policy decisions. General sentiments suggest that this is a good thing. However, the field does not yet have empirical evidence to back up its efforts and hold itself accountable to this societal investment. We currently are making so many claims that it is not clear what the expectations are or what the relationship is between desired outcomes for specific cases and desired outcomes for programs. Do programs seek the aggregate of case outcomes, or is something qualitatively different expected because of the institutional investment?

We do not have clear or effective measures for all the claims we are making. Even though researchers and evaluators have carefully thought about outcome measures for some claims, we are not implementing enough evaluations, particularly at the program level, to assess whether those outcomes are being achieved. Should legislators voting on appropriations for conflict resolution programs believe that decisions will be better, that greater financial savings will be made, that democracy will be enhanced? How will they know? And, if they do not know, will other needs surpass the use of mediation as a priority for limited tax dollars?

More attention has been given to success measures for individual cases than to those for conflict resolution programs, at least for environmental and public policy disputes. However, the challenge for researchers in the public policy arena is somewhat different from the challenge faced by program evaluators who handle cases otherwise addressed by the judicial system, where concepts of and measures for standards of justice are more clearly established. Measuring success may be more difficult in the public policy context, where the issues are so varied, so many decisionmaking forums are available, and the stakeholders are so numerous that our processes could address a wide array of social, political, environmental, and economic goals. Evaluation research in the court system is never easy, but at least some connections between theories of effects and outcomes have been made. Consequently, there has been far more empirical research on the long-term effects of these programs. Court systems typically collect data on the outcomes both of cases handled by traditional means and cases referred to as alternative dispute resolution (ADR). Therefore, comparisons between court program outcomes and ADR program outcomes may be possible.

Fourth, additional thought about expected outcomes may focus more attention on the development of improved measures for outcome research and thus contribute to the quality and usefulness of research being conducted. Without some shared concept of the bases for evaluating the effects of mediation and other collaborative processes, researchers will be less effective at defining clear metrics and sophisticated research methods. Without a shared rationale for what data to collect, researchers and practitioners alike will also be less able to recreate the comprehensive set of data necessary to describe trends in the field, such as was begun in the early 1980s. And without comprehensive data on at least some measures, the field will be left with ad hoc, anecdotal responses to important evaluation questions. As a result, the potential effects of the field will not be realized because we will have no case to make to a society beset with competing demands for social investment.

Finally, broader awareness of the different assumptions being made about expected outcomes may increase the understanding and acceptance of the evaluative research being conducted. Whether or not practitioners and researchers agree with the conclusions in any particular study, the field needs more evaluative research. Some conclusions will stand the test of time; others will not. Attacks on evaluative research, when based principally on disagreements over whether the "right" outcome goals were selected, are counterproductive to fostering a reflective culture in the field that encourages the very inquiry that will lead to conclusions that stand the test of time. Either a more commonly shared set of assumptions about outcomes or greater clarity about and willingness to respect differences among them should open up more opportunities to learn from one another.

What Claims Are Made about Environmental Conflict Resolution?

Almost 30 years ago, the most commonly asked questions had to do with what public issues would benefit from mediation, which parties would be interested, and how it would work. Early pioneers were careful not to be promoters, as their intent was to ask about—not assume—the usefulness and appropriateness of mediation techniques for environmental disputes. Environmental disputes were on the rise, largely because of the success of advocacy strategies. Given the early mediators' strong concerns for issues of power and justice, most did not want to push mediation if it was not in the best interest of all involved. Thus, the field focused on questions oriented toward looking for appropriate cases and exploring innovative approaches. It was a period of experimentation.

Seventeen years ago, the questions addressed in my book *Resolving Environmental Disputes: A Decade of Experience* were largely descriptive:

—How many environmental and natural resources cases have been mediated?
—Has the use of mediation for these issues grown over time?
—What was the intended objective of the mediation?
—What issues were involved?
—Who were the parties? How many parties were involved?
—Where were the cases?
—What kind of decisions were being made (e.g., site-specific or policy)?

Some evaluative questions were asked:

—How many agreements were reached, and in what percentage of the cases?
—How many of those agreements were implemented?
—How long did a case take?
—What did the mediation cost?
—What factors may increase or decrease the likelihood of success in reaching the intended objective?

Today, a more extensive variety of claims can be found in the literature about outcomes to be expected from mediating environmental and other public policy disputes. Generally, these desired outcomes fall into three categories: substance, process, and relationships. Attributes both researchers and parties commonly say they use to measure whether a specific mediation case was successful are listed below for each category of outcomes.

Substantive results are important. Some look concretely to the articulation of new and more creative options, or the achievement or implementation of agreements. Some argue that conflict resolution processes ought to improve the quality of the decisions reached. Often this case is made in the general game theory language of joint gains—each party has the opportunity to achieve outcomes that are better than they would have achieved without the mediated negotiations. Others make a more specific point, arguing that inclusive, consensual solutions to environmental problems should seek to be more environmentally sound and ecologically sustainable.

Attributes of the *process* are also discussed in the literature. Many continue to seek processes that are "cheaper and faster" than litigation and expect this of conflict resolution processes such as mediation. Others also assert that the criteria of fairness, accessibility, representation of all interests, consistency with applicable laws, adequacy of information and analysis, and transparency to which the justice system is commonly held should be applied to other forms of dispute resolution such as mediation.

Relationships and community values are also considered important benefits to be sought from consensus-building processes such as mediation. Some claim that consensus-building processes should and can improve relationships or teach individuals how to handle their differences better in the future. Another view some scholars and practitioners promote argues that mediation transforms individuals by promoting individual growth and moral development. They say that this ought to be the goal, more than problem solving per se. This value placed on individual process learning is also extended to aspirations for social learning. Others, both neutrals and researchers, add claims that conflict resolution processes can and should contribute to social change and social justice as disputants learn new skills, legitimize their opponents' interests, and transform relationships between parties. Some environmental and public policy conflict resolution promoters argue that conflict resolution processes can empower individuals and groups to reclaim their voices, increase their self-esteem, democratize their neighborhoods, involve the disenfranchised in policy deliberations, and create new social or governing structures.

What Barriers Exist to Advancing Current Thinking and What Next Steps Would Be Useful?

Greater clarity about intended outcomes will not be easy to achieve. There are many challenges. First, many claims about the possible effects of mediation are *attractive*. Second, many assumptions about its benefits are *implicit* and, therefore, not easily shared and discussed. Third, it is conceptually difficult to establish empirical *measures* for whether some of the desired outcomes have been achieved. Fourth, the *context* within which mediation is applied differs, potentially affecting the most appropriate approach to the current challenges. Few forums currently exist to encourage dialogue about and significant progress on these issues taken as a whole.

The positive intent behind these claims is evident. It is hard to argue that one should not try to achieve any one of them. It is also likely that some of these outcomes actually are being achieved some of the time—and that none of them is being achieved all of the time. Correspondingly, it is likely that the insights that come from good evaluation would allow more of these outcomes to be achieved more of the time. Unfortunately, as noted above, inherent trade-offs may make it difficult, if not impossible, to achieve all of them.

As attractive as they are, with so little systematic evidence, it is difficult to accept the general truth of so many claims any longer.

One useful first step would be to organize a roundtable or series of discussions, involving mediators, researchers, parties, and program sponsors, to help better articulate why each of these outcome claims may be important (to whom and under what circumstances), whether trade-offs are perceived, and what possible distinctions or connections exist between outcomes expected in individual cases and for program investments more generally. Such a dialogue might clarify the issues in new ways such that progress can be made to overcome current barriers.

This effort should be approached much like a mediator would approach other topics about which people have diverse perspectives. Generally, this would involve initial conversations to assess the interest in such a process and to propose a design that would have widespread engagement. It would be useful to ask practitioners, researchers, parties, and funders about whether they indeed consider these barriers problematic, what products would make discussions useful, what issues they face from their perspectives, and who they think should be involved.

At this stage, it is not clear what the best initial focus of a dialogue should be. It is likely that we already know and can build on a full list of claims and aspirations for the mediation of public issues, but that assumption needs to be confirmed. In planning such discussions, an explicit choice also needs to be made about whether to combine the scales of evaluation for cases and programs. Dialogue needs to start, at least, with achieving greater understanding of the assumptions that underlie each of the outcome claims and whether any patterns emerge in those assumptions.

It is not clear how we should tackle the question of whether to narrow our collective assertions about outcome claims and, if so, how to select among them. Identifying whether and what kinds of trade-offs exist in the choices mediators and parties make in specific cases (and program sponsors in their policies and procedures) to achieve different desired outcomes would seem to be valuable. This could be either a part of the dialogue suggested above or a separate second stage.

The issue of context may emerge at this point. Do parties seek different kinds of outcomes in different circumstances? If so, which contextual circumstances are most significant? Power relations between the parties? The parties' incentives to negotiate? The parties' knowledge base about the issues in dispute? The current forum in which the issue is being decided? The number of parties affected? How well organized the parties are? The preferences and personalities of the individuals involved? It may be that the question of context in relation to expected outcomes can only be clarified but not answered through discussions, and research may be needed. However, again, understanding the assumptions being made about contextual variables may help illuminate next steps in the thought process.

In the end, the best basis for choosing among current claims for mediation and other consensus-building processes may be to test them with real data and then to make only those claims that can be supported. Conceptual discussions about what outcomes ought to be claimed will need to be followed by rigorous thought to define those concepts so that they are measurable, for without measures, we do not know what data to collect. More work has been done in this regard for measuring success in individual cases, with some initial work having only just begun for research on program results. This should be built upon.

A comprehensive effort at systematic data collection would certainly benefit from the shared conceptual rationale that could emerge from the dialogue and improved measurement methods suggested above. The architecture—or at least starting points—for such a data collection effort exist. However, this work needs to be framed as an effort around which practitioners, parties, and researchers can collaborate. Such an effort will help researchers, practitioners, program managers, and parties learn and synthesize their knowledge about how to evaluate long-term program effects and thus benefit the whole field.

In sum, 30 years ago, the mediation of controversial environmental decisions did not exist as an explicit option for government agencies and other parties responsible for and affected by these disputes. Evaluation of the use of mediation and other consensus-building processes to address environmental issues remains too infrequent. With important exceptions, case study research continues to be the most typical approach to evaluating outcomes and impact. However, we now have streams of cases handled by environmental conflict resolution programs and institutionalized into agencies and regulations. Several thousands of cases have almost certainly been mediated. However, despite 30 years of impressive development in the field, more rhetoric than empirical research can be found concerning the long-term effect of environmental and public policy conflict resolution programs.

We still do not know what the parties or public institutions are "buying" when they invest their time and money in mediation and other consensus-building processes. Legislatures, governmental program managers, private parties, and foundations allocate resources to conflict resolution programs based on a multitude of attractive claims. Conceptual discussions have taken place, but only a few efforts have been undertaken in the field to define those concepts so that they are measurable, and then to frame those concepts into testable propositions. Few, if any, program evaluations have focused on consistent concepts to measure long-term effects, nor are they widely available. Because so few evaluation projects have been done of effects, and those are not easily comparable, the field cannot yet answer legitimate questions about the benefits of environmental conflict resolution. Increased dialogue on what ought to be the core expectations of mediation processes would be an important step to advance the thinking and then to provide the foundation for more systematic evaluation attempts in this field. This book is an important and valuable step in such a dialogue. As such, this book is a major contribution both to the scholarship and the practice of environmental conflict resolution.

GAIL BINGHAM
President
RESOLVE, Inc.
Washington, DC

RESOLVE is a nonprofit organization with expertise in alternative dispute resolution and consensus building in public policy. Founded in 1977, RESOLVE has offices in Washington, DC; Denver, CO; and Portland, OR.

Acknowledgements

Most of the chapters in *The Promise and Performance of Environmental Conflict Resolution* were first presented at a research conference. They have been revised to reflect developments in the field and to provide constructive guidance for effective change. Without the conference, this book never would have been developed.

The conference, "Evaluating Environmental and Public Policy Dispute Resolution Programs and Policies," was held in 2001 at the Syracuse University Greenberg House in Washington, DC, and funded by the William and Flora Hewlett Foundation; Dean John Palmer of Syracuse University's Maxwell School; Vice President for Research and Computing Ben Ware of Syracuse University; the Campbell Public Affairs Institute, the Maxwell School; and the Indiana Conflict Resolution Institute, School of Public and Environmental Affairs at Indiana University Bloomington. All of the participants were experts in dispute resolution but with very different areas of expertise: psychology, law, anthropology, economics, sociology, planning, political science, public administration, public policy, business, and education. Some of the experts were mediators, facilitators, arbitrators, and negotiators. Some were from the public sector, others were from the private sector, and still others were from the nonprofit sector. Some were academics, some were practitioners, and some were "pracademics"—individuals who bridge both worlds.

The group was united, however, by one common interest: a desire to make environmental and public policy dispute resolution programs and policies better, more robust, and more effective. The discussion topics ranged from audits to efficiency reviews to stakeholder satisfaction surveys. Cost–benefit analysis, case studies, quasiexperimental designs, and meta-analyses were placed under the microscope.

We are grateful to the conference participants who did not write chapters for this book but whose comments and insights while participating in the two-day conference were invaluable in creating the final product: the late Senator Daniel Patrick Moynihan (Syracuse University), Sarah Connick (University of California, Berkeley), Michael DeGuzman Nobleza (Syracuse University), Michael Elliot (University of Georgia), Juliette Falkner (U.S. Department of the Interior), Judith E. Innes (University of California, Berkeley), Dale Keyes (U.S. Institute for Environmental Conflict Resolution), Harry Manasewich (Massachusetts Office of Dispute Resolution), John Murray (Syracuse University), Evan Ringquist (Indiana University), Lee Scharf (U.S. Environmental Protection Agency), and Sue Senecah (The State University of New York).

We thank our mentors at the Hewlett Foundation, Melanie Greenberg and Terry Amsler, for their faith in our abilities; the staff of the Indiana Conflict Resolution Institute for their dedication and hard work, especially Tina Nabatchi; the staff of the Campbell Public Affairs Institute for their expert assistance, including Bethany Walawender, Kelley Coleman, Willow Jacobson, Heather Balent, Mark Jesionowski, Paul Alexander, Lisa Daughtry-Weiss, Neelakshi Medhi, and Elliott Stiles; Alyssa Colonna for her excellent editing; and the staff of the Greenberg House, particularly Dugan Gillis, for providing a delightful conference setting and staff assistance.

We also thank reviewers Lawrence Susskind, Evan Ringquist, and our anonymous reviewers, as well as Don Reisman and Rebecca Henderson at RFF Press, for helpful suggestions for improvement, and Betsy Kulamer and Paula Bérard for their design and copyediting expertise. Finally, we are grateful to the contributors for their consistently excellent work and for their enthusiasm. We sincerely hope this book will inform the debate about environmental conflict resolution programs and policies. Our intent is to honor and build upon the solid work of our predecessors while serving as catalysts for positive and effective change to help fulfill the promise of environmental conflict resolution.

ROSEMARY O'LEARY
Syracuse, New York

LISA B. BINGHAM
Bloomington, Indiana

This book is dedicated to

Daniel Patrick Moynihan,

*scholar and senator, orator and author,
evaluator and peacemaker,
whose intellectual and political leadership
helped to shape national and international policy
on the major issues of our time.*

Contributors

THOMAS C. BEIERLE is a fellow at Resources for the Future (RFF). With his coauthor Jerry Cayford, Beierle published an evaluation of 239 cases of public participation in the RFF Press book *Democracy in Practice: Public Participation in Environmental Decisions.*

FRANCES STOKES BERRY is professor and director of the master's in public administration program at the Askew School of Public Administration and Policy, Florida State University, and has consulted on numerous projects with state agencies. Her areas of research and teaching include state and local government management reform, state policy, intergovernmental relations, strategic management, program evaluation, and policy innovation.

LISA B. BINGHAM is Keller-Runden Professor of Public Service and director of the Indiana Conflict Resolution Institute at the Indiana University School of Public and Environmental Affairs, in Bloomington, Indiana. Bingham cofounded the Indiana Conflict Resolution Institute, which conducts applied research and program evaluation on mediation, arbitration, and other forms of dispute resolution.

JULIANA E. BIRKHOFF is a conflict resolution scholar and an experienced mediator, facilitator, and trainer. She is a senior mediator at RESOLVE, Inc., an evaluator for a U.S. Environmental Protection Agency (EPA) community-based collaborative pilot project in Cleveland, and the project director of the

National Environmental Case Database project, funded by the William and Flora Hewlett Foundation.

METTE BROGDEN is an anthropologist and the program manager of the Environmental and Public Policy Conflict Resolution Program at the University of Arizona's Udall Center for Studies in Public Policy. She has facilitated the Arizona Common Ground Roundtable, a statewide policy dialogue among ranchers, conservationists, public agency personnel, researchers, and sports enthusiasts concerning land use and open space.

MARCIA CATON CAMPBELL is an assistant professor in the Department of Urban and Regional Planning at the University of Wisconsin–Madison. She recently began an evaluation of Wisconsin's waste facility siting program, which tracks the state-mandated negotiations that must take place between applicants and host communities before hazardous and solid waste facilities can be sited.

CHRISTINE CARLSON is an adjunct professor in the Conflict Resolution Program of Antioch University's McGregor School and coexecutive director of the Policy Consensus Initiative, which works with state leaders to establish and strengthen the use of consensus building and conflict resolution. Her publications include *A Practical Guide to Consensus.*

JERRY CAYFORD is a research associate at RFF. He works on risk perception and on the issues raised by patents on biotechnology. A philosopher by training, he is the coauthor (with Thomas C. Beierle) of *Democracy in Practice: Public Participation in Environmental Decisions.*

CARY COGLIANESE is associate professor of public policy at Harvard University's John F. Kennedy School of Government and chair of the Regulatory Policy Program at the school's Center for Business and Government. His interdisciplinary research focuses on issues of regulation and administrative law, particularly focusing on alternative means of designing regulatory processes and the role of disputing in regulatory policymaking. He is coeditor of the RFF Press book, *Regulating from the Inside: Can Environmental Management Systems Achieve Policy Goals?*

BONNIE G. COLBY is professor of agricultural and resource economics at the University of Arizona. Her research focuses on the economics of natural resource policy and disputes over water, public lands, and environmental regulation. She has provided invited testimony on these matters to state legislators and the U.S. Congress.

AYSIN DEDEKORKUT is a research assistant at the Florida Conflict Resolution Consortium and a doctoral student in urban and regional planning at Florida State University, with research interests in collaborative processes in water resource planning. She is on leave from the Izmir Institute of Technology in Turkey.

TAMRA PEARSON D'ESTRÉE is the Henry R. Luce Professor of Conflict Resolution at the University of Denver. Her research includes work on social identity, intergroup relations, and conflict resolution processes, as well as evaluation research. She and Bonnie Colby have coauthored *Braving the Currents: Evaluating Conflict Resolution in the River Basins of the American West* (forthcoming). d'Estrée has facilitated interactive problem-solving workshops in various intercommunal contexts, including the Middle East, Ethiopia, and U.S. intertribal disputes.

KIRK EMERSON is the director of the U.S. Institute for Environmental Conflict Resolution. Prior to her appointment, Emerson developed and coordinated the environmental conflict resolution program at the University of Arizona's Udall Center for Studies in Public Policy, where she conducted research, taught, and directed several conflict management and public participation projects involving water resources, endangered species, and western range policies.

DAVID FAIRMAN is a vice president at the Consensus Building Institute in Cambridge, Massachusetts, where he mediates and facilitates the resolution of public policy conflicts, designs and delivers negotiation training and capacity building programs for public and private agencies worldwide, and evaluates and designs dispute resolution systems and programs. He also directs the Workable Peace project, which teaches intergroup conflict management skills through high schools in the United States, Israel, and Palestine.

DANIEL J. FIORINO is the director of the Performance Incentives Division in the Office of Policy, Economics, and Innovation at EPA headquarters in Washington, DC, and he serves as the program manager for EPA's National Environmental Performance Track. He is the author of *Making Environmental Policy* and coauthor of *Managing for the Environment*.

BARBARA GRAY is a professor of organizational behavior and director of the Center for Research in Conflict and Negotiation at Pennsylvania State University. As a mediator and trainer, she has worked for the U.S. Department of Energy, the U.S. Fish and Wildlife Service, the Federal Highway Administration, the Pennsylvania Department of Environmental Protection, the Pennsylvania House of Representatives, and many other public and private organizations.

SANDA KAUFMAN is professor of planning and public administration at Cleveland State University's Levin College of Urban Affairs. Kaufman's area of expertise is negotiation and intervention in public, organizational, and environmental conflicts.

LISA A. KLOPPENBERG is dean of the University of Dayton School of Law. Formerly, she was founding director of the Appropriate Dispute Resolution Program at the University of Oregon School of Law. She is author of *Playing It Safe: How the Supreme Court Sidesteps Hard Cases and Stunts the Development of Law.*

WILLIAM LEACH is a researcher and lecturer in the Department of Environmental Science and Policy at the University of California, Davis, where he has taught courses in policy analysis and research methods. His research focuses on water quality regulation and collaborative watershed management.

KEM LOWRY is professor and chair of the Department of Urban and Regional Planning and director of the Program on Conflict Resolution, University of Hawaii. His research and practice focuses on planning and environmental management, conflict resolution, community participation, and evaluation.

TINA NABATCHI is a doctoral candidate in the public affairs program at Indiana University's School of Public and Environmental Affairs. Her research interests include public management, public policy, and law, particularly in relation to conflict resolution and sustainable development administration.

ROSEMARY O'LEARY is professor of public administration and director of the Ph.D. program at the Maxwell School of Citizenship and Public Affairs at Syracuse University. Previously, O'Leary was cofounder and codirector of the Indiana Conflict Resolution Institute. She has won six national awards and one international award for her research.

SUSAN SUMMERS RAINES is an assistant professor at Kennesaw State University in Atlanta, where she teaches courses in conflict resolution. She also works as a professional mediator of both domestic and international disputes, with an emphasis on environmental conflict resolution.

ANDY ROWE is vice president of GHK International in Toronto, Canada, which works on North American and international development assignments. He has undertaken evaluation consultancies in health and human services, resource policy and programs, regional and community development, and gender and disability issues.

PAUL SABATIER is a political scientist in the Department of Environmental Science and Policy and codirector of the Center for Environmental Conflict Analysis at University of California, Davis. He is best known for his work on policy implementation and the development of the Advocacy Coalition Framework of policy change and Lake Tahoe.

JOHN STEPHENS directs the University of North Carolina Institute of Government's public dispute resolution program and is also a teacher and advisor for the Natural Resources Leadership Institute at North Carolina State University. He has consulted with the United Methodist Church and United Church of Christ on faith-based disputes over homosexuality and church policy.

BRUCE STIFTEL is professor of urban and regional planning at Florida State University and faculty associate of the Florida Conflict Resolution Consortium. His research interests include dispute resolution, citizen participation, and the heuristic uses of decision science.

The Promise and Performance of
ENVIRONMENTAL CONFLICT RESOLUTION

PART I

Introduction

I magine that you are the administrator of the U.S. Environmental Protection Agency and you have just approved a $41 million program to contract for facilitators, convenors, and mediators to assist in the resolution of environmental disputes (as did EPA Administrator Carol Browner in 1999). You have been advised by alternative dispute resolution (ADR) advocates that environmental conflict resolution (ECR) saves money and time. You have been told that participants to ECR processes are generally satisfied with outcomes. You also have been counseled that ADR can serve as a catalyst for individual empowerment, personal transformation, and positive relationships among disputing parties.

But how will you *really* know whether your $41 million investment is a success? Congress has passed the Government Performance and Results Act, mandating that you demonstrate accountability in your annual performance reports. How will you do this? Furthermore, what will you show to concerned taxpayers who are interested in seeing public funds spent as judiciously as possible? You're all for more transparency in government, but how can that transparency be implemented with ECR, which requires discretion and sometimes confidentiality? Are there any lessons from the world of research that might be useful here? Is there any proof of claims both for and against ECR? And what about the environment? Does ECR protect the environment?

These are the types of questions that are raised in Part I, with the goal of driving home the need for new approaches to environmental conflict resolution. In Chapter 1, Kirk Emerson, Tina Nabatchi, Rosemary O'Leary, and

1

John Stephens provide a useful primer on ECR, as well as an overview of the challenges involved in ECR research. A central question presented is, how can we build more systematic knowledge about the ECR field that can also inform our thinking about governance? After highlighting the major methodological problems, as well as the major conceptual issues, the authors close by discussing opportunities for research on ECR.

In Chapter 2, after reviewing the different perspectives both supporting and questioning ADR in general, Juliana Birkhoff and Kem Lowry argue that the current evaluation agenda promoted by researchers and funded by foundations and government agencies is inadequate. The authors maintain that although this agenda is helping to enhance the legitimacy of mediation as a genuine dispute resolution alternative and to persuade skeptical policymakers of its potential usefulness, it inadequately serves the needs of mediators and the disputants they serve. The primary challenge for ECR lies in providing feedback to mediators so that they may learn from their practice. Another important challenge is ensuring that ECR practices are accountable to the disputants they serve. A balanced and well-rounded agenda that recognizes the multiple purposes and audiences of ECR is needed.

Together, these two chapters serve as valuable analyses of the state of the field of ECR and set the stage for the chapters that follow.

1

The Challenges of Environmental Conflict Resolution

KIRK EMERSON, TINA NABATCHI,
ROSEMARY O'LEARY, AND JOHN STEPHENS

The way humans and their institutions handle conflict is changing. The alternative [or appropriate] dispute resolution (ADR) movement has grown from a handful of mediators working in community mediation centers to institutionalized programs in courts, public agencies, nonprofit organizations, and corporations. Interest has grown dramatically in consensus building, facilitation, mediation, and other forms of resolving conflict through assisted negotiation and voluntary settlement. A framework supporting public agency ADR exists in federal statutes and in a growing number of state statutes. Some people speculate that we are seeing a generational shift from command and control to less authoritarian forms of organization and decisionmaking, from vertical forms of hierarchy to horizontal forms of hierarchy.

Environmental conflict resolution (ECR) reflects this pronounced growth. In 1998, building on more than two decades of ECR experience, Congress created a new federal agency, the U.S. Institute for Environmental Conflict Resolution, to address the complex intergovernmental relationships and the myriad disputes involving the environment.[1] Hundreds of environmental conflicts have been addressed through ECR processes, yet there is relatively little systematic research on how ADR processes and programs, when applied to environmental concerns, compare to traditional ways of doing business.

For example, do conflict resolution processes yield better environmental policies? Is mediation a better way to resolve natural resource disputes? Is ECR faster, cheaper, and better, as much of the literature claims? What do we know about how these processes work? Most importantly, how can we build more systematic knowledge about and for the ECR field that can also inform our thinking about governance and the future of public administration and planning?

These questions inspired and guided this book. The best thinkers in the field were asked to examine issues at the cutting edge of ECR. Their ideas, which form the chapters in this book, span the categories of public participation, participant satisfaction, intractable conflicts, relationship change, evaluation of facilitators, state agency ECR, court-annexed ECR, federal agency ECR, outcome measures, and the evaluation of ECR programs and policies. When considered in their entirety, the ideas and insights of this book reflect the promise of environmental conflict resolution. Let us begin with a primer on ECR and an examination of the challenges of ECR research.

Framing the Issues: An ECR Primer

Environmental conflicts are fundamental and ongoing differences among parties concerning values and behavior as they relate to the environment.[2] More specifically, environmental conflicts are actual or potential disputes involving the environment, natural resources, public lands, or all three.[3] They usually involve multiple parties who are engaged in a decisionmaking process and disagree about issues traceable to an action or policy that has potential environmental effects.[4] The range of environmental conflict is large; however, disputes can be classified as upstream, midstream, or downstream. Upstream environmental conflicts involve planning or policymaking. For example, they may include the creation and implementation of governmental policy at the national, regional, state, or local level, such as environmental, natural resource, health, or safety policy. Midstream environmental conflicts involve administrative permitting, for example, the granting of environmental permits. Downstream environmental conflicts are about compliance and enforcement. They can involve the ways that lands are used, the allocation or distribution of natural resources, and the siting of industrial or other large facilities. Environmental conflicts can also involve the prevention, cleanup, and consequences of pollution.

Environmental disputes are public disputes in several respects. Not only do they address issues of the proverbial "public interest," but they also tend to have a high public profile and involve the public sector (government agencies) as parties to the dispute. Gail Bingham's study of 160 disputes discussed in the foreword found that government was a party in more than 80% of the cases, and intergovernmental conflicts involving more than one governmental entity among the parties was the most frequent category of dispute. Thirteen years after Bingham's study, Lawrence Susskind and others studied 100 disputes for the Lincoln Institute of Land Policy. They found that government officials initiated 78% of the cases.[5] Most importantly, environmental disputes are carried out within a context of a larger set of overlapping systems of dispute resolution. Even site-specific disputes exist within a context of one or more public dispute resolution systems, with respect to local development permitting, the public designation of a natural wildlife corridor, or standard setting for grazing permits.

Environmental conflicts can also be generally categorized by the scope of the dispute. For example, policy-level disputes pertain to classes of resources, locations, or situations, whereas site-specific disputes involve particular natural resources, locations, or situations.[6] A policy-level dispute is an upstream dispute, whereas a site-specific dispute is considered a downstream dispute.

Conflict emerges from differences in values and world views, conflicting interests, and the uncertainty that surrounds various courses of action.[7] In addition, popular attitudes, political culture, technology, laws, political interests, economics, and religion (especially as related to Native American culture) can influence environmental conflicts.[8]

Given this breadth of contextual issues and potential philosophical differences, hosts of possible government, public, and private interests have a stake in an environmental conflict. Elected and appointed government officials at the local, county, state, federal, or all of these levels are usually involved in environmental conflicts because many of these conflicts arise from the creation or implementation of legislation and policies. Often these government officials represent different agencies (for example, the U.S. Department of the Interior and the U.S. Environmental Protection Agency), different departments or subdivisions within an agency (for example, the Bureau of Land Management and the Fish and Wildlife Service within the U.S. Department of the Interior), or even different branches of government (for example, officials in Congress and officials from an administrative agency such as the U.S. Department of Agriculture). The numerous public interests represented in environmental conflicts include community residents, interest groups, and public interest law firms. Finally, private interests also play a large role in environmental conflicts. Industry, commercial, and other business people are often involved in environmental conflicts, such as those that involve siting facilities, working on pollution abatement issues, or granting various permits. Frequently these government, public, and private interests also require and use the services of scientific, research, and technical consultants, adding to the number of stakeholders involved in the conflict.

Why ECR?

The terms environmental conflict resolution (ECR) and environmental dispute resolution (EDR) refer to various ADR techniques as applied to environmental conflicts. Certain characteristics of environmental conflicts add to their complexity, making the application of ADR techniques more difficult. Some of these characteristics include multiple forums for decisionmaking; interorganizational, as opposed to interpersonal, conflicts; multiple parties; multiple issues; technical complexity and scientific uncertainty; unequal power and resources; and public and political arenas for problem solving.[9]

More specifically, ECR consists of a set of techniques, processes, and roles that enable parties in a dispute to reach agreement, usually with the help of one or more third-party neutrals, as discussed in more detail below. Despite the variance in ECR techniques and processes, researchers have identified five characteristics[10] shared by all forms of ECR (with the exception of binding arbitration):

1. Participation is usually voluntary for all participants.
2. The parties or their representatives must be able to participate directly in the process.
3. Any and all participants must have the option to withdraw from the ECR process and seek a resolution through a more formal process, such as litigation.
4. The third-party neutral must not have independent, formal authority to impose an outcome but rather should help the parties reach their own agreement.
5. The parties must agree to the outcome or resolution of the dispute. The purpose of the process is to help parties reach their own solutions, which requires their consent to the decision or recommendation.[11]

These characteristics of ECR processes are clearly derived from the philosophy of the ADR movement, which stands in contrast to traditional, adversarial methods of dispute resolution, especially litigation. Unlike traditional litigation, in which a judge or jury makes a final determination or issues a judgement, ECR techniques use various forms of assisted negotiation to help the parties reach a mutually satisfactory agreement on their own terms. Advocates of ECR generally find fault with traditional modes of environmental policymaking and dispute resolution and specifically point to failures in the legislative and administrative arenas and to the drawbacks of litigation.[12]

Advocates of ECR suggest several problems that subvert the ideal of approaching environmental conflicts through the legislative process. First, it is difficult for all of the interests affected by environmental decisions to be heard. Many environmental and other interest groups cannot effectively participate in the legislative arena because they lack adequate financial and human resources to engage in lobbying. The innate controversy surrounding environmental policies often precludes a viable consensus among legislators, which results in vague and ambiguous legislation.

The legislative arena's failure to address conflicts effectively sets the stage for conflicts to reemerge in the administrative arena. As agencies try to interpret and implement vague policies, controversies about specific actions or projects flare. Just as in the legislative arena, it is difficult for groups to become involved in administrative decisionmaking processes. Some parties are deliberately ignored or left out of processes, and some parties, even if invited to the table, lack the financial or human resources necessary to participate effectively.[13] Most agencies, at least at the federal level, must receive public comments or hold hearings where concerned parties can voice their preferences. However, critics suggest that these procedures only give the appearance of participation and that comments and testimony are not actually considered in policy implementation.[14]

The failure of the legislative and administrative arenas to effectively address environmental conflicts often means that litigation will ensue. However, advocates of ECR produce two primary criticisms of litigation as a dispute resolution process for these conflicts.[15] First, litigation does not allow for adequate public participation in important environmental decisions. The costs of litigation are often prohibitive to interest groups, especially those groups that are small or represent local interests. The process of litigation is also extremely time-consuming, often taking months for cases to come to trial. After accounting for appeals time, the litigation process can take years. These delays inherent in litigation are costly to all the parties involved. Second, litigation is ineffective for resolving the issues at stake in environmental disputes. Court decisions frequently fail to resolve the basic issues in dispute between the parties because the courts are often limited in their ability to address the substantive dimensions of environmental conflicts; thus, they render decisions only on procedural grounds.[16] Many underlying controversies therefore remain unresolved, and hence, more lawsuits emerge in the future.

In short, there are many drawbacks to dealing with environmental controversies in the legislature, bureaucracy, and courts. Participation is expensive and time-consuming, with limited access for many groups. The legislature tends to produce vague policies that do not resolve basic controversies and conflicts. Conflicts then reemerge in the administrative arena, where unilateral, top-down decisionmaking results in choices that lack political legitimacy. As a result, legal challenges to the decisions frequently arise. The litigative approach, however, is not designed to resolve differences, but rather to decide issues.[17] With its adversarial nature and win–lose approach, litigation fails to resolve controversies and instead encourages losing parties to continue the conflict. These criticisms of legislation, administration, and litigation strengthen the argument that various ECR techniques can result in better informed, more legitimate and enduring decisions for parties involved in environmental conflicts. ECR assistance may also be useful in settling disputes arising out of past events.[18]

Advocates of ECR suggest that the informal negotiation techniques of ECR have several advantages, including (a) less risk for the parties than the risk associated with win-all or lose-all litigation; (b) reduction in court costs, legal fees, inflationary delays, and other conflict-related expenses; (c) an increase in the efficiency of the outcome, such that all disputants or stakeholders prefer it over all other feasible outcomes; and (d) the increased likelihood of achieving a stable agreement or an agreement that all parties honor for at least several years.[19] In short, when all parties are at the table, there is a better chance that all the relevant issues will be raised and that the parties will be better situated to make efficient trades and reach decisions that effectively address the substantive nature of the dispute.[20]

The participatory nature of ECR techniques and processes also promotes a sense of procedural justice. Procedural justice, a commonly used framework in ADR research, suggests that participants' satisfaction with an ADR process is a function of their opportunities to control and participate in the process, present views, and receive fair treatment from the mediator.[21] When participants sense that they have received procedural justice, the perceived legitimacy of the decisions and outcomes increases, which reduces the likelihood that participants will challenge them in the future. By contrast, when a decision is forced on stakeholders, they have a natural tendency to reject it.[22] Finally, the participatory nature of ECR compensates for the lack of public access to negotiation sessions in the legislative, administrative, and judicial systems and promotes the idea of citizen participation in policy- and decisionmaking, a strongly heralded value in this era of administrative reforms.

ECR techniques have been used successfully in both site-specific and policy-level disputes.[23] Specifically, ECR has been successfully used for land-use disputes involving commercial development, housing, facility siting, and transportation; natural resource use or management issues involving fisheries, timber, and mining; water resource issues such as water quality, flood protection, and water use; air quality issues such as odor, acid rain, and air pollution; issues related to toxics such as chemical regulation, asbestos removal, and waste cleanup policies; and others.[24]

By far the most frequent use of ECR has been with site-specific conflicts over resource allocation. Early notable disputes were over the location of a dam in the Snoqualmie River Valley in the state of Washington and the Storm King dispute over building a hydroelectric power plant on the Hudson River in New York state. Typically, these land-use disputes involve competing interests over the future control or use of the site. The resolution has distributive effects—parties share in dividing a resource pie. There have also been many "upstream" policy dialogues among conflicting stakeholders, mediated by third-party facilitators, to develop public policy for subsequent legislation and administrative action (for example, the National Coal Policy Project or President George W. Bush's "no net loss" wetland policy).

When an ECR case arises, the approaches used to address the dispute often vary with the type of conflict. For example, environmental cases involving pollution, land use, natural resource use or distribution, environmental permitting, facility or infrastructure siting disputes, or compliance and enforcement may require different ECR techniques than public policy cases involving the creation or implementation of national, regional, state, or local government policy. Likewise, a complex environmental or public policy case—defined by the U.S. Institute for Environmental Conflict Resolution as one involving at least 100 case hours; multiple issues; and multiple parties, at least one of whom is a government entity—would likely necessitate a different ECR set of techniques than a more simple case.[25] Generally, the neutral party's goal is to develop a process of agreement building, dispense recommendations on the issues in controversy, and facilitate collaborative processes among the multiple parties to address the issues in controversy. The next section of this chapter briefly explains some of the most commonly used ECR techniques and processes, including when and where they might be used.

ECR Processes

We can arrange ECR processes along a continuum from less formal, interest-based negotiation or consensus-based techniques to more formal, adjudicatory arrangements. In consensus-based processes, disputing parties work together to develop a mutually acceptable agreement. Consensus-based techniques require that everyone, not just a majority, agree with the outcome or decision, unless all parties agree to define consensus as a majority decision (as provided for in the Negotiated Rulemaking Acts of 1990 and 1996). Moving along the continuum toward quasijudicial processes, the goal of the processes shifts to making a determination about the issues in controversy, and, as a result, the nature of the techniques used also changes. Consensus-based techniques are generally used for upstream disputes, whereas quasijudicial processes are generally used for downstream disputes.

ECR processes—perhaps with the exception of binding arbitration—are based on the idea of negotiation. Negotiation is simply bargaining—a process of discussion and give-and-take among disputants who want to find a solution to a common problem. It can be relatively cooperative, as when both sides seek a solution that is mutually beneficial (commonly called interest-based or principled negotiation), or it can be confrontational (commonly called win–lose or adversarial negotiation), in which each side seeks to prevail over the other.[26] A discussion of some of the more common ECR techniques, categorized as consensus-based or quasijudicial processes, follows.[27]

Consensus-Based Processes

Consensus-based, or consensus-building, processes describe collaborative decisionmaking techniques in which a third-party neutral such as a facilitator or mediator assists diverse or competing interest groups to reach an agreement on an environmental conflict.[28] Consensus-building processes are typically used to foster dialogue, clarify areas of agreement and disagreement, improve the information on which a decision is based, and resolve controversial issues in ways that all interests find acceptable. Consensus building, usually used for upstream disputes, typically involves informal but structured face-to-face interaction among representatives of stakeholder groups who hold different viewpoints. The goals are to promote early participation by the affected stakeholders; produce sensible and stable policies or decisions that have a strong, broad base of support; and reduce the likelihood of subsequent disagreements or legal challenges. The most common consensus-based processes used in environmental conflicts are conflict assessment, convening, facilitation, mediation, conciliation, negotiated rulemaking, and policy dialogues.

Conflict Assessment

Conflict assessment is valuable as a first step in ECR processes.[29] The goal of conflict assessment is to help identify the issues in controversy, the affected stakeholders, and the appropriate forms of ECR for handling the conflict. The assessment process typically involves multiple steps, beginning with conferring among potential stakeholders to evaluate the causes of the conflict and to identify the entities and individuals who would be substantively affected by the conflict's outcome. The next steps are to assess the stakeholders' interests and needs, identify a preliminary set of relevant issues to be discussed, and evaluate the feasibility of using consensus-building or other collaborative processes to address the issues in dispute. Other prospective stakeholders are then informed and educated about the possible consensus-building ECR processes so that they can determine whether they want to participate. The final step of the conflict assessment process is to select the ECR technique most appropriate for the conflict. In this step, often called process design, a neutral party recommends or assists in developing a process for addressing a particular controversy or dispute. The process designer typically interviews representatives of the interested or affected groups to understand their perceptions about the conflict, interests in the conflict, and suggestions about useful ways to handle it. With this information, the designer then reports to the parties with recommendations or a plan for handling the dispute.

Facilitation

Facilitation is a collaborative process in which a neutral party assists a group of stakeholders in constructively discussing the issues in controversy. The facilitator typically works with participants before and during these discussions to ensure that the appropriate people are at the table, to help the parties set and enforce ground rules and agendas, to assist parties in effectively communicating, and to help the participants keep on track in working toward their goals. Facilitation may work in any number of situations, ranging from scientific seminars to management meetings to public forums. For example, the Project XL Program at the U.S. Environmental Protection Agency uses facilitation processes to increase public participation by affected stakeholders in some of the agency's policymaking activities. The Office of Surface Mining Reclamation and Enforcement in the U.S. Department of the Interior also uses facilitation before each proposed rulemaking to expand conflict resolution options and reduce the future potential use of litigation.

Mediation

Mediation is a form of facilitated negotiation in which a skilled, impartial third party with neither decisionmaking authority nor the power to impose a settlement assists the parties in reaching a voluntary, mutually agreeable resolution to all or some of the disputed issues.[30] (Whereas facilitation bears many similarities to mediation, the neutral or facilitator usually plays a less active role than a mediator and, unlike a mediator, often does not see "resolution" as a goal of his or her work.) Mediation is one of the oldest forms of conflict resolution and one of the most common forms of ECR. A mediator works with the disputing parties to help them improve communication and analysis of the conflict, identify interests, and explore possibilities for a mutually agreeable resolution that meets all of the disputants' interests or needs. The mediator lacks power to impose any solution but rather must assist the process in ways acceptable to the parties and help the disputants design a solution themselves. Typically this involves supervising the bargaining, helping the disputants find areas of common ground and understand their alternatives, offering possible solutions, and helping parties draft a final settlement agreement. Mediation usually occurs in the context of a specific dispute involving a limited number of parties, but mediation procedures are also employed to develop broad policies or regulatory mandates and may involve dozens of participants representing a variety of interests. Mediation is most often a voluntary process, but court orders or statutes mandate its use in some jurisdictions.

In most ECR cases, a mediator will use a facilitative, directive, or evaluative mediation style, or a combination of styles. In facilitative mediation, mediators are less likely to provide direct advice, propose solutions, or predict outcomes. They are more likely to assist the parties in identifying and merging their interests by establishing an atmosphere that allows parties to communicate more effectively about their interests, options, and realistic alternatives.[31] Conversely, in directive mediation, the mediator is engaged more directly in diagnosing the problem and persuading the parties to reach a mutually acceptable agreement. In its most extreme form, directive mediation resembles settlement judging.[32] In evaluative mediation, the mediator may give the parties more guidance on the merits of the dispute[33] to help the parties understand the strengths and weaknesses of their cases, assess the likely outcome if the case were to go to trial, and suggest appropriate grounds for settling. Evaluative mediation is similar to early neutral evaluation in the context of litigation. In terms of reaching effective settlements, research suggests that mediation increases settlement opportunities because it allows more room to consider participants' underlying needs, concerns, and interests.[34]

Conciliation

Conciliation involves efforts by a third party to improve the relationship between two or more disputants. Generally, the third party works with the disputants to correct misunderstandings, reduce fear and distrust, and improve communication among the parties in conflict. Conciliation is not a common ECR technique; however, it usually prepares the disputants for a future ECR process.

Negotiated Rulemaking

Negotiated rulemaking, also known as regulatory negotiation or "reg-neg," is another form of ECR that involves efforts by regulatory agencies to design environmental regulations by first negotiating with interested stakeholders.[35] In this multiparty consensus-building process, a balanced negotiating committee, composed of representatives of the rulemaking agency and other stakeholders who might be affected by or have an interest in the new rule, seeks to reach agreement on the substance of the proposed agency rule, policy, or standard. The intent of negotiated rulemaking is to avoid litigation that may arise to challenge the new rule by generating agreement among the affected interests so that they abide by the decision and its implementation. Federal law requires a thorough conflict assessment before the use of reg-neg and the involvement of a skilled, neutral mediator or facilitator during this type of consensus-building ECR process. Reg-neg is an example of a new and potentially significant development that represents the formal institutionalization of ECR within governmental policy and decisionmaking processes.[36] The U.S. Environmental Protection Agency was among the first to use negotiated rulemaking and did so before the passage of the Negotiated Rulemaking Act. Since then, many federal and state government agencies have found it to be a useful tool with the potential to reduce future litigation.

Policy Dialogues

Policy dialogues are a relatively new form of ECR. They are generally used to address complex environmental conflicts or public policy disputes. In the policy dialogue process, representatives of groups with divergent views or interests are assembled to generate discussion and improve communication and mutual understanding. The goal is to explore the issues in controversy to see if general recommendations can be developed and to try to reach agreement on the policy standard or guidelines the government will propose. Unlike the other consensus-based ECR processes discussed above, policy dialogues usually do not seek to achieve a full, specific agreement that would bind all participating interests. Rather, participants in a policy dialogue may seek to assess the potential for developing a full consensus resolution later or may put forward general, nonbinding recommendations or broad policy preferences for an agency (or other governmental entity) to consider in its subsequent decisionmaking.

For example, the Electric Power Research Institute (EPRI) convened the National Review Group (NRG) to discuss relevant energy policies. NRG is composed of more than 30 organizations, including the Federal Energy Regulatory Commission (FERC), hydroelectric power operators, government and public agencies, tribes, and public interest groups. The objectives of the group are to tackle tough issues and, if possible, identify solutions, develop preferred practices, track and share experiences, improve dialogue, foster exchange among stakeholder groups, and coordinate with and add value to other administrative relicensing reform efforts. The NRG completed a report, "Hydro Licensing Forum: Relicensing Strategies," in December 2000, which provides some common-sense solutions encountered by stakeholder group relicensing efforts and is intended to be used as a base from which stakeholders can address issues in their own relicensing processes.[37]

Quasijudicial Processes

Quasijudicial processes supply the disputing parties with an expert opinion about the merits of their cases, furnish more information about their best alternative to a negotiated agreement, and provide the disputants with a loop back to negotiation. In theory, the parties can engage more effectively in settlement negotiations with the additional information they gain in these processes. Some of the most common quasijudicial processes used in environmental conflicts are early neutral evaluation, minitrials, summary jury trials, settlement judges, fact finding, and arbitration.

Early Neutral Evaluation

Early neutral evaluation is a quasijudicial process in which a third-party neutral, often someone with specifically relevant legal, substantive, or technical expertise, hears informal evidence and arguments from all of the parties involved in the dispute and issues a nonbinding report advising them about the strengths and weaknesses of their cases.[38] The report may also evaluate the likely reaction of a judge or jury if settlement is not reached, provide guidance about an appropriate range of outcomes, and assist the parties in narrowing the areas of disagreement or in identifying information that may enhance the chances of settlement.

For example, the Dispute Resolution Service (DRS) at FERC has established an enforcement hotline. The hotline receives informal complaints and questions from the public, and DRS staff attempt to resolve disputes before a formal proceeding is instituted. Hotline assistance can range from clarifying procedural questions, to preventing natural gas pipelines from cutting off transportation services, to resolving tariff and other contract disputes, to assisting landowners in obtaining restoration of property as a result of pipeline construction, to helping wholesale marketers of electricity obtain transmission services. The attorney assigned to an inquiry gathers information and consults with other staff and then conducts a type of early neutral evaluation. Any findings of the hotline are nonbinding and advisory only.

Minitrials and Summary Jury Trials

Minitrials and summary jury trials are commonly used to resolve litigation over complex environmental issues. In a minitrial, parties are generally represented by counsel and an agent with the authority to agree to a settlement or decision, for example, a CEO or agency official. Abbreviated versions of the evidence and arguments are presented, after which the decisionmaking representatives attempt to negotiate a settlement.[39] In a summary jury trial, the disputing parties call a jury and present short versions of the evidence and arguments. The jury deliberates and makes findings of fact and liability when appropriate, which are then released by the judge.[40] The parties are not bound by the jury's findings but rather use the information to assist with settlement negotiations. Minitrials and summary jury trials are alike in that they both serve as a loop back to future negotiations.

Settlement Judges

A settlement judge process is also similar to minitrials and summary jury trials; however, this process is used for litigation that has already reached administrative adjudication. In a settlement judge process, a judge who is different from the presiding judge in the case acts as a mediator or neutral evaluator and meets both separately and jointly with the parties.[41] If the settlement judge's efforts do not produce full agreement, the case returns to the presiding judge. A settlement judge often plays a more authoritative role than a private mediator by sometimes providing parties with specific legal or substantive information and recommendations.

This settlement judge process is used at FERC, where administrative law judges (ALJs) may act as settlement judges to reduce delays in the settlement process, provide structure, control the pace of negotiations, and give a report to the commission or the chief ALJ on the likelihood of settlement. The settlement judge meets with all parties separately or collectively, looks for common ground, and crafts or suggests solutions. The settlement judge may express an opinion based on precedent and the judge's experience as to how the case may be decided by a presiding judge or the commission if the case proceeds to litigation.

Fact Finding

Fact finding is closely related to nonbinding arbitration. In this process, a neutral, called a "fact finder," receives information and listens to arguments presented by the disputants. The fact finder, who may conduct additional research to investigate the issues in dispute, evaluates the evidence and submits a report that contains findings of fact and sometimes recommendations based on those findings.[42] Typically, the fact-finding process is informal, and the recommendations are nonbinding. Occasionally, disputants use this process to literally define the facts of a case so that all parties can use them in subsequent negotiations. In such cases, the disputed facts usually involve highly technical scientific or engineering issues and require the fact finder to have subject matter expertise.

Arbitration

In arbitration, the disputants present their cases to an impartial third party, who then issues an opinion. An arbitration decision can be binding or nonbinding. In binding arbitration, the opinion is final and subject to limited judicial review. In nonbinding arbitration, the opinion is advisory and can be rejected; however, it can also serve as a loop back to negotiation.

The Challenges of ECR Research[43]

As the body of ECR research has grown, so too have the commentaries on the state of the scholarship and the critiques of the research. The early and primary focus in ECR research has been on the "micro" level of analysis, where individual disputes are described and evaluated, and consideration is given to the various factors and ingredients that contribute to the mechanics of successful mediation (e.g., the role of the mediator, the ripeness of the dispute, or the number of parties at the table). To the extent that there has been theory building, it has focused on informing practice with prescriptions for improved techniques, processes, and conditions. Much of this work has been done, as others point out, by mediator–scholars who are ideologically and professionally committed to the success of ECR. There is a lot of "boosterism" in this literature, but one can also observe that theory about practice has begun to develop through a cumulative, self-conscious process based on reflective experience in the field and principally through inductive, rather than deductive, reasoning.

Throughout these critiques, evidence of methodological and conceptual obstacles that have posed challenges for ECR research and have contributed to the lack of theory building and evidence-based research emerge. They are highlighted here.

Methodological Issues

Single, Small-*n* Descriptive Case Studies. The field is dominated by single- or small-*n* descriptive case studies, making generalizable theory building difficult. Only three large-*N* studies were found in Rosemary O'Leary's 1994 review of the state of research in the field.[44]

Practitioner–Scholar Bias. Theory is primarily understood as theory of and out of practice. Prescriptions abound, based largely on the longstanding experience of mediators.[45] Whereas professional wisdom is essential to building the field, there is also a need for careful and creative research derived from more detached, objective investigation.

Diversity and Uniqueness of Applications. The diversity of applications, topics, and conditions makes cross-case comparisons and controlling for particular variables difficult. After all, these are custom-designed processes by definition, with few exceptions.[46]

Inaccessible Data. Access to data is problematic; processes are often confidential; mediators pledge confidentiality to their clients; records are rarely kept of meetings (certainly not to the extent of more formal fact finding); and opportunities for direct observation are limited.

Retrospective Reporting. As a consequence of the above, reporting on experience in mediation or facilitated group processes is often retrospective and subject to recall bias. Many of these methodological limitations can be overcome, and conditions are changing. In particular, we now have more than 20 years of experience in ECR, hence more disputes to compare and analyze, and a plethora of documented case studies that can yield useful systematic evidence. The potential for comparison has improved as well. For example, case-series analysis across the scores of these individual case studies may now be possible. The potential for more careful, strategic sampling now exists. We also have the opportunity to look back on past settlements and find out more about implementation and compliance patterns, whether the dispute was later transferred to other forums or transformed into other issues, how the relationship among the parties changed over time, and what the nature of subsequent conflict experiences has been. Therefore, the potential for longitudinal analysis is riper today, and opportunities are greater for overcoming these methodological limitations.

Conceptual Issues

Ideology of Mediation. There is a strong presumption that ADR, mediation in particular, is good and right. This has been fostered directly by the practitioners, scholars, and trainers themselves, having been born from the supportive origins of the field—from labor relations (collective bargaining), urban planning (exploring enhanced citizen participation and involvement), the legal reform movement (pushing for more efficient processes and greater access to the courts), and the legal process theory school studying procedural justice. From multiple sectors, then, has come the conclusion that the conventional adversarial system for resolving public disputes is inadequate or inappropriate for many contemporary environmental and natural resource conflicts. This presumption has resulted in empirical research that has tended to be less critical, and perhaps even celebratory.

Comparison to Litigation. The tendency to compare mediation and other ADR techniques to litigation has continued. Definitionally, ADR has meant alternatives to adversarial dispute processing within the formal legal system. However, increasingly, this comparison is seen as faulty—not only because mediation does not always measure up (particularly on efficiency grounds), but also because ECR is now situated in several different institutional contexts (e.g., administrative processes, legislative policymaking, and community-based collaborations).

Disputes as Discrete Phenomena. The majority of ECR research addresses disputes individually as discrete, almost static phenomena, rather than as a moment or condition in a more dynamic conflict process, or alternatively, as an expression of some larger pattern of social conflict. This perspective has tended to focus the research on micro-level variables, where the unit of analysis is an individual dispute. It neglects the frequent occurrence of patterns of disputes moving through different processes over time or simultaneously.[47] Moreover, practitioners and scholars have tended to view settlement as a fixed point at the end of a linear process, rather than as a point in the trough of a wavelike process. In this wave-form, the settlement occurs after the peak point of tension, just past the crest in a diminishing sine wave. Conflict often will recur and the wave will recrest; however, the earlier agreement functions to allow the parties to handle the conflict differently and more productively should the conflict recur.[48] The emphasis of this view is on examining ECR as a time-extended phenomenon and looking for evidence that people who participate in ECR processes think or behave differently over time from those who do not. In short, more systematic analysis addressing some of the critical questions raised by legal scholars, political scientists, and public administration scholars is neglected in most empirical studies.

Limited Use of Research from Other Fields. There has been limited use of the scholarship in other fields. See Box 1-1 concerning the variety of journals in which ECR research may be found. As Kolb and Rubin point out, there is a substantial literature on ADR and mediation in communication, social psychology, sociology, anthropology, and family therapy, among many other

---

Box 1-1. ECR Journals: An Excerpt by Kirk Emerson

The growing literature on ECR is spread across a wide range of journals and fields, being multidisciplinary, though as yet, rarely interdisciplinary. Most of the praxis articles are found in the journals of negotiation and mediation, such as *Negotiation Journal, Mediation Quarterly* (now called *Conflict Resolution Quarterly*), *Journal of Dispute Resolution, Ohio State Journal on Dispute Resolution,* and AAA's *Arbitration Journal* (now called *Dispute Resolution Journal*). Articles based on and about practice dominate the literature here.

ECR research that emphasizes environmental regulation and policy is found in the *EIS Review, Environmental Forum, Journal of Environmental Management,* and *Journal of Environmental Economics and Management.* Where natural resources and land use are the focus, look to the *Natural Resources Journal* and the *Journal of the American Planning Association.*

Where public administration is the focus, research and commentary are found in the *Policy Studies Journal, Policy Studies Review, Public*

fields that can inform theory in ECR.[49] For example, a relatively long and not unrelated research area in legal sociology and social psychology deals with procedural justice. John Stephens makes a compelling argument that the field of ECR as a subset of the field of public policy conflict resolution can learn much from studying the literature of public participation (see Box 1-2).

Push toward Settlement. The tendency to focus on settlement has obscured such issues as the distributive justice of the outcomes or the quality of the resolution. Bingham, Susskind and Ozawa, Cormick, and Godshalk all provide thoughtful criteria for judging the success of an ECR outcome (such as a signed agreement; all parties involved; positive public perception of agreement; technical, economic, and political efficacy; contingency plans; and links to formal governmental decisionmaking),[50] but the predominant measure of success in the research has been settlement. An exception is the Buckle and Thomas-Buckle study, which is one of the few empirical studies with a respectable number of cases to analyze.[51] They specifically addressed failed mediations and suggested that we change our definition of success, or at least our expectations of the mediators. After reviewing 81 cases from a firm that handled ECR cases, they found that 90% had not reached agreement (contrary to Bingham's 78% success rate).[52] Of the 8 cases that made it

Management, Policy Sciences, Journal of Policy Analysis and Management, and *Public Administration Review.* Political science coverage has been slim, with a few pieces in *The Journal of Politics* and *Western Political Science Quarterly.*

Where public and environmental law are the focus, a number of law reviews cover the topic, including *Environmental Law Reporter, Administrative Law Journal, Administrative Law Review, Boston College Environmental Affairs Law Review, Ecology Law Quarterly, Harvard Law Review,* and a variety of Yale journals, as well as those from Georgetown, Duke, Temple, and others.

For more critical perspectives on public and environmental dispute resolution, you can find coverage in *Law and Contemporary Society, Law and Policy, Journal of Social Issues,* and *Law and Society Review.* For classic references in social psychology, see *American Behavioral Scientist, Journal of Applied Behavioral Science,* and *Psychological Science.*

Source: From Kirk Emerson, "The Challenges of Environmental Conflict Resolution Research," paper presented at the Syracuse University–Indiana University Greenberg House Conference, "Evaluating Environmental Conflict Resolution," March 2001.

Box 1-2. The Contribution of Evaluation of Public Participation for Public Policy Dispute Resolution: An Excerpt by John Stephens

Public participation (PP) as a professional field of practice[a] has much in common with public policy dispute resolution (PPDR). However, PPDR evaluation has been largely separate from PP evaluation. This separation is a detriment to achieving a stronger perspective on PPDR as a whole and for the goals and methods of its evaluation. I argue that it is necessary to integrate the two for evaluation purposes.

PP and PPDR share an overarching purpose and values. For example, the general purpose of both fields is to help reach better decisions, and PP and PPDR largely agree on what "better" means. They say better decisions have greater legitimacy, based on stakeholders' involvement, open-minded consideration of a variety of information and perspectives, the opportunity for new proposals to be raised and considered, and the ultimate decisionmaker's acknowledgement of the participants' input or direction. PP and PPDR depart from one another on what constitutes a superior decision. The two fields are not technocratic; the highest goal is not to seek the best science, economic, environmental, or political calculus as the arbiter of decision quality. PP and PPDR eschew a confrontational debate format, where some arguments triumph over others.[b] Even when an outcome is disappointing or detrimental to one or more interests, PP and PPDR focus on the legitimacy of an open, inclusive, reasonable, and integrative process....

A core conceptual category for both PPDR and PP is that of a third party, separate from decisionmakers or stakeholders, who plays a productive role in designing and managing a process that pursues a variety of goals beneficial to the decisionmakers and stakeholders. The third party is "a protector of the process...."[c] The other core commonality is an explicit invitation by the decisionmaker to be influenced by the views, ideas, and needs of stakeholders and interested parties....[d]

[They both] ...

— Focus on an issue or limited list of issues.
— Have stakeholders participate.
— Provide information and assess the meaning of data.
— Share and understand different, often conflicting views, on the desired outcomes, preferred methods, and decision criteria.

—Create a forum that promotes mutual education, shows respect for stakeholders, minimizes conflict, and considers alternative policies or methods.

—Leave the members of the participation–negotiation with an experience of significant value, and possibly greater benefit, than pursuing their goals through other means (e.g., litigation, legislative lobbying, or demonstrations).

... Furthermore, PP and PPDR exhibit similar practices. The nature of the work focuses on reaching a wide audience of potential or actual stakeholders and engaging their questions, information, views, and reactions to proposals in a productive manner....

Source: From John Stephens, "Learning from Your Neighbor: The Contribution of Evaluation of Public Participation for Public Policy Dispute Resolution," paper presented at the Syracuse University–Indiana University Greenberg House Conference, "Evaluating Environmental Conflict Resolution," March 2001.

a. For example, the International Association for Public Participation website, www.iap2.org (accessed January 17, 2003).

b. For PPDR, even where the issue is one of compliance, enforcement, or compensation, the focus is on negotiation.

c. October 1998 Society for Professionals in Dispute Resolution [SPIDR] Annual Conference session, "Intersection of Mediation and Public Participation," Portland, OR. Author's transcription and Lucy Moore, "SPIDR Annual Conference, Portland, October 1998, Brown Bag Session: Intersection of Mediation and Public Participation," unpublished.

d. Or, more often for PPDR, there is the creation of a temporary body (e.g., a task force, forum, or some other gathering of negotiators) with "strong influence," if not formal, binding authority. The body's goal is to reach an agreement that government authorities will adopt or implement.

past the first meeting, 6 did reach agreement. However, the researchers found from case files, survey responses, and interviews with participants in all 81 disputes, that the mediators had been appreciated and were helpful as teachers of negotiation skills, even if they did not end up facilitating a settlement.

Models of Social Conflict. There is one additional conceptual issue that merits more elaboration: the underlying models of social conflict being used by mediators, designers of ECR systems, and researchers. There are at least five models represented in the literature: pluralistic bargaining, utilitarian decision theory, jurisprudential models, deep political conflict, and republican deliberation.

In the pluralistic bargaining model (the prevalent model, promoted by Fisher, Ury, and Patton and Ury, Brett, and Goldberg),[53] the process is geared to balancing power and outcomes among contending interests. Another interest-based approach, the utilitarian decision theory model,[54] is geared toward optimization, seeking to maximize the net benefits for some aggregation of interests (the social welfare approach). The jurisprudential model is a rights-based approach of the judicial system,[55] where the rule of law is in play. Judgements are sought based on principles, as opposed to settlements based on interests. The deep political conflict model assumes that conflicts are inevitable and often run deep and that there are structural inequalities and imbalances in power in society that require political solutions.[56] Finally, the model of republication deliberation, or civic discourse, the originating model of the conflict resolution–ADR movement, is one where disputes are seen to be transformed by and transforming of the common public interest, where consensus is possible and shared interests and values are identifiable. In this republican model, espoused by John Forester,[57] among others, party preferences are not predetermined, to be summed and divided at the table, but are indeed inputs to a process that constructs new meaning and values and is synergistic and creative (the latest incarnation of this school being Bush and Folger's *The Promise of Mediation*).[58]

The underlying models of social conflict have important implications for both practice and theory. In practice, the model or models of social conflict that parties and neutrals presume to be operating in a given conflict should be critical to the nature and quality of the ECR process and its expected outcomes. Two important questions for practice are, how do the underlying views of practitioners and participants affect the disputes? and, are some models appropriate for different kinds or stages of disputes? In theory, the models of social conflict used by researchers will affect not only how they frame environmental disputes, but also their expectations and preferences concerning appropriate outcomes. In other words, these views influence the way that ECR is evaluated. Thus, two important questions for theory are, how do the underlying views of researchers affect their evaluations? and, if a different view were used in the same research effort with the same data, would a different set of conclusions be reached? This challenge is discussed in greater depth by Juliana E. Birkhoff and Kem Lowry in the next chapter of this book.

Conclusions

As some of these methodological and conceptual obstacles are overcome, we most likely will see a new generation of ECR research that is much more systematic and rigorous, with the potential for more useful and integrative theory building.

If we can translate some of the important conceptual questions being asked in the critical literature into more rigorous research designs and make better use of the data, then we will move scholarship and the field forward. If we can make better use of the findings in a variety of other disciplines and produce truly interdisciplinary approaches to ECR research, the field will benefit. Herein lies the promise of environmental conflict resolution.

We envision this book as a meandering stream with the interdisciplinary issues and questions highlighted above intertwined at every bend. This chapter described the ECR stream by presenting both a primer on ECR and background on the challenges of conducting ECR research. The following chapter continues this description by highlighting some of the most salient ECR research, from a wide variety of perspectives, to date. The chapters in Part II of this book focus on upstream uses of ECR in the form of public participation. The ideas presented in the chapters that comprise Part III form a midstream analysis of ECR by focusing on intractable conflicts, relationship change, the use of frame elicitation to evaluate ECR, and the evaluation of facilitators. The chapters included in Parts IV and V are downstream examinations of ECR: Part IV presents ideas about, and concrete examples of, ECR program evaluation in state agencies, courts, and federal agencies, and Part V examines new ideas about ECR outcomes. Finally, the concluding chapter examines the common themes among the diverse authors with a view to the future of ECR.

Notes

1. The 1998 Environmental Policy and Conflict Resolution Act (P.L. 105–156).

2. Alissa J. Stern and Tim Hicks, *The Process of Business–Environmental Collaborations: Partnering for Sustainability* (Westport, CT: Quorum Books, 2000), p. 196.

3. U.S. Institute for Environmental Conflict Resolution (USIECR), www.ecr.gov (accessed January 17, 2003).

4. J. Walton Blackburn and Willa Marie Bruce, *Mediating Environmental Conflicts: Theory and Practice* (Westport, CT: Quorum Books, 1995), pp. 1–2.

5. Lawrence Susskind, Ole Amundsen, Masahiro Matsuura, Marshall Kaplan, and David Lampe, *Using Assisted Negotiation To Settle Land Use Disputes* (Cambridge, MA: Lincoln Institute of Land Policy, 1999).

6. Gail Bingham, *Resolving Environmental Disputes: A Decade of Experience* (Washington, DC: Conservation Foundation, 1986); Rosemary O'Leary, Robert F. Durant, Daniel J. Fiorino, and Paul S. Weiland, *Managing for the Environment: Understanding the Legal, Organizational, and Policy Challenges* (San Francisco: Jossey-Bass, 1999), p. 203.

7. O'Leary and others, *Managing for the Environment.*

8. Scott Mernitz, *Mediation of Environmental Disputes: A Sourcebook* (New York: Praeger, 1980).

9. Blackburn and Bruce, *Mediating Environmental Conflicts*, pp. 18–19.

10. O'Leary and others, *Managing for the Environment.*

11. Ibid.

12. For a more detailed discussion, see Douglas Amy, *The Politics of Environmental Mediation* (Columbia University Press, 1987).

13. Ibid.

14. Ibid.

15. Ibid.

16. Laura M. Lake, "Judicial Review: From Procedure to Substance," in Laura M. Lake, ed., *Environmental Mediation: The Search for Consensus* (Boulder, CO: Westview Press, 1981), pp. 32–57.

17. Amy, *The Politics of Environmental Mediation.*

18. Melvin Aron Eisenberg, "Private Ordering through Negotiation: Dispute-Settlement and Rulemaking," *Harvard Law Review*, vol. 89 (1976), pp. 637–681.

19. Lawrence Susskind, Lawrence Bacow, and Michael Wheeler, *Resolving Environmental Regulatory Disputes* (Cambridge, MA: Schenkman Publishing Company, 1983), p. 2.

20. Lawrence S. Bacow and Michael Wheeler, *Environmental Dispute Resolution* (Plenum Press, 1984), p. 360.

21. A.E. Lind and T.R. Tyler, *The Social Psychology of Procedural Justice* (Plenum Press, 1988).

22. Amy, *The Politics of Environmental Mediation.*

23. O'Leary and others, *Managing for the Environment.*

24. Bingham, *Resolving Environmental Disputes.*

25. USIECR website.

26. Conflict Resolution Information Sources (CRInfo) website, www.crinfo.org (accessed January 17, 2003).

27. Definitions for the following terms can be found at the USIECR website and the Conflict Resolution Information Sources (CRInfo) website. Other sources are cited as relevant.

28. For more information on consensus building, see Barbara Gray, *Collaborating* (San Francisco: Jossey-Bass, 1987); Lawrence Susskind and Jeffrey Cruikshank, *Breaking the Impasse: Consensual Approaches to Resolving Public Disputes* (Basic Books, 1987); Judith E. Innes, "Evaluating Consensus Building," in Lawrence Susskind, Sarah McKearnan, and Jennifer Thomas-Larmer, eds., *The Consensus Building Handbook: A Comprehensive Guide to Reaching Agreement* (Thousand Oaks, CA: Sage Publications, 1999); Cary Coglianese, "Is Consensus an Appropriate Basis for Regulatory Policy?" in Eric Orts and Kurt Deketelaere, eds., *Environmental Contracts: Comparative Approaches to Regulatory Innovation in the United States* (Kluwer Academic Publishers, 2001). See also the Policy Consensus Initiative website, www.policyconsensus.org (accessed January 17, 2003).

29. See Gerald W. Cormick, "Strategic Issues in Structuring Multi-Party Public Policy Negotiations," *Negotiation Journal*, vol. 5 (April 1989), pp. 125–132; Gerald W. Cormick, "The Theory and Practice of Environmental Mediation," *The Environmental Professional*, vol. 2 (1980), pp. 24–33; Gerald W. Cormick and L.K. Patton, "Environmental Mediation: Defining the Process through Experience," in Laura M. Lake, ed., *Environmental Mediation: The Search for Consensus* (Boulder, CO: Westview Press, 1981), pp. 79–97; Lawrence Susskind and Jennifer Thomas-Larmer, "Conducting a Conflict Assessment," in Lawrence Susskind, Sarah McKearnan, and Jennifer Thomas-Larmer, eds., *The Consensus Building Handbook: A Comprehensive Guide to Reaching Agreement* (Thousand Oaks, CA: Sage Publications, 1999). See also Chapter 5 in this volume by Marcia Caton Campbell.

30. See Christopher Moore, *The Mediation Process, 2nd ed.* (San Francisco: Jossey-Bass, 1996); James A. Wall and A. Lynn, "Mediation: A Current Review," *Journal of Conflict Resolution,* vol. 37, no. 1 (1993), pp. 160–194; James A. Wall, John B. Stark, and Rhetta L. Standifer, "Mediation: A Current Review and Theory Development," *Journal of Conflict Resolution,* vol. 45, no. 3 (2001), pp. 370–391.

31. Roger Fisher, William Ury, and Bruce Patton, *Getting to Yes, 2nd ed.* (Penguin Books, 1991); E.A. Waldman, "Identifying the Role of Social Norms in Mediation: A Multiple Model Approach," *Hastings Law Journal,* vol. 48, no. 4 (1997), pp. 703–769; E.A. Waldman, "The Evaluative–Facilitative Debate in Mediation: Applying the Lens of Therapeutic Jurisprudence," *Marquette Law Review,* vol. 82, no. 1 (1998), pp. 155–170.

32. Adrienne E. Eaton and Jeffrey H. Keefe, eds., *1999 Industrial Relations Research Association Research Volume: Employment Dispute Resolution and Worker Rights in the Changing Workplace* (Champaign, IL: Industrial Relations Research Association, 1999), pp. 95–135.

33. Waldman, "The Evaluative–Facilitative Debate in Mediation."

34. Fisher, Ury, and Patton, *Getting to Yes.*

35. See Cornelius M. Kerwin, *Rulemaking: How Government Agencies Write Law and Make Policy* (Washington, DC: Congressional Quarterly Press, 1994); Cornelius M. Kerwin, "Negotiated Rulemaking," in Phillip J. Cooper and Chester A. Newland, eds., *Handbook of Public Law and Administration* (San Francisco: Jossey-Bass, 1997), pp. 225–236; Cornelius M. Kerwin and Laura Langbein, *An Evaluation of Negotiated Rulemaking at the Environmental Protection Agency, Phase I* (Washington, DC: Administrative Conference of the United States, 1995); Laura Langbein and Cornelius M. Kerwin, "Regulatory Negotiation versus Conventional Rule Making: Claims, Counterclaims, and Empirical Evidence," *Journal of Public Administration Research and Theory,* vol. 10 (2000), pp. 599–632; Cary Coglianese, "Assessing Consensus: The Promise and Performance of Negotiated Rulemaking," *Duke Law Journal,* vol. 46 (1997), p. 1255; Phillip J. Harter, "Negotiating Regulations: A Cure for Malaise," *Georgetown Law Journal,* vol. 71, no. 1 (1982), pp. 100–118; Phillip J. Harter, "Assessing the Assessors: The Actual Performance of Negotiated Rulemaking," *New York University Environmental Law Journal,* vol. 9 (2000), p. 32; Lawrence Susskind and Gerard McMahon, "The Theory and Practice of Negotiated Rulemaking," *Yale Journal of Regulation,* vol. 3 (1985), p. 133.

36. Amy, *The Politics of Environmental Mediation;* Kerwin, "Negotiated Rulemaking."

37. The report is located on the EPRI website, www.epri.com/targetHigh.asp?program=149&objid=232146 (accessed January 17, 2003).

38. Lisa B. Bingham, "Alternative Dispute Resolution in Public Administration," in Phillip J. Cooper and Chester A. Newland, eds., *Handbook of Public Law and Administration* (San Francisco: Jossey-Bass, 1997), pp. 546–566.

39. Ibid.

40. Ibid.

41. Ibid.

42. Ibid.

43. This section is drawn primarily from Kirk Emerson, "The Challenges of Environmental Conflict Resolution Research," presented at the National Conference of the American Society for Public Administration in Seattle, Washington, 1998.

44. See Bingham, *Resolving Environmental Disputes*; Leonard G. Buckle and Suzann R. Thomas-Buckle, "Placing Environmental Mediation in Context: Lessons from 'Failed' Mediations," *Environmental Impact Assessment Review,* vol. 6 (March 1986), pp. 55–70; Philip J. Harter, "Negotiated Rulemaking: An Overview," *Environmental Law Reporter,* vol. 17 (July 1987), pp. 245–247; Philip J. Harter, "Points on a Continuum: Dispute Resolution Procedures and the Administrative Process," *Administrative Law Journal,* vol. 1 (Summer 1987), pp. 141–211.

45. See Lawrence Susskind, "Environmental Mediation and the Accountability Problem," *Vermont Law Review,* vol. 6, no. 1 (Spring 1981), pp. 1–47; Susan L. Carpenter and W.J.D. Kennedy, *Managing Public Disputes* (San Francisco: Jossey-Bass, 1988); Gerald W. Cormick, "The Myth, the Reality and the Future of Environmental Mediation," *Environment,* vol. 24, no. 7 (1982), p. 14; Moore, *The Mediation Process.*

46. Jack D. Kartez and Peter Bowman, "Quick Deals and Raw Deals: A Perspective on Abuses of Public ADR Principles in Texas Resource Conflicts," *Environmental Impact Assessment Review,* vol. 13 (1993), pp. 319–330.

47. Melanie J. Rowland, "Bargaining for Life: Protecting Biodiversity through Mediated Agreements," *Environmental Law,* vol. 22, no. 2 (1992), pp. 503–527.

48. Chris Honeyman, "The Wrong Mental Image of Settlement," *Negotiation Journal,* vol. 17 (2001), pp. 25–32.

49. Deborah M. Kolb and Jeffrey Z. Rubin, "Mediation through a Disciplinary Prism," paper presented at the Conference on Research on Negotiation in Organizations, Northwestern University, March 31–April 2, 1989.

50. See Bingham, *Resolving Environmental Disputes*; Lawrence Susskind and Connie Ozawa, "Mediating Public Disputes: Obstacles and Possibilities," *Journal of Social Issues,* vol. 41 (Summer 1985), pp. 145–159; Gerald W. Cormick, "Environmental Conflict, Community Mobilization, and the 'Public Good,'" *Studies in Law, Politics, and Society Annual,* vol. 12 (1992), pp. 309–329; David R. Godshalk, "Negotiating Intergovernmental Development Policy Conflicts," *American Planning Association Journal,* vol. 58, no. 3 (1992), pp. 368–378.

51. Buckle and Thomas-Buckle, "Placing Environmental Mediation in Context."

52. Bingham, *Resolving Environmental Disputes.*

53. See Fisher, Ury, and Patton, *Getting to Yes*; William L. Ury, Jeanne M. Brett, and Stephen B. Goldberg, *Getting Disputes Resolved: Designing Systems To Cut the Costs of Conflict* (Program on Negotiation at Harvard Law School, 1993).

54. Howard Raiffa, *The Art and Science of Negotiation* (Cambridge, MA: Harvard University Press, 1982).

55. Owen Fiss, "Against Settlement," *Yale Law Journal,* vol. 93, no. 6 (1984), pp. 1073–1090.

56. See Amy, *The Politics of Environmental Mediation*; Douglas Amy, "Environmental Dispute Resolution: The Promise and the Pitfalls," in Norman J. Vig and Michael E. Kraft, eds., *Environmental Policy in the 1990s* (Washington, DC: Congressional Quarterly Press, 1990), pp. 211–234; James E. Crowfoot and Julia M. Wondolleck, *Environmental Disputes: Community Involvement in Conflict Resolution* (Washington, DC: Island Press, 1990).

57. John Forester, "Envisioning the Politics of Public Sector Dispute Resolution," *Studies in Law, Politics, and Society,* vol. 12 (1992), pp. 247–286.

2

Whose Reality
Counts?

Juliana E. Birkhoff and Kem Lowry

The growing interest in alternative dispute resolution generally and environmental mediation in particular has provoked both enthusiasm and
skepticism. Mediation enthusiasts see not only a more effective way to
address complex disputes, but also the possibilities of more deliberative politics. Critics argue that some of the claims made for mediation are overblown
or unsubstantiated. Evaluation research has become one of the principal
vehicles for testing the claims about environmental mediation and other collaborative problem-solving processes, for enlarging our understanding of
practice-related issues, and for ensuring accountability. In this chapter, we
outline some of the promises and criticisms of environmental mediation. We
then explore several questions about the dominant evaluative purposes and
research strategies reflected in current evaluative studies.

Background

In 1973, Gerald Cormick and Jane McCarthy began discussions with 10 interest groups in Washington state concerned with flood control, recreation, and
agricultural viability in the Snoqualmie River Basin. In 1974, the parties signed
a mediation agreement. Although the roots of environmental mediation and
collaborative problem solving go back farther, many trace the beginning of
environmental mediation to this case. By 1986, approximately 200 environmental and public policy cases had been mediated.[1] There have been no calculations of the number of mediation cases since 1986, but growth has continued.

In the 1970s and 1980s, collaborative problem-solving and consensus-building efforts focused on clear disputes. These were usually cases in litigation or at administrative agency impasse. They often involved cleaning up hazardous sites, developing policies or regulations, or negotiating about controversial developments and facility siting. The current range of uses for collaborative approaches includes those uses as well as developing plans, recommendations, and policies; designing permits and operating requirements; managing natural resources; solving problems; and resolving disputes.[2,3]

In the 1970s and 1980s, a few sole practitioners and dispute resolution nonprofits provided most of the services, often with funding from foundations. In 2003, hundreds[4] of dispute resolution professionals work in local, state, and federal government; in universities and for-profit consulting firms; as well as in nonprofit organizations and as sole practitioners.

Established programs with mission statements, strategic plans, personnel policies, and budgets now provide collaborative problem-solving and consensus-building services. There have been debates about the ethics, effectiveness, and goals of environmental and public policy mediation since its inception, but institutionalization has highlighted questions of accountability and long-term societal and environmental outcomes. As local, state, and federal governments allocate taxpayer dollars to an increasing array of providers, projects, and programs, the quest for better evaluation and research has increased. The increasing demand for rigorous applied and theoretical research highlights the proliferating lists of benefits and critiques of environmental collaborative problem-solving and consensus-seeking processes. Researchers and program evaluators operationalize these proposed benefits and detriments into program and process goals, performance criteria, outcome measures, program objectives, and indicators for studies of environmental collaborative problem-solving and consensus-seeking processes. We begin our exploration of the current terrain of applied research with an overview of the advocates' benefits and opponents' critiques.

The Promise of Environmental Dispute Resolution

Mediators and mediation proponents have never been reticent in claiming wide benefits and effects for the process. These claims about the benefits, outcomes, and effects cluster on the individual, relationship, social, and institutional or ecological levels. Some of these claims have held up under scrutiny, some have not been researched adequately, and some seem too good to be true.

Individual Satisfaction and Met Needs

As with most other mediation processes, proponents of environmental collaborative problem solving and consensus seeking explain that the processes can uniquely satisfy individuals' interests, wants, and needs.[5] Scholars and practitioners claim that environmental mediation results in higher levels of satisfaction.[6] Proponents explain that the processes allow participation by all parties—those who are involved in the dispute, those who have a role in implementation, and those who can block implementation of the solution. Therefore, they can clearly explain their preferences, advocate for them in an open negotiation process, and make trade-offs that are sensible to them. Proponents emphasize that these features ensure that parties will be satisfied with the results.

Another argument at the individual level for environmental collaboration is increased efficiency and cost savings. Mediation is generally asserted to be far less expensive than litigation.[7] Several authors have also argued that environmental mediation takes less time than conventional processes do.[8]

Because the consensus-based process is flexible and the parties design it, the solutions, too, can be more carefully crafted, creative, and "adaptive." Advocates explain that administrative and rulemaking bodies cannot take the time—nor do they have the knowledge—to hammer out the tailored details of environmental plans or solutions for specific regions or species.[9]

Some scholars and practitioners refer to procedural justice theories and research to explain why individuals should be, and are, more satisfied in collaborative and consensus-based processes.[10] This research demonstrates that parties feel the process is fair when they believe that:

— they are well represented or have control over the process;
— they are being treated as someone else would be;
— they are treated with dignity, respect, and in accordance with their rights;
— the mediator is impartial, honest, and fair;
— decisionmakers have enough information to make decisions and problems are publicly discussed; and
— there is a chance to correct or modify a decision once it has been made.[11]

Mediators and mediation proponents point to design features of the collaborative problem-solving and consensus-building processes to illustrate how procedural justice elements are incorporated into collaborative processes and thus, why stakeholders are satisfied with the results of the processes.

Individual Empowerment and Capacity Building

Some mediators claim that dispute resolution processes teach individuals how to handle their disputes better. Many proponents explain that participation in collaborative problem solving or a consensus-based process develops individuals' skills in negotiation, communication, active listening, group process, and coalition building.[12] Still other proponents explain that mediation "empowers" the parties in some way.[13] To these proponents, empowerment means that by participating in the process, individuals learn how to articulate their interests, listen to others, and negotiate well for themselves.[14] Participating in the process is thus a modeling exercise in which the model the mediator provides and the positive interaction among parties implicitly teach the individuals involved good process skills. Some mediators provide a day of negotiation training at the beginning of a process. Furthermore, almost all mediators coach stakeholders on how to frame proposals and bargain productively.

Mediators also explain that collaborative problem-solving processes empower parties through greater access to key decisionmakers and important information. Parties have more opportunities to influence or control meeting agendas. Finally, they maintain that public processes empower parties by enhancing parties' credibility, legitimacy, or confidence.[15]

Personal Transformation

A popular ideology that scholars and practitioners promote argues that mediation transforms individuals by promoting individual growth and moral development.[16] According to Lon Fuller, the central strength of mediation is "its capacity to reorient the parties toward each other, not by imposing rules on them, but by helping them to achieve a new and shared perception of their relationship, a perception that will redirect their attitudes and dispositions toward one another."[17] Whereas environmental mediation proponents are not as explicit about the role of mediation in helping to transform individuals, many mediators believe in the redemptive nature of collaborative problem-solving and consensus-seeking processes.[18]

Advocates of the personal transformation benefits of collaborative problem-solving and consensus-building processes tell compelling stories of stakeholders who experience personal growth during the process. They also explain that during the process, participants gradually recognize and appreciate the other party's situation. Participants begin to value the nature of human diversity.

Relationship-Level Outcomes and Benefits

Mediators frequently explain that the mediation process helps parties to be more open, honest, and trusting.[19] The mediator structures a transparent and fair communication process designed to encourage parties to engage in meaningful dialogue. The process of jointly working through differences leads to rational relationships.[20] Mediators describe how collaborative processes break down barriers between adversaries. Parties can identify with each other after increased and regular contact, as well as other rituals.[21] Collaborative processes help develop new networks and relationships among the parties.[22]

The development of the "social capital" concept has helped provide a theoretical basis for mediators' observations.[23] Social capital is the ability of individuals to draw upon rich relationship networks to facilitate coordination and cooperation. Scholars and practitioners explain that the exchange of information and perspectives and the improved communication, understanding, and focus on problem solving develop norms, trust, and networks among the participants.[24] The focus on social capital connects the benefits that collaborative processes offer individuals with larger social or institutional benefits.

Social-Level Outcomes and Benefits

Mediators claim that dispute resolution processes contribute to social change and social justice as disputants learn new skills, legitimize their opponents' interests, and transform relationships between parties.[25] Environmental and public policy dispute resolution promoters also argue that dispute resolution processes empower individuals and groups to reclaim their voices, increase their self-esteem, democratize their neighborhoods, involve the disenfranchised in policy deliberations, and create new social or governing structures.[26] Participation in collaborative processes can translate to greater resources and efficacy for groups in future negotiations with agencies and organizations. Furthermore, because the mediator creates and controls a fair communication and negotiation space, parties participate fully and thus gain legitimacy and respect.[27] Some scholars and mediators believe that the process challenges established power relationships. New agreements can produce structural or institutional change that institutionalizes new patterns of interaction or legitimizes broader participation.[28]

Frank Dukes summarizes the individual benefits of collaborative problem-solving and consensus-seeking processes and connects them to broader social goals. Dukes explains that collaborative processes can respond to disintegrating communities and relationships, citizen alienation, and societies' inability to solve problems and resolve conflicts. Collaborative problem-solving and consensus-seeking processes sustain and create spaces for authentic public dialogue and civic consciousness.[29] They also strengthen and institutionalize meaningful public participation in public and agency planning and decision-making. Ultimately, these processes strengthen democracy and communicative action and build social capital.[30] Innes asserts that

> Consensus building processes, whether or not they result in an agreement, typically produce numerous secondary consequences that are sometimes more important than any agreement. For example, consensus building can result in new relationships and trust among stakeholders who were either in conflict or simply not in communication.[31]

We explore how these claims and benefits are examined later in the chapter.

Ecological-Level Outcomes and Benefits

Finally, a few scholars and mediators argue that inclusive, consensual solutions to environmental problems are more environmentally sound and ecologically sustainable. These arguments are based on the linkage between fair and inclusive processes with better environmental outcomes. Wider public participation in planning, management, and problem solving can incorporate more ways of knowing. Collaborative processes involve more parties; with more people involved, more extensive information about the areas, species, histories, and risks can be integrated into problem analysis and solution generation. Because more parties are involved, the solutions can be more creative and synergistic. Finally, proponents explain that with more bargaining latitude and release from standard operating procedures negotiators are freer to invent better agreements. Collaborative processes can produce better agreements that have more adaptive management ideas.[32]

Criticisms of Environmental Dispute Resolution

At the same time that environmental and public policy mediation programs were becoming institutionalized, legal, social, and political theorists criticized specific cases, specific dispute resolution programs, and the collaborative problem-solving and consensus-seeking process. There is a long list of critiques about the effectiveness and fairness of collaborative problem-solving processes. These criticisms have shaped applied and theoretical research concerns and projects.

Douglas Amy's scathing critique of environmental mediation created a stir in the dispute resolution community when it was published in 1987. Amy argued that the use of dispute resolution in public policy disputes was fundamentally inequitable. Most of the concerns he raised focus on power. He argued that powerful actors control access to the legislative, administrative, and judicial processes, as well as technical expertise, alternatives to the negotiated process, and the way mediators and parties think of their options.[33]

A longstanding criticism argues that alternative policymaking and decisionmaking processes weaken attempts to fundamentally reform governance and judicial processes. Similarly, many critics focus on dispute resolution's delegitimizing effect on the roles of administrative and regulatory structures.[34] Other critics argue that collaboration, like other policymaking reforms, is a fad and cannot take hold in Western systems with Western approaches to governance.[35]

Some analysts have criticized processes for leading to compromise, with science and information the focus of the compromise. The Science Advisory Boards of the U.S. Environmental Protection Agency focused on criticisms that collaborative processes do not integrate scientific and technical information.[36]

Alex Conley and Ann Moote summarized the criticisms in their literature review on collaboration. They explain that critics argue that collaborative processes

— delegitimize conflict;
— produce lowest common denominator outcomes;
— often include members with unequal resources such as time, money, information, and negotiation training;
— address issues such as national forest management and grazing on public lands through local collaboration instead of through national dialogue;
— consist of stakeholders whose roles may not be well defined;
— exclude urban-based environmental groups;
— disempower national and local majorities when using consensus-based approaches;
— may circumvent the authorities whose role it is to manage resources; and
— co-opt environmental advocates.[37]

These criticisms fuel practitioners' efforts to create processes that are more transparent and researchers' efforts to explore how programs stack up against these criticisms.

The Changing Terrain of Evaluation

The use of collaborative and consensus-based strategies has grown, but research and program evaluation have not increased at the same rate. The first applied research of collaborative problem-solving and consensus-seeking processes was in the form of case studies. The researchers attempted to accurately describe what happened in specific cases with different experimental processes. Later, collections of case studies, most written by different authors with different theoretical lenses, were compared for general lessons. Some more explicit comparative case study research projects were conducted in other areas, for example, regulatory negotiation and land use conflicts.[38] Gail Bingham's descriptive research with analysis of success factors was the first study to empirically examine a number of disparate cases.[39] By the 1990s, some comparative studies had been conducted on streams of cases—for example, Superfund cases and enforcement cases.[40] We describe some of the findings from these program evaluations and applied research projects below.

In 2002, researchers, program managers, legislatures, practitioners, funders, and sponsors are asking broader questions, and the sense of urgency is higher. Although practitioners have been able to convince stakeholders to use processes with several engaging case histories, gatekeepers and stakeholders press for detailed answers about benefits. Legislatures are asking for empirical proof about the benefits. More people are asking broader questions stemming from claims that are more extensive and from wider criticisms.

Some research, principally evaluative research, has addressed these and other questions about the efficacy of environmental mediation. Interest in using the results of evaluations and applied research to increase the use and legitimacy of environmental mediation has greatly increased. The emerging significance of environmental mediation (and other collaborative processes) and the potential usefulness of evaluation research as a way to address many of the questions about these processes has led gatekeepers, funders, program managers, and practitioners to dramatically increase their attention to evaluation and applied research. This interest is manifest in the proliferation of articles, papers, conferences, and, perhaps most importantly, funding. Evaluation debates have focused primarily on the evaluation criteria and methods for generating valid data about mediation and other collaborative processes.

Purposes of Evaluation

Lost—or at least buried—in these debates are other questions we regard as critical: what purposes should evaluation serve? Whose purposes are they? How effectively do current evaluations of collaborative processes address the multiple needs of evaluation clients?

Influential clients—foundations and government agencies—shape the evaluative agenda by the types of evaluation and other research they encourage and fund. In environmental collaboration and dispute resolution, influential clients are encouraging research organized around several broad purposes: making program and project judgements, building theory, learning from performance, and ensuring accountability.

Making Program and Project Judgements. How effective are programs and projects in terms of the goals they set for themselves? How valid are the more general claims that program proponents make about the efficiency, effectiveness, satisfaction, and durability of outcomes for collaborative processes such as environmental mediation?

Building Theory. How can evaluation contribute to practice theory? For example, what strategies for incorporating technical information into complex processes are regarded by participants as most effective? How can evaluation contribute to general social and political theory about the role of informal deliberative processes? For example, should collaborative processes be thought of as informal problem melioration activities, or is transformation of relationships among disputants the desired goal? What attributes of collaborative processes enhance their effectiveness? Which impede their effectiveness?

Whereas evaluation is funded and encouraged primarily to make program judgements and to contribute to theory building, broadly construed, two other purposes of evaluation are also important: learning from performance and ensuring accountability.

Learning from Performance. To what extent does evaluation contribute to improved practice? How should evaluation be organized to provide practice-related evaluative information to individual mediators or program officials?

Ensuring Accountability. To whom should mediation program managers and mediators be accountable? To what standards of collaborative process design and mediator practice should programs and mediators be held accountable?

These evaluative purposes are not necessarily mutually exclusive, but in practice the judgemental and theory-building purposes—and the audiences they serve—dominate the published research, conference discussions, and proposals for foundation funding. Performance-oriented learning and accountability to practitioners, disputants, and the public receive relatively less emphasis in the evaluative literature.

Some of the research associated with each of these evaluative purposes is outlined below. Our purpose is not to undertake a comprehensive review of the research associated with each purpose, but rather to provide a flavor of the type of research questions raised and the audiences served by each.

Making Program and Project Judgements

Evaluative research is comparative. Findings are compared to baseline conditions, to nonprogram controls, or to explicit or implicit standards or goals. In the case of environmental mediation, efforts have been made to contrast attributes of environmental mediation efforts with the costs, time spent, disputant satisfaction, and quality of outcomes of the judicial or administrative forums in which environmental cases would otherwise be addressed.

Generally, the results of early research suggest that environmental mediation is at least moderately more effective in terms of cost, process efficiency, and disputant satisfaction. In the evaluative research, disputants have reported that environmental mediation has been either moderately or very cost-effective.[41] Whereas there are methodological problems in calculating actual time spent in mediation, particularly when mediation occurs in cases that began as lawsuits, disputants perceive time savings over court.[42]

Disputant satisfaction with the mediation process and outcome is another criterion suggested as a basis for evaluating the relative effectiveness of environmental mediation. One study found that more than 80% of disputants were satisfied with the process and outcome.[43] Sipe and Stiftel reported that 66% of participants in 19 mediated environmental enforcement cases were either moderately or very satisfied with the mediation process in which they were engaged, and 80% believed that the mediated outcome met some or all of their objectives.[44] Andrew's analysis of mediations involving waste management found that 65% of disputants were satisfied with the mediation process, and 61% were satisfied with the outcome.[45]

One of the principal claims of mediation literature is that the flexibility and informality of the process allow disputants to surface their concerns, to explore options for meeting their interests, and to craft agreements that optimize joint gains among the parties.[46] Because mediated agreements are voluntary, many researchers have chosen to use the frequency of agreement in mediated processes as a rough surrogate indicator of success, on the assumption that parties would not make agreements that they did not perceive to be better than their alternatives. Bingham found that agreements were reached in 78% of the 133 environmental conflicts she analyzed.[47] Buckle and Thomas-Buckle reported a 75% agreement rate in mediated processes that went beyond the initial session.[48] Sipe and Stiftel reported settlement rates of 74% in the environmental cases they examined.[49] In his analysis, Andrew found settlement rates of 81%.[50]

This evaluative research tends to support claims that mediation is cheaper, more efficient, more satisfying to disputants, and more likely to result in agreements. Nevertheless, this judgemental purpose continues to be a primary focus for evaluation of environmental mediation efforts for two reasons. First, critics argue that too many claims about environmental mediation have not been addressed or have been inadequately evaluated. They argue that evidence is nonexistent or inconclusive that mediators are neutral, that collaborative processes are sufficiently inclusive, or that negotiated agreements are demonstrably better.[51] Second, the list of claims about the potential benefits of environmental mediation keeps expanding. As we outlined above, practitioners and scholars have moved from a focus on individual and case-specific benefits of collaborative processes to a focus on broader and deeper societal and environmental outcomes and advantages. These and other claims have not been systematically evaluated.

Building Theory

Evaluative writing about environmental mediation—including both research on specific projects and programs and more general essays or studies on evaluative themes—contributes to two types of theory, practice theory and normative theory.

Whereas evaluations are frequently conducted to make judgements about effectiveness, cost, efficiency, and the like or to identify potential program improvements, the results are also used to influence thinking about practice-related issues. They may be used to illuminate concepts, to identify key practice issues, to generate lessons, and to contribute to the identification of best practices.

Evaluation research may lead to the generation of lessons and best practices in some fields, but in environmental mediation the relationship is reversed. The lists of lessons and best practices are far more extensive than the research supporting them. The literature on representation issues in environmental mediation is a case in point. The principle that effective environmental mediation requires the participation of all parties potentially affected in a dispute finds strong support in both the theoretical and best practice literature. According to Kubsak and Silverman, for example, environmental mediation obtains input from "all major interests."[52] Likewise, Croce argues that mediation involves "meaningful participation by all those who have a stake in the outcome."[53] This stance is reinforced in several lists of best practices. For example, *A Practical Guide to Consensus*, a useful handbook developed by the Policy Consensus Initiative, identifies the following best practices regarding participation:

— All necessary interests are represented or at least approve of the process.
— Participants usually represent stakeholder groups or interests and not just themselves.
— Participation is voluntary.
— Participants share responsibility for both process and outcome.
— All participants must be able to participate effectively.[54]

There is a high level of consensus about the importance of involving all key stakeholders, but surprisingly little research explores representation issues in depth. Bingham found that "there is no evidence among the cases studied to indicate that a large number of parties in a dispute resolution process make(s) it more difficult to reach an agreement."[55] However, she found that representation and involvement of all parties present a more complex set of questions. She notes that full participation is difficult because parties may not be organized, because they participate at different points in a potentially lengthy process, or because there are simply so many parties that including them all can be difficult.[56] She concludes that

It is difficult to evaluate how much impact the selection of participants has had on the success or failure of the various mediation attempts examined for this report. ... In the simplest terms, looking at the 29 cases in which the parties failed to agree, there appears to be no case in which parties who were included in the negotiation failed to reach an agreement because other parties had been left out. Still, in some cases in which agreements were reached, there does seem to be evidence that special efforts had to be made to satisfy the concerns of groups not directly involved in the process, and … there sometimes were problems during implementation because unrepresented parties objected after an agreement was reached.[57]

Other research confirms the finding that not all parties participate. In his research on waste management cases, Andrew asked disputants whether the relevant parties participated. He found that all relevant parties participated in only six (or 26%) of the processes he examined.[58] The effect of nonpartic-ipation on the outcomes of the processes he examined is not clear.

Issues about representation in environmental mediation are part of a longer list of potentially important research questions that have not been subjected to rigorous empirical analysis. In addition to questions about par-ticipation, much still needs to be learned about issues of power, assertions of rights, incorporating technical information, issues of mediator neutrality, and a host of other questions.

As O'Leary puts it,

> … [D]espite the plethora of literature touting the advantages of envi-ronmental mediation (and, at times, the disadvantages), the empirical foundations for most of the conclusions are quite weak. While there are some strong conceptual works, few scholars have studied environ-mental mediation through one or more of the standard empirical methods: theoretically informed case studies, comparative cases anal-ysis, surveys, interviews, and statistical analyses of quantitative data. Given the paucity of empirically based research, it must be concluded that much of our "knowledge" concerning environmental mediation is based primarily on thoughtful speculation or wisdom, with few data (broadly defined) to support it.[59]

Recognizing how evaluation could contribute to theory building, founda-tion and government agencies are supporting a variety of multicase, multi-program empirical studies that address some of these questions and contrib-ute to practice theory. The Policy Consensus Initiative and the U.S. Institute for Environmental Conflict Resolution are working with five state offices of dispute resolution on a comprehensive program evaluation. The National Academy of Sciences is supporting a broader study of citizen participation and collaboration, and the National Science Foundation has funded a study of watershed partnerships. The William and Flora Hewlett Foundation has supported a number of projects to gather new data, develop analytic frame-works, and share data and theoretical frameworks.

Whereas some critics have emphasized the need to test more rigorously the assumptions of good practice in environmental mediation, others seek to connect the evaluation of environmental mediation to larger debates about political theory. Forester is supportive of the emphasis in conventional evaluation on explicit standards and criteria for assessing environmental mediation, but he argues that this conception of evaluation is too limited:

> We need a good deal more: an ongoing debate, reflecting careful research, that will formulate and clarify those "standards and criteria" that we are to take as the very foundation of the evaluation efforts.... Just what is to be evaluated, what is to be asked, is too often presumed by social scientists and self-limitingly considered the sole province of (beware:) "theorists."[60]

He adds

> We must also ask how, in particular, cultural, historical, and institutional settings, public dispute-resolution processes should work: what qualities of participation mediators should encourage, what qualities of outcomes mediators should seek, what qualities of commitment, virtue, and judgment [sic] mediators should embody? These are the questions that political theorists, particularly those interested in democratic politics and practices, aspire to ask and address, if never answer for all time.[61]

The "ongoing debate" Forester calls for is occurring. A rich array of criteria is being suggested that operationalizes the broadest claims for collaborative problem solving and consensus seeking.[62] In arguing for a broader evaluative framework, Innes has suggested that criteria be "grounded significantly in the idea that consensus building is a method for creating a self-organizing, adaptive system for a complex, changing, and uncertain context."[63] She adds that "the ideas underlying the criteria have also been influenced by the works of social theorist Jurgen Habermas, who has proposed a concept of communicative rationality to guide the processes of dialogue and collective learning that is remarkably parallel to the best practices of consensus building."[64]

Like many scholars, Innes distinguishes between process and outcome criteria. Several of her proposed process criteria are similar to conventional best practices for mediation processes:

— The consensus-building process includes representatives of all relevant and significantly different interests.
— It follows the principles of civil discourse.
— It adapts and incorporates high-quality information.[65]

The rest of her candidate criteria hold mediation and other collaborative processes accountable to standards (and expectations) that go well beyond conventional process criteria:

— The process is driven by a purpose that is practical and shared by the group.
— It is self-organizing.
— It encourages participants to challenge assumptions.
— It keeps participants at the table, interested and learning.
— It seeks consensus only after discussions fully explore the issues and interests and significant effort was made to find creative responses to differences.

Likewise, her proposed outcome criteria include "high-quality agreements" but go well beyond that. "Desirable outcomes," she argues, "mirror the outcomes of a complex, adaptive system."[66] She asserts that a "process that achieves many of these outcomes is probably better than one that achieves few, but the outcomes may vary in importance depending on the situation. The relative importance of each criterion in each case ultimately will have to be judged by those affected by the results."[67] In addition to "high-quality agreements," good outcomes include the following:

— It ended stalemate.
— It compared favorably with other planning or decision methods in terms of costs and benefits.
— It produced feasible proposals from political, economic, and social perspectives.
— It produced creative ideas for action.
— Stakeholders gained knowledge and understanding.
— The process created new personal and working relationships and social and political capital among participants.
— It produced information and analyses that stakeholders accept as accurate.
— Learning and knowledge produced within the consensus process were shared by others beyond the immediate group.
— It had second-order effects beyond agreements or attitudes developed in the process, such as changes in behaviors and actions, spin-off partnerships, collaborative activities, new practices, or even new institutions.
— It resulted in practices and institutions that were both flexible and networked, which permitted a community to respond more creatively to change and conflict.
— It produced outcomes that were regarded as just.
— The outcomes seemed to serve the common good or public interest.
— The outcomes contributed to the sustainability of natural and social systems.[68]

These criteria—and others like them—can be read as minimum standards to which environmental mediations should be held accountable. They can also be read as outlining a normative case for more inclusive, more deliberative politics.

As practitioners (and as academics), we share concerns about perceived gaps between the claims of environmental mediation and the evidence to support them. We also share broad aspirations for the potential of more deliberative politics through collaborative processes. What tends to be lost or subsumed in the emphasis on legitimacy and theory building is an evaluative orientation toward improving mediators' performance and ensuring their accountability to a variety of stakeholders.

Learning from Performance

The current evaluative emphasis on creating legitimacy and building theory at the expense of practice and accountability is sometimes explained as a gap between the theory community and the practice community. Deborah Kolb has suggested that it is less a gap between these communities than it is a difference in perception about what is useful and interesting in research. As a result, researchers investigate quite different questions than practitioners have in mind.[69]

What does it mean to organize evaluation in ways that are more responsive to practitioners' needs? What types of questions should be addressed? What data-gathering strategies would be most appropriate? What types of interpretive actions are needed?

Mediators seeking to learn from practice have to make sense of particular experiences through the lens of their training, their exposure to conceptions of best practice, their general practice experience, and whatever evaluative data are available. Most mediators have been through training in which mediation is presented as a sequence of activities (e.g., opening statements, initial caucus) and a set of skills (e.g., active listening). Training may include typical obstacles or dilemmas encountered by mediators (e.g., power imbalances among the parties, "difficult" people) and lists of best practices for handling such obstacles. Once they have begun to practice, they encounter situations in which the familiar sequence of mediation activities is not appropriate, dilemmas and obstacles are unfamiliar, or the skill set they have developed seems inadequate. Mediators have the difficult task of making sense of these experiences. How do they decide what new skills they need? What process lessons should be drawn from experience?

Confronted with self-doubts and questions about what processes or skills might have been more effective, mediators have a limited number of tools to facilitate making sense of their experiences. Conventional evaluative tools are often of limited utility. Postmediation assessments completed by the participants may be available to the mediator. These assessments indicate such things as participants' satisfaction with the mediator and how well the mediator facilitated communications. Such surveys are a source of validation when collaborative processes go well, but are of more limited usefulness as a diagnostic when processes seem to falter. They may provide clues about needed skills and process design, but making sense of such clues requires a great deal of interpretive skill.

Program evaluations may also be periodically available. Surveys of disputants that are aggregated across multiple disputes provide more data about disputant satisfaction, perceptions of effectiveness, efficiency, and the like. Even when evaluative surveys are supplemented by interviews with disputants, findings are likely to be aggregated at the program level. Individual mediators have to make their own judgements about whether and how program-level findings apply to their individual performances.

Making such judgements is subject to a number of potential errors. Two of the most common are "competency traps" and causal misattribution. Competency traps can occur when mediators (or other clinical practitioners) are "successful" *in spite of* rather than *because of* the process choices they made.[70] After excluding some "difficult" people from a process and successfully reaching an agreement among the remaining participants, for example, a mediator may conclude that inclusiveness is not that important.

Causal misattribution occurs when mediators misunderstand or improperly specify the causal processes that led to a perceived success or failure. A single text negotiation that co-mediators regard as a brilliant tactical move on their part may be experienced by disputants as an after-the-fact summation of tacit agreements already made by the parties. A facilitator leading a collaborative process may decide that he had erred in not designing an inclusive agenda development process for the meeting because during the meeting stakeholders objected to the agenda and how it was developed. However, the stakeholders could explain that this was the first meeting in which they felt engaged in the process and took responsibility for their issues.

What is needed is a different approach to learning about performance. At the core of such an approach are conceptions about how people learn and how to learn from practice in particular. Some theories of individual learning are based on the idea that learning activities that emphasize the social and interactive elements of learning from practice are most relevant. Constructivist learning theories are built on the belief that all knowledge is based on experience, and people arrive at meanings by continuously seeking order in their experiences.[71] Humanist learning theories emphasize that learners do best when there is a focus on problem solving.[72] Constructivist and humanist conceptions of learning both emphasize the social dimension of learning. They both assume that a substantial amount of learning occurs as people exchange information and make sense of experience collectively. Participants in "communities of practice" such as a mediation program create interpretations and explanations that are broadly shared.

Practitioners need information that helps them reflect on the effects of the tactical choices they make.[73] Cases and the stories practitioners tell about cases are the primary material from which information about tactics and practices can be gleaned.[74] Some of the activities associated with "action learning" and "action research" are relevant and are briefly noted here. In particular, we note *asking questions and dialogue and reflection as key learning activities. Both are based on constructivist conceptions of learning, in which mediators work collaboratively to learn from practice.*

Asking questions is a means of identifying key practice-related dilemmas, issues, and problems. It is also a way of focusing inquiry that is more detailed. A variety of types of questions can help stimulate dialogue.[75]

A brief excerpt generated from interviews with about 25 environmental and public policy mediators provides a flavor of useful practice-related questions:

— What helps parties start building relationships and trust?
— What questions can or do mediators ask that help build a consensus-building process effectively?
— What models exist for information exchange?
— What is known about how to sequence or link mediation processes with more formal decisionmaking procedures?
— When politics is involved in a dispute that also has a strong technical component, what are effective ways to integrate the two?
— What increases the likelihood that constituencies will approve or ratify agreements?

Dialogue and reflection are other key learning activities. Dialogue and reflection on surprises, key questions, critical mediation incidents, and mediator interventions can be powerful stimuli to learning. The reflection focuses both on what is happening and on the assumptions behind what participants thought was happening. Dialogue and reflection can help identify key assumptions that practitioners make about their interventions, and it can allow mediators to surface "undiscussable" sensitive issues and to develop shared understandings.[76]

Because mediators know and learn in cases and through problem solving, evaluation activities have to be connected to particular cases and the surprises, questions, or dilemmas of those cases. Team meetings, case consultations, observations, and dialogues are evaluation activities borrowed from other clinical practices that are used by some conflict resolution organizations (such as Antioch University, Institute for Conflict Analysis, and RESOLVE, Inc.). Case consultations impose the discipline of reciprocal reflection in action. Observations allow practitioners to learn in the moment how a tactic or strategy looks to an observer. In the dialogue, both the observer and the observed can discuss assumptions and learn. These meetings and workshops are the occasion for practitioners to pose questions and practice dilemmas, examine assumptions, and explore options. We see these and related performance-learning practices not as substitutes for conventional evaluation, but as valuable supplementary activities.

Ensuring Accountability

Ensuring accountability, like learning from performance, is notably lacking as a primary theme in current efforts to evaluate environmental mediation programs. Accountability was once a dominant theme in evaluation. In its narrowest sense, it referred to the procedures for ensuring that project or program fund expenditures complied with prescribed rules. A second conception of accountability assumes that programs should be implemented in ways consistent with their design.

Whereas we regard these conceptions of accountability as valid, we are making a case for a somewhat broader conception of accountability. At the core of this conception of evaluation are several questions:

— To whom are mediation program managers and environmental mediators accountable?
— For what are they accountable?
— How should accountability be assessed?

To whom are mediators accountable? The responsibility of environmental mediation program managers to funding agencies or other sponsoring organizations is obvious and requires little elaboration. We also assume that mediators (and program managers) are accountable to stakeholders in a dispute—including stakeholders who are not direct participants in the process. Some mediation advocates argue for a broader political accountability on the part of mediators.[77] They suggest that mediators are accountable for inclusive, deliberative politics. We think responsibility is best served by careful mediator attention to the processes and outcomes of the specific disputes in which they are involved.

For what is the mediator accountable? The mediator is accountable for process and outcome quality. Process quality involves a number of elements, including trustworthiness of mediators, disputant feelings of control over the presentation of arguments, civility among disputants, perceived competence of mediators, perceived process fairness, and respect.[78] Mediators are also accountable for the efficient use of disputants' time and resources. Outcome quality also involves multiple elements. It includes stakeholder satisfaction with the outcome, perceived fairness, durability, and the degree to which agreements set good precedents.[79] Mediators should also be accountable for helping stakeholders incorporate the best available technical and cultural information, if that is relevant.

Mediation can lead to improved relationships among disputants. Indeed, some have argued that mediation should result in *transformed* relationships—new ways of thinking and feeling about adversaries.[80] Mediation and other collaborative processes may lead to increased community capacity to manage environmental and public policy disputes.[81] Such increases in community capacity are part of what has been called "social capital." To what extent should mediators be held accountable for transformed relationships and increases in social capital? Here lies one of the fault lines in the field. A few academics have boldly asserted that transformation and building social capital are outcomes for which mediators should be held accountable, and some practitioners have drawn the conclusion that such expectations are being created.

How should accountability be assessed? In 1999, a workshop sponsored by the Indiana Conflict Resolution Institute, the Policy Consensus Initiative, and the U.S. Institute for Environmental Conflict was held in Tucson, Arizona. The workshop brought together researchers, program managers, and practitioners to develop an approach to evaluating government agency environmental dispute resolution programs. Some of the tensions between academic observers and practitioners were evident at the workshop.

The participants disagreed about how to assess outcomes of individual cases. Some researchers argued that in the end, programs should advance sustainable development and build social capital. Practitioners argued that they should not be responsible for social or political outcomes beyond the case that they were mediating. Program managers agreed that their programs were accountable for more than just agreements. They acknowledged that they were responsible for using the best information available to design quality processes and systems. They also believed deeply in their obligation to the participants in the process. However, they were troubled by the prospect of using evaluative information to develop judgements about the effects of their mediation activities on larger social processes.

For example, the Massachusetts Office of Dispute Resolution (MODR) is charged by Massachusetts statute to aid the three branches of state government, municipalities, and other public institutions in the resolution of disputes. Program managers developed mediation, training, and dispute system design programs to carry out their mandate. The program managers, staff, and mediators in the Massachusetts program are accountable to the legislature to aid specified stakeholders to resolve those specific disputes. MODR program managers argued that it would be unreasonable to hold them accountable for something they never set out to accomplish, such as "sustainable natural and social systems" or "flexible and networked practices and institutions."

The discussions in this workshop were civil and engaging. Nevertheless, the gaps between the research interests of academics and the practical information needs of program managers were obvious. Program managers need information that would help mediators reflect on the quality of their process designs and mediation activities. They also need information that would help mediators be more accountable to the disputants they serve, but those practical needs can get lost in the creation of more elaborate research agendas designed to increase the visibility and legitimacy of environmental mediation.

Conclusions

The current evaluation agenda promoted by researchers and funded by foundations and government agencies is providing a more nuanced perspective on the strengths and limitations of collaborative processes for resolving public disputes, including environmental disputes. The agenda is helping to enhance the legitimacy of mediation as a genuine dispute resolution alternative and to persuade skeptical policymakers of its potential usefulness. However, as useful as it is, this is an evaluative research agenda that inadequately serves the needs of mediators and disputants. Too little evaluative research helps mediators learn from their practice. Too little is designed to ensure that mediation practices are accountable to the disputants they serve. This is not an issue of choosing one agenda over another. Rather, it reflects a problem of not recognizing the multiple purposes—and audiences—that a more well-rounded research agenda might serve.

Notes

1. Gail Bingham, *Resolving Environmental Disputes: A Decade of Experience* (Washington, DC: Conservation Foundation, 1986).

2. Scholars distinguish between *disputes* and *conflicts*. Conflicts involve many parties and extend over time. Disputes involve specific and bounded issues that parties pursue. Conflict takes place within social structures at particular points in time and is conditioned by historical patterns. Disputes are dissensions or controversies over specific, bounded issues. A conflict is an ongoing relationship or state of affairs that is larger, more complex, and overladen with historical and situational elements. Disputes may be incidents in a larger conflict.

3. Much of the confusion in the field comes from unclear definitions about processes. We describe the overarching category of nonadversarial, alternative approaches to handling environmental issues as "collaborative problem solving and consensus building." Under this overarching category is a wide range of processes used in earlier or later stages of difference. Used in earlier stages of the development of difference, consensus-building techniques originated in the planning and citizen participation fields. The goals of these consensus-building, facilitation, joint problem-solving, and collaborative planning processes are different from the goals of environmental mediation. Used in later stages of dispute development, environmental mediation processes originated in the labor management field. Policy dialogues and regulatory negotiation are yet other processes with different goals and thus different design

attributes. They are often compared to other governmental decisionmaking processes. It is important for researchers, practitioners, and policymakers to carefully distinguish the attributes and goals of the process they are discussing.

4. There are approximately 375 members of the Environmental and Public Policy Section of the Association of Conflict Resolution and 191 practitioners on the U.S. Institute for Environmental Conflict Resolution roster.

5. Larry Susskind and Connie Ozawa, "Mediated Negotiation in the Public Sector: The Planner as Mediator," *Journal of Planning Education and Research,* vol. 4, no. 1 (1984); see also Susan Silbey and Austin Sarat, "Dispute Processing in Law and Legal Scholarship: From Institutional Critique to the Reconstruction of the Juridical Subject," *Denver University Law Review,* vol. 66, no. 3 (1989) for critique of needs.

6. Jay Folberg and Alison Taylor, *Mediation: A Comprehensive Guide to Resolving Conflicts without Litigation* (San Francisco: Jossey-Bass, 1984).

7. Larry Susskind and Connie Ozawa, "Mediating Negotiation in the Public Sector: Mediator Accountability and the Public Interest Problem," *American Behavioral Scientist,* vol. 27, no. 2 (1983).

8. Scott Mernitz, *Mediation of Environmental Disputes: A Sourcebook* (Praeger, 1980); Susskind and Ozawa, "Mediating Negotiation in the Public Sector"; Jane McCarthy and Alice Shorett, *Negotiating Settlements: A Guide to Environmental Mediation* (New York: American Arbitration Association, 1984); Thomas Gunton and Sarah Flynn, "Resolving Environmental Conflict: The Role of Mediation and Negotiation," *Environment,* vol. 21, no. 3 (1992).

9. Daniel Fiorino, "Citizen Participation and Environmental Risk: A Survey of Institutional Mechanisms," *Science, Technology and Human Values,* vol. 15, no. 2 (1990), pp. 226–243.

10. E.A. Lind and Tom Tyler, *The Social Psychology of Procedural Justice* (Plenum, 1988); Tom Tyler, "What Is Procedural Justice? Criteria Used by Citizens To Assess the Fairness of Legal Procedures," *Law and Society Review,* vol. 22, no. 2 (1988).

11. Lind and Tyler, *The Social Psychology of Procedural Justice;* Tom Tyler, "The Quality of Dispute Resolution Processes and Outcomes: Measurement Problems and Possibilities," *Denver University Law Review,* vol. 66, no. 3 (1988).

12. James E. Crowfoot and Julia M. Wondolleck, *Environmental Disputes: Community Involvement in Conflict Resolution* (Washington, DC: Island Press, 1990).

13. Folberg and Taylor, *Mediation;* Albie Davis and Richard A. Salem, "Dealing with Power Imbalances in the Mediation of Interpersonal Disputes," *Mediation Quarterly,* vol. 6 (1984); Bernard Mayer, "The Dynamics of Power in Mediation and Negotiation," *Mediation Quarterly,* vol. 16 (1987); Edward W. Schwerin, *Mediation, Citizen Empowerment, and Transformational Politics* (Westport, CT: Praeger, 1995).

14. Sally E. Merry, "Defining 'Success' in the Neighborhood Justice Movement," in R. Tomasic and M.M. Feeley, eds., *Neighborhood Justice: Assessment of an Emerging Idea* (New York: Longman, 1982), pp. 172–193; Jennifer Beer, *Peacemaking in Your Neighborhood: Mediator's Handbook* (Philadelphia: Friends Suburban Project, 1987); Sara Cobb, "Empowerment and Mediation: A Narrative Perspective," *Negotiation Journal,* vol. 9, no. 3 (1993), pp. 245–255; N. Milner and Sally E. Merry, *The Possibilities of Popular Justice* (University of Michigan Press, 1993).

15. Larry Susskind and J. Cruikshank, *Breaking the Impasse* (Basic Books, 1987); Crowfoot and Wondolleck, *Environmental Disputes.*

16. R.A. Baruch Bush and J. Folger, *The Promise of Mediation: Responding to Conflict through Empowerment and Recognition* (San Francisco: Jossey-Bass, 1994).

17. Lon Fuller, "Mediation—Its Forms and Functions," in Lon L. Fuller and Kenneth Winston, eds., *The Principles of Social Order, revised ed.* (New York: International Specialized Book Services, 2002), p. 144.

18. Heidi Burgess and Guy Burgess, "Constructive Confrontation: A Transformative Approach to Intractable Conflicts," *Mediation Quarterly,* vol. 13, no. 4 (1996); Chris Maser, *Resolving Environmental Conflict: Toward Sustainable Community Development* (Delray Beach, FL: St. Lucie Press, 1996); Barnett Pearce and Stephen Littlejohn, *Moral Conflict: When Social Worlds Collide* (Thousand Oaks, CA: Sage Publications, 1997); see also Chapter 5 in this volume by Marcia Caton-Campbell.

19. John Murray and Michael Nobleza, "Building Trust: Lessons from the Nile River Basin," paper presented at the Syracuse University–Indiana University Greenberg House Conference, "Evaluating Environmental Conflict Resolution," March 2001.

20. Jane E. McCarthy, "Resolving Environmental Conflicts," *Environmental Science and Technology,* vol. 10 (1976).

21. James L. Creighton, *Public Involvement Manual: Involving the Public in Water and Power Resource Discussions* (Government Printing Office, 1980); Michelle LeBaron and Rosemary Romero, "Exploring Symbolic Successes in Public Policy Dispute Resolution" (Washington, DC: RESOLVE), no. 28 (1997).

22. Susan Carpenter and W.J.D. Kennedy, *Managing Public Disputes: A Practical Guide to Handling Conflict and Reaching Agreements* (San Francisco: Jossey-Bass, 1989); Crowfoot and Wondolleck, *Environmental Disputes.*

23. Robert Putnam, "Bowling Alone: America's Declining Social Capital," *Journal of Democracy,* vol. 6 (1995).

24. Susskind and Ozawa, "Mediating Negotiation in the Public Sector"; Franklin Dukes, *Public Disputes: Transforming Community and Governance* (New York: St. Martin's Press, 1996); Connie Ozawa and Susan Podziba, "Social Learning and the Building of Social Capital," no. 28 (Washington, DC: RESOLVE, 1997).

25. J. Laue and G. Cormick, "The Ethics of Intervention in Community Disputes," *The Ethics of Social Intervention* (Washington, DC: Halsted Press, 1978).

26. Merry, "Defining 'Success' in the Neighborhood Justice Movement"; Beer, *Peacemaking in Your Neighborhood;* Cobb, "Empowerment and Mediation"; Milner and Merry, *The Possibilities of Popular Justice;* Dukes, *Public Disputes.*

27. Crowfoot and Wondolleck, *Environmental Disputes,* pp. 254–259.

28. Barbara Gray, *Collaborating: Finding Common Ground for Multiparty Problems* (San Francisco: Jossey-Bass, 1989).

29. Dukes, *Public Disputes,* p. 172.

30. Ibid.

31. Judith E. Innes, "Evaluating Consensus Building," in Lawrence Susskind, Sarah McKearnan, and Jennifer Thomas-Larmer, eds., *The Consensus Building Handbook: A Comprehensive Guide to Reaching Agreement* (Thousand Oaks, CA: Sage Publications, 1999), p. 635.

32. Lawrence S. Bacow and Michael Wheeler, *Environmental Dispute Resolution* (Plenum Press, 1984); Tamra Pearson d'Estrée and Bonnie Colby, *Guidebook for Analyzing Success in Environmental Conflict Resolution Cases* (Fairfax, VA: Institute for Conflict Analysis and Resolution, George Mason University, 2000); Allisa J. Stern,

The Process of Business/Environmental Collaborations: Partnering for Sustainability (Westport, CT: Quorum, 2000).

33. Douglas Amy, *The Politics of Environmental Mediation* (Columbia University, 1987).

34. Robert Ellickson, *Order without Law: How Neighbors Settle Disputes* (Harvard University Press, 1991); Gary Coglianese, "The Limits of Consensus," *Environment,* vol. 41, no. 3 (1999).

35. Barry Rabe, "The Politics of Environmental Dispute Resolution," *Policy Studies Journal,* vol. 16 (1988), pp. 585–601.

36. U.S. Environmental Protection Agency website, www.epa.gov/science1/fiscal01.htm (accessed January 17, 2003).

37. Alex Conley and Ann Moote, "Collaborative Conservation in Theory and Practice: A Literature Review" (Udall Center for Studies in Public Policy, University of Arizona, February 2001).

38. Cornelius Kerwin and Laura Langbein, "An Evaluation of Negotiated Rulemaking at the Environmental Protection Agency, Phase I" (Washington DC: Administrative Conference of the United States, 1995).

39. Bingham, *Resolving Environmental Disputes.*

40. Neil Sipe and Bruce Stiftel, "Mediating Environmental Enforcement Disputes," *Environmental Impact Assessment Review,* vol. 15 (1995); Rosemary O'Leary and Susan Summers Raines, "Lessons Learned from Two Decades of Alternative Dispute Resolution Programs and Processes at the U.S. Environmental Protection Agency," *Public Administration Review,* vol. 61, no. 6 (2001), pp. 661–671.

41. Leonard G. Buckle and Suzann R. Thomas-Buckle, "Placing Environmental Mediation in Context: Lessons from 'Failed' Mediations," *Environmental Impact Assessment Review,* vol. 6 (March 1986), pp. 55–70; Sipe and Stiftel, "Mediating Environmental Enforcement Disputes"; John S. Andrew, "Examining the Claims of Environmental ADR," *Journal of Planning Education and Research,* vol. 21 (2001).

42. Buckle and Thomas-Buckle, "Placing Environmental Mediation in Context"; Andrew, "Examining the Claims of Environmental ADR."

43. R.F. Cook, J.A. Roehl, and D. Sheppard, *Neighborhood Justice Field Test: Final Evaluation Report* (Government Printing Office, 1980).

44. Sipe and Stiftel, "Mediating Environmental Enforcement Disputes."

45. Andrew, "Examining the Claims of Environmental ADR."

46. Susskind and Cruikshank, *Breaking the Impasse*; Kenneth Kressel and Dean Pruitt, "Conclusion: A Research Perspective on the Mediation of Social Conflict," in Kenneth Kressel and Dean Pruitt, eds., *Mediation and Research* (San Francisco: Jossey-Bass, 1989).

47. Bingham, *Resolving Environmental Disputes.*

48. Buckle and Thomas-Buckle, "Placing Environmental Mediation in Context."

49. Sipe and Stiftel, "Mediating Environmental Enforcement Disputes."

50. Andrew, "Examining the Claims of Environmental ADR."

51. See, e.g., Amy, *The Politics of Environmental Mediation.*

52. Nancy Kubsak and Gary Silverman, "Environmental Mediation," *American Business Law Journal,* vol. 26 (1988).

53. Cynthia Croce, "Negotiation Instead of Confrontation," *EPA Journal,* vol. 11 (1985).

54. *A Practical Guide to Consensus* (Santa Fe, NM: Policy Consensus Initiative, 1999).

55. Bingham, *Resolving Environmental Disputes,* p. 162.

56. Ibid.

57. Ibid, p. 165.

58. Andrew, "Examining the Claims of Environmental ADR."

59. Rosemary O'Leary, "Environmental Mediation: What Do We Know and How Do We Know It?" in J.W. Blackburn and W.M. Bruce, eds., *Mediating Environmental Conflicts: Theory and Practice* (Westport, CT: Quorum Books, 1995), p. 32.

60. John Forester, *The Deliberative Practitioner* (MIT Press, 1999), p. 192.

61. Ibid.

62. Innes, "Evaluating Consensus Building"; d'Estrée and Colby, *Guidebook for Analyzing Success in Environmental Conflict Resolution Cases.*

63. Innes, "Evaluating Consensus Building," p. 647.

64. Ibid.

65. Innes, "Evaluating Consensus Building."

66. Ibid, p. 650.

67. Ibid, p. 651.

68. Ibid, pp. 651–654.

69. Deborah Kolb cited in Carolyn Roth, "Overcoming the Research/Practice/ Evaluation Gap," *Forum,* vol. 32 (1997).

70. James G. March, *The Pursuit of Organizational Intelligence* (Oxford, U.K.: Blackwell, 1999).

71. Hallie Preskill and Rosalie Torres, *Evaluative Inquiry for Learning in Organizations* (Thousand Oaks, CA: Sage Publications, 1999).

72. P. Friere, *Pedagogy for the Oppressed* (New York: Seabury, 1970); C.R. Rogers, *Freedom To Learn* (Columbus, OH: Merrill, 1969).

73. Donald A. Schon, *The Reflective Practitioner* (New York: Basic Books, 1983).

74. Ibid.; Forester, *The Deliberative Practitioner.*

75. Garvin quoted in Preskill and Torres, *Evaluative Inquiry for Learning in Organizations.*

76. Preskill and Torres, *Evaluative Inquiry for Learning in Organizations.*

77. Innes, "Evaluating Consensus Building."

78. Lind and Tyler, *The Social Psychology of Procedural Justice*; Tyler, "What Is Procedural Justice?"; d'Estrée and Colby, *Guidebook for Analyzing Success in Environmental Conflict Resolution Cases.*

79. Susskind and Cruikshank, *Breaking the Impasse.*

80. Bush and Folger, *The Promise of Mediation.*

81. Innes, "Evaluating Consensus Building."

PART II

Upstream Environmental Conflict Resolution

In the 1980s, selenium poisoning of wildlife at Kesterson National Wildlife Refuge in California's San Joaquin Valley focused national attention on the problem of trace element contamination from irrigation drainage. As a result, in the 1990s the U.S. Department of the Interior (DOI) initiated the National Water Quality Irrigation Program to investigate similar wetland sites throughout the West associated with federal water projects. As the enormity of the task grew more daunting, DOI called on the National Academy of Sciences (NAS) for advice on the scientific, legal, political, economic, and public involvement issues involved in these sites.

The public involvement challenges facing DOI involved public participation as well as conflict resolution. The NAS panel of experts pointed out that effective public involvement efforts can

— help to identify areas of concern and vulnerability among potentially affected populations, interest groups, and stakeholders;
— create opportunities to generate innovative action alternatives that are more likely to meet the criteria of effectiveness as well as public acceptability; and
— increase the potential for cooperation among management agencies and affected parties.[1]

Whereas formal public hearings, written comment processes, and other conventional approaches generally satisfy legal and administrative mandates for public involvement, such formal procedures provide few real opportunities for open communication, learning, and participation of the affected public in the planning process. Participants in traditional public involvement processes are generally dissatisfied with the process, pessimistic about the degree to which their input is considered seriously, and frustrated by the inability to tell whether or how their input has affected management actions.

To avoid this situation, the NAS committee recommended public involvement that creates opportunities for personal interactions that emphasize two-way communication between and among agency participants and potentially affected or concerned stakeholders. Such interactive approaches can take a variety of forms, including conflict resolution activities, such as facilitated discussions and mediation. This section focuses on the evaluation of these forms of interactive approaches.

Thomas C. Beierle and Jerry Cayford evaluate dispute resolution as a method of public participation in Chapter 3. The authors draw on data from 239 published case studies of public involvement in environmental decision-making. Among their findings is the conclusion that decisions made through dispute resolution are sometimes revisited or rejected in the implementation phase when a broader set of actors and issues comes into play. Additional findings are presented.

Cary Coglianese asks whether satisfaction is an appropriate basis for evaluating the overall public value of different participation structures. In Chapter 4, Coglianese begins by elaborating on the use of satisfaction as a policy evaluation criterion. Next, the author explains two conceptual limitations of using participant satisfaction: it does not necessarily equal good public policy, and it is inherently limited because it excludes those who did not participate. The author closes by detailing a series of problems in applying, measuring, and interpreting participant satisfaction.

When examined as a whole, these two chapters bring together cutting-edge research concerning public participation and conflict resolution. The authors raise provocative issues. Challenging public policy agendas are presented.

Note

1. "Review of the Department of Interior's National Irrigation Water Quality Program: Planning and Remediation" (National Academy of Sciences Committee on Planning and Remediation for Irrigation-Induced Water Quality Problems, 1996).

3

Dispute Resolution as a Method of Public Participation

Thomas C. Beierle and Jerry Cayford

Over the past 30 years, public participation has moved to center stage in the play of influences that determine how the environment will be protected and managed. In doing so, it has evolved considerably. More intensive approaches to participation have joined traditional public hearings and public comment procedures. Loosely termed "dispute resolution," these processes emphasize face-to-face deliberation, problem solving, and consensus building among a relatively small group of participants selected to represent the wider public. The purpose of dispute resolution processes goes far beyond the traditional role of public participation in ensuring a minimum level of government accountability. Increasingly, dispute resolution is used as a strategy for improving the quality of decisions and dealing with the mistrust and conflict (and the resulting litigation) that are endemic to environmental policy.

Here we evaluate dispute resolution as a method of public participation. Methods for engaging the public in decisionmaking range along a continuum from informal consultations with individual citizens to highly formalized processes for seeking agreement among organized nongovernmental interest groups. Dispute resolution lies at the more formal and intensive end of the continuum. Examining dispute resolution as a form of public participation can provide insights not available from the more common practice of evaluating dispute resolution just as an alternative to litigation.[1]

To compare dispute resolution with other forms of public participation, we evaluate a large number of actual cases based on a set of public participation's social goals. These social goals embody many of the expectations and aspirations for how increased involvement of the public can improve environmental decisionmaking.[2] They are the following:

— incorporating public values into decisions,
— increasing the substantive quality of decisions,
— resolving conflict among competing interests,
— building trust in institutions, and
— educating and informing the public.

The thousands of cases in which the public has become involved in environmental policy decisions over the past three decades have produced many hundreds of documents describing what happened in the cases. We identified case studies from an extensive review of journals, books, dissertations, conference proceedings, and government reports.[3] We analyzed 239 cases of public involvement in environmental decisionmaking, coding each case for more than 100 attributes describing its context, process, and outcomes. Using the social goals as measures of success, we sought to understand what participation accomplished and what determined success.[4]

Data from the cases can be used to compare dispute resolution with other, less intensive, forms of public participation. Our definition of dispute resolution is similar to that provided by Gail Bingham's *Resolving Environmental Disputes*: a "variety of approaches that allow the parties to meet face-to-face to reach a mutually acceptable resolution of the issues in a dispute or potentially controversial situation ... [and] that involve some form of consensus building, joint problem solving, or negotiation."[5] Included in the dispute resolution category are negotiations, mediations, policy dialogues, and other small-group processes where participants explicitly seek consensus. The less intensive processes to which we compare dispute resolution include public meetings and hearings, public workshops, and advisory committees in which participants are not explicitly seeking consensus.

The comparison between dispute resolution and other forms of participation reveals an interesting tension: dispute resolution cases are clearly more effective in achieving the social goals of public participation, but only among the small group of participants. Dispute resolution processes are much less effective in using outreach to spread the benefits of education and trust formation. Dispute resolution processes may do a poor job of representing the values and interests of the wider public because participants are not representative of the wider public in terms of education, income, ethnicity, and the like, and they often put little effort into fostering broader public input into the participatory process. Dispute resolution processes therefore limit whose values are heard, whose conflicts are resolved, and whose priority issues are addressed.

The results suggest that using dispute resolution as a form of public participation entails a trade-off between success in achieving the social goals and the social significance of that achievement. In practical terms, the trade-off between success and significance means that decisions made through dispute resolution are sometimes revisited or rejected in the implementation phase when a broader set of actors and issues comes into play. In normative terms, the trade-off means that dispute resolution alone is often inadequate to make decisions on issues that engage broad public interests.

Evaluating Public Participation

The criteria we use for evaluating public participation derive from the increasingly high expectations for what public participation can accomplish in the modern environmental management system. As participation has become integral to the substance of environmental decisionmaking, governments are also using participation to achieve a variety of social goals that traditional approaches to decisionmaking have failed to meet. The range of social goals reflects the reality that public participation is not only expected to keep government accountable; it also aims to help agencies make good decisions, help resolve longstanding problems with conflict and mistrust, and build capacity for dealing with future problems. These new demands translate into the five social goals listed previously, which we use as criteria for evaluating the public participation case studies.

The goal of incorporating public values into decisions is fundamental to democracy and has been the driving force behind challenges to managerial, expert-led models of decisionmaking. Risk perception and communication literatures, for example, outline dramatic differences in the ways that lay citizens and experts view risks.[6] Their values, assumptions, and preferences differ, which implies that direct public participation would better capture the public interest than traditional bureaucratic processes alone would.[7]

The second goal, increasing the substantive quality of decisions, recognizes the public as a valuable source of knowledge and ideas for making decisions.[8] Citizens can improve the substantive quality of decisions in a number of ways, such as by identifying relevant information, discovering mistakes, or generating alternative solutions that satisfy a wider range of interests.

The third goal is resolving conflict among competing interests. The environmental regulatory system in the United States was born of conflict between environmental and industrial interests, and conflict has persisted as the system has matured. Court battles consume substantial amounts of money and energy, and environmental problems go unsolved. One of the principal arguments for dispute resolution is that collaborative decisionmaking is more likely than adversarial processes to result in lasting and more satisfying decisions.[9]

In addition to resolving conflict, public involvement can create opportunities for building trust in institutions. This fourth goal addresses the dramatic decline in public trust of government over the past 30 years.[10] It recognizes that such a loss of trust is a legitimate reaction to government mismanagement and that a healthy dose of skepticism is important for ensuring government accountability. However, as trust in the institutions responsible for solving complex environmental problems decreases, the institutions' ability to solve those problems becomes seriously limited.[11] Research suggests that agencies can try to rebuild trust through greater public involvement in and influence over decisionmaking.[12]

The goal of educating and informing the public addresses the need to build capacity by increasing public understanding of environmental problems. Education here should be interpreted as something more than science lessons. It is a more fundamental education that integrates information about the problem at hand with participants' intuition, experience, and local knowledge to develop the practical ability to contribute to collective environmental solutions.

In each case, we measured success according to how well the participants in the process achieved these five social goals. For example, if there were 10 participants, we examined how well conflict was resolved among those 10 people. If there were 100 participants, we examined how well conflict was resolved among this larger group, and so forth. To understand the significance of achieving the social goals, however, it is important to expand our vision beyond the actual, direct participants and ask how broadly through society the benefits of achieving the goals were distributed. Indeed, the social significance of meeting the goals largely depends on whether the goals are being met for only a small, select group of participants or for a broader swath of the affected public.

To gather information on how well public participation has met the social goals, we collected data from a large body of case studies on public participation. We screened more than 1,800 case studies, and ultimately identified 239 cases for intensive investigation.[13] These we studied using a "case survey" methodology. A case survey is analogous to a normal closed-end survey, except that a reader-analyst asks a standard set of questions of a written case study rather than of a person.[14] By coding the answers to these questions, we compiled data on more than 100 attributes of each case. Coded attributes included the type of environmental issue discussed in the case, characteristics of the participants, important features of the participatory process and its context, and the process's outcomes. Each attribute was assigned a score—usually low, medium, or high—based on a standard template. Each attribute score was given one of three weight-of-evidence indices, ranging from "solid evidence" to "best-informed guess." (The analysis excluded data with the lowest weight of evidence.) A written entry describing the attribute in qualitative terms also accompanied the scores.

The 239 case studies covered a wide variety of planning, management, and implementation efforts carried out by environmental and natural resource agencies at many levels of government. The cases described public participation processes ranging from informal public meetings to intensive multiparty negotiations. About one-half of the cases involved participation that could be considered dispute resolution. All cases took place in the United States during the past 30 years.

The case survey approach is not new to the public participation literature, but much refinement of the methodology comes from the business and management literature.[15] Despite its appeal and track record, the case survey approach is still somewhat experimental, and accordingly, there are a few important caveats to mention about the method and its application. The quality of the data used in a case survey is only as good as the quality of the case studies from which the data come. Because different authors report on different aspects of a process, there are inevitable data gaps, and no case has data for every attribute. (Throughout the rest of the discussion, reported percentages are relative to the total number of cases with relevant data.) Finally, there is always the threat of a biased sample of cases. This analysis accounted for some of these problems by drawing on enough cases to overcome problems with data gaps and by dropping scores with the lowest weight-of-evidence index. An analysis of bias found that, although there was some bias toward successful cases, it did not change any of our major conclusions.[16]

Public Participation's Qualified Success

Figure 3-1 shows the degree to which the variety of public participation processes employed in the 239 cases—including dispute resolution—achieved the social goals.[17] As is apparent from the figure, the results are much more positive than negative: high-scoring cases represented more than one-half of all cases coded for almost all the goals. As a group, the cases performed best on educating participants and worst on building trust in agencies, with results for incorporating public values, improving decision quality, and resolving conflict goals falling in between. The results on the five social goals paint a positive picture of public involvement in environmental decisionmaking, as seen through the case study record.

The story could end here, with public participation demonstrating its capacity to achieve a number of desirable social goals. But how significant has that achievement actually been? That is, how broadly have the benefits of participation been felt among the wider public as well as among the participants themselves? We can expand the scope of analysis beyond the actual participants by considering a second set of evaluation criteria:

— socioeconomic representativeness,
— the extent of consultation with the wider public,
— the extent to which parties were left out of the process,

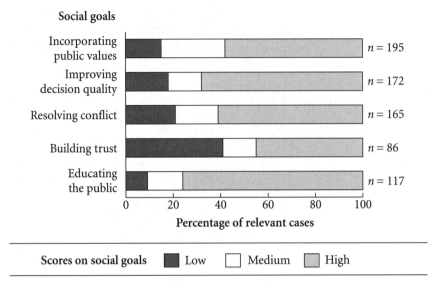

FIGURE 3-1. Scores on Social Goals

Note: n = the number of cases scored.

— the extent to which issues were avoided during the process, and
— the extent of educational outreach to the wider public.

As shown in Figure 3-2, the second set of evaluation criteria paints a much less positive picture of the 239 public participation cases. In only one-quarter of cases were participants socioeconomically representative of the public they sought to represent. Perhaps more importantly, in fewer than one-half the cases was there a concerted effort to consult the wider public outside the often small group of actual participants. Fewer than one-half of the cases involved all the parties with an interest in a particular decision, and only slightly more than one-half dealt with all the relevant issues. Educational outreach to the wider public was successful in only about one-quarter of the cases.

The results in Figure 3-2 are significant qualifications to the successes illustrated in Figure 3-1. Although almost 60% of the cases successfully incorporated public values into decisions, in many cases only the participants' values, not the wider public's, were represented. About 60% of the cases were also successful in reducing conflict, but many did so by excluding certain parties or leaving particular issues off the table. Although many cases were successful in educating participants, limited outreach meant that little information made its way to the wider public. Likewise, participants reported increased trust in lead agencies in 45% of cases, but the lack of outreach meant that little of this trust formation filtered out to the wider public, where widespread mistrust of government resides.

Qualifications to social goals

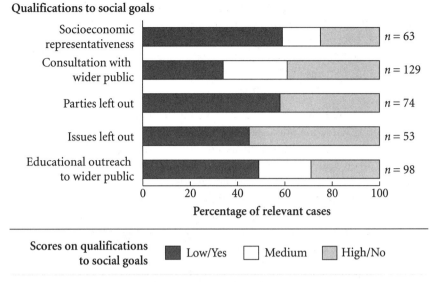

FIGURE 3-2. Scores on Qualifications to Social Goals

Notes: n = the number of cases scored. The scores for "parties left out" and "issues left out" were coded as yes or no, where "yes" corresponds to a low score and "no" corresponds to a high score.

Dispute Resolution and the Significance of Success

The pattern of successful public participation—and the qualifications to its success—is largely attributable to dispute resolution cases. Such cases clearly achieved the social goals more successfully among participants than did other forms of public participation, but they were less successful in engaging or representing the wider public.

Before comparing the success of dispute resolution processes with the success of other types of participation, it is useful to describe each type in detail. About one-half of the cases in the data set can be described as dispute resolution cases. They included negotiations, mediations, and other consensus-based processes. In these cases, parties either formulated agreements that bound their organizations to particular courses of action or recommended policies that they all could agree to support. Examples of dispute resolution cases in the dataset included the following:

— regulatory negotiations on a variety of topics, from reformulated gasoline to wood-burning stoves;

— mediations on the siting and operation of industrial facilities, hazardous waste cleanups, or habitat management plans;

— consensus-based resource management groups organizing around rivers or watersheds;

— policy dialogues on topics ranging from oyster harvesting in the Chesapeake Bay to broad national environmental policy goals; and

— advisory committees seeking consensus on the cleanup of U.S. Department of Energy facilities, requirements for facilities under the U.S. Environmental Protection Agency's regulatory flexibility programs, or risk priorities for states and counties.

Approximately one-half of the cases involve other, less intensive, types of participation. These included public consultations, public hearings and meetings, workshops, and nonconsensus advisory committees. These processes primarily involved information exchange and did not seek agreements among parties. Agencies informed citizens about their activities, and citizens provided input and opinions on agency policy. Agencies had an implicit obligation to review the information obtained through these processes, but in most cases, there was little commitment to sharing significant decisionmaking authority with the public. Examples from the dataset included these cases:

— informal public consultations on transportation planning in Boulder, Colorado, and flood control near San Francisco;

— public hearings on Army Corps of Engineers projects, statewide water quality planning, and hunting and fishing policy;

— public workshops on coastal land management in Oregon and power plant siting in California; and

— advisory committees providing input on demand-side management policies for public utilities.

Figure 3-3 compares the success of dispute resolution processes with that of other forms of public participation. Across each of the five goals, a much higher percentage of dispute resolution cases were successful (that is, they scored high on the achievement of the social goals).[18]

The dispute resolution cases, however, performed worse on three of the measures concerned with engaging the wider public than did other public participation, as shown in Figure 3-4. Participants in dispute resolution processes were much less likely to be socioeconomically representative of the wider public. They were also much less likely to seek input from the wider public through surveys, consultations, or supplementary public meetings. Educational outreach was also somewhat worse in the dispute resolution cases.[19] Outreach was particularly poor in the most intensive dispute resolution cases, those involving negotiations and mediations. Only 15% of these cases had any effective outreach. In some of these negotiations and mediations, outreach was actively discouraged to protect sensitive negotiations, but in most cases it simply was not considered important.

We can attribute the overall tendency of the public participation cases to resolve conflict by leaving out participants or ignoring issues (see Figure 3-2)

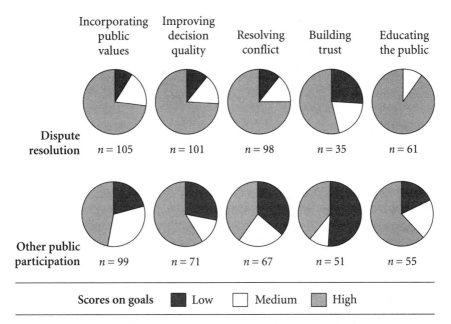

FIGURE 3-3. Social Goals—Dispute Resolution vs. Other Forms of Public Participation
Note: n = the number of cases scored.

to dispute resolution cases as well. (There was insufficient data to determine whether the less intensive participation processes had a better record.) Reaching a consensus is much easier when all the parties (or those still at the table when it is finally time to agree) already think alike or when the issues on which they do not agree are tabled for future discussions. Indeed, the exclusion of certain groups, the departure of dissenting parties, or the avoidance of particular issues ultimately made consensus possible—or at least easier—in 36% of the 73 dispute resolution cases in which conflict was resolved.

Dispute Resolution's Virtues and Vices

Among the dispute resolution cases, the high degree of effectiveness in influencing decisions, resolving conflict, building trust, and educating a small group of participants often came at the expense of engaging or adequately representing the wider public. The narrow success of dispute resolution cases is clear, but the wider significance of that success is not. Interestingly, the very nature of dispute resolution is what leads to this tension. In short, dispute resolution's virtues lead to its vices.

Dispute resolution's first virtue is the effective use of deliberation to identify common values, generate ideas, and solve problems. Intensive delibera-

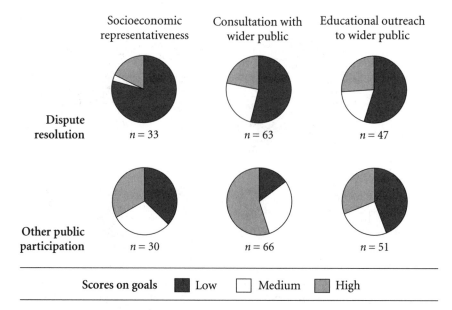

FIGURE 3-4. Qualifications to Social Goals—Dispute Resolution vs. Other Forms of Public Participation

Note: n = the number of cases scored.

tions are feasible only among relatively small groups of people. Limiting participation to a small group entails a selection process. Whereas less intensive processes are often open to all interested individuals or have relatively loose selection criteria, dispute resolution processes typically have much stricter and more exclusive processes for selecting participants.

The second virtue of dispute resolution is the high-level capacity of those selected to participate. The participants in dispute resolution cases are typically professional representatives of organized interest groups or other entities; they speak for the views of those they represent and make commitments on their behalf. These participants are far more likely to have experience in participatory processes, to have technical training in the issues at hand, and to have more experience in interest-group politics. However, there is a trade-off between capacity and representativeness. High-capacity nongovernmental actors are simply not like the average citizen in terms of income, education, type of employment, or other socioeconomic characteristics.

Finally, dispute resolution processes seek to forge agreements, either on a set of recommendations for a lead agency or for actions that the participants themselves commit to take. The need to come to agreement, which usually translates into developing consensus, can lead to a number of characteristic problems. We have already mentioned the tendency to exclude particularly conflictive actors or contentious issues. However, the need to seek consensus can also limit outreach. Negotiation theory has long acknowledged that outside intervention—whether in the form of public pressure, press coverage, political involvement, or otherwise—can dramatically complicate the negotiating dynamic. There is pressure, then, to keep issues among the select few negotiators and out of the public view. Thus, the need to forge an agreement through sensitive negotiations can be a barrier to sharing information with the wider public.

Implications

Dispute resolution's trade-off of wider significance for narrower success has both normative and practical implications. On the normative side, project planners need to assess what kind of participation is most appropriate for a given decision. Whether to use an effective problem-solving group that sacrifices engagement of the wider public or a broadly representative process that may lack somewhat in capacity will largely depend on the goals of the public participation effort and the nature of the problem under discussion. Project planners need to ask, "Does this issue involve a broad public interest or a relatively narrow set of private interests? Are problems with conflict or mistrust pervasive in a community, or can they be dealt with through interaction among a handful of organized interest groups? Are there larger public values and interests at stake that cannot be represented by a select few participants?"

The implications of choosing dispute resolution are not just normative, however. Using dispute resolution inappropriately can come back to haunt projects in the implementation phase. Implementation opens the door to wider political forces, giving those not involved in decisionmaking an opportunity to make their views known and to support or thwart the decisions made. Two case studies illustrate this point.

A classic case of a dispute resolution process that was internally effective but failed to garner the wider support necessary for implementation is the U.S. Environmental Protection Agency's reformulated gasoline regulatory negotiation. The aim of the negotiation was to bring business, environmental, and public interest groups together with regulators to develop rules on the use of reformulated gasoline to reduce urban smog, pursuant to the 1990 Clean Air Act Amendments. Even though many of the parties were traditional adversaries, over the course of the negotiations they were able to find creative solutions to their differences. They agreed on rules that were arguably more cost-effective and more satisfying to the range of interests involved than would have been likely otherwise.[20] However, the regulations ultimately generated a massive outcry from the people who were never consulted but would have to pay for the regulations: the driving public.

> An EPA hearing [in Milwaukee] resulted in the expression of tremendous anger by the consuming public against the regulation imposed by the national government. The general perception was that consumers were the victims of a rigged inside game that excluded the "small guy," leaving the American motoring public to foot the bill for urban smog.[21]

The same pattern can play out locally as well, as when a dispute resolution process was convened to deal with deer management near Rochester, New York. The Citizen Task Force on Deer Management, made up of 11 stakeholders selected by the state conservation agency, arrived at a consensus on recommendations for controlling deer populations. The state's Department of Environmental Conservation ultimately adopted the group's recommendations, which it made with minimal outside input. According to the author of the case study, the press conference at which the recommendations were presented drew protesters, and "before the plan was implemented, those opposed entered a court injunction, which delayed implementation. ... Later, two households in support of the deer management plan were threatened by fake pipe bombs on their property."[22] Pipe bombs, fake or not, are undoubtedly good indications that the dispute resolution work of the task force did not resolve the broader conflict in the community.

Conclusions

The reformulated gasoline regulatory negotiation and the Citizen Task Force on Deer Management illustrate a larger pattern among cases of dispute resolution. Although dispute resolution approaches are clearly more effective for achieving the social goals of public participation, that achievement is often largely confined to the actual participants. Dispute resolution frequently succeeds in making decisions that are responsive to participants' values and are substantively robust. It resolves conflict among participants, builds trust between participants and agencies, and increases participants' knowledge about the environment. Dispute resolution is less successful than other forms of public participation, however, in engaging or representing the wider public. Participants in dispute resolution cases are likely to be unrepresentative of those they are supposed to represent. Efforts to draw wide public values into decisionmaking or to provide educational outreach are often limited. In many cases, participants in dispute resolution reach their goal of consensus partly by excluding certain parties and leaving out particularly contentious issues.

We can draw four lessons for those who design dispute resolution processes and those who evaluate them. *The first lesson deals with the decision of whether to use dispute resolution in a particular case.* Such a choice should depend on the nature of the issue under discussion and the specific objectives of the public participation effort. A partial list of criteria for deciding that dispute resolution is the appropriate form of public participation to use would include the following:

— The issues are complex enough to warrant a small-group, deliberative process.
— The issues involve mostly private interests rather than a broader public interest.
— Accountable representatives of the private interests can be gathered together in a small group.
— Implementation of the agreement depends mainly on the participants and the organizations they represent.

It may be true that relatively few environmental issues share these criteria, which leads to a second lesson—and a challenge—for practitioners. In cases that do not meet the above criteria, *dispute resolution* may still be appropriate, but it *should not be the only approach to public participation.* Practitioners need to find ways to combine the problem-solving capabilities of dispute resolution processes with the broad public involvement often found in less intensive participatory processes. One approach would be emphasizing the responsibility of participants in dispute resolution processes to communicate with, and be accountable to, their broader constituencies. However, because the affected public may not be organized into formal constituencies, most situations probably require the combination of different forms of participation, so that participants in one process can analyze, and ideally ratify, decisions made by those in another.

The *third lesson deals with the scope of analysis,* and it is directed more toward evaluators than toward practitioners. The analysis presented here has demonstrated the benefits of looking at dispute resolution not just as an alternative to litigation, but as a form of public participation. By broadening the scope of analysis, we can identify failings and opportunities that would not be visible otherwise. In addition, the examination of dispute resolution needs to look not only at the actions of actual participants, but also at who is not participating and how the benefits of the process extend, or not, to this larger group. The examination also needs to extend beyond the process itself and look at the extent to which parties actually implement the agreements. Only by including the nonparticipating public and the record of implementation in the scope of analysis can we see how choices made in the dispute resolution process ultimately play out.

The final lesson encompasses the overall success of public participation. The various public participation processes can, and often do, achieve demonstrable benefits for environmental management. These benefits go beyond the traditional goal of accountability and even beyond the procedural benefits of introducing greater democracy into public decisionmaking. The benefits are precisely those that have eluded other, less participatory decisionmaking processes in the past: incorporating public values, improving the quality of decisions, resolving conflict, building trust, and educating and informing the public. That some processes do not achieve these goals, or that the benefits of achieving them do not extend broadly, are not indictments of public participation. They are merely challenges to think creatively about how to design and undertake effective participation efforts in the future.

Notes

1. J. Clarence Davies, "Environmental ADR and Public Participation," *Valparaiso University Law Review,* vol. 34, no. 2 (2000), pp. 389–401.

2. For a more complete discussion of public participation's social goals, see Thomas C. Beierle, "Using Social Goals To Evaluate Public Participation in Environmental Decisions," *Policy Studies Review,* vol. 16, no. 3 (1999), pp. 75–103.

3. We should note that these cases cover only part of what might be considered under a broad definition of "public participation." We do not, for example, examine voting, lobbying, citizen suits, market choices, protest, or other methods by which citizens can influence public policy.

4. The complete results of the research are contained in Thomas C. Beierle and Jerry Cayford, *Democracy in Practice: Public Participation in Environmental Decisions* (Washington, DC: RFF Press, 2002).

5. Gail Bingham, *Resolving Environmental Disputes: A Decade of Experience* (Washington, DC: Conservation Foundation, 1986), p. xv.

6. For a discussion of public and expert risk perception, see Paul Slovic, "Perceptions of Risk: Reflections on the Psychometric Paradigm," in Sheldon Krimsky and Alonzo Plough, eds., *Social Theories of Risk* (Westport, CT: Praeger, 1992).

7. For an argument about the need to address differences in values through participatory processes, see Paul C. Stern and Harvey V. Fineberg, eds., *Understanding Risk: Informing Decisions in a Democratic Society* (Washington, DC: National Academy Press, 1996).

8. Daniel J. Fiorino, "Citizen Participation and Environmental Risk: A Survey of Institutional Mechanisms," *Science, Technology and Human Values,* vol. 15, no. 2 (1996), pp. 226–243; Carolyn Raffensperger, "Guess Who's Coming to Dinner: The Scientist and the Public Making Good Environmental Decisions," *Human Ecology Review,* vol. 5, no. 1 (1998); Stern and Fineberg, *Understanding Risk.*

9. Lawrence Susskind and Jeffrey Cruikshank, *Breaking the Impasse: Consensual Approaches to Resolving Public Disputes* (New York: Basic Books, 1987).

10. For evidence on the decline in trust, see Pew Research Center, *Deconstructing Trust: How Americans View Government* (Washington, DC: Pew Research Center, 1998).

11. For a discussion of the challenges of administering environmental programs in the presence of low public trust, see William D. Ruckelshaus, *Trust in Government:*

A Prescription for Restoration (Washington, DC: National Academy of Public Administration, 1996).

12. Paul Slovic, "Perceived Risk, Trust, and Democracy," *Risk Analysis,* vol. 13, no. 6 (1993), pp. 675–682; Mark Schneider, Paul Teske, and Melissa Marschall, "Institutional Arrangements and the Creation of Social Capital: The Effects of Public School Choice," *American Political Science Review,* vol. 91, no. 1 (1997), pp. 82–93.

13. Case studies were screened based on the following criteria: dealt with public involvement in environmental decisionmaking, generally at the administrative level; occurred in the United States; occurred since 1970; had an identifiable lead (or otherwise interested) government agency; described a discrete mechanism (or set of mechanisms) used to engage the public; described participation of nongovernmental citizens other than regulated parties; and contained sufficient information on context, process, and outcomes.

14. A number of sources describe the case survey methodology, including William A. Lucas, *The Case Survey Method: Aggregating Case Experience,* Report No. R–1515–RC (Santa Monica, CA: RAND, 1974); Robert K. Yin and Karen A. Heald, "Using the Case Survey Method To Analyze Policy Studies," *Administrative Science Quarterly,* vol. 20 (1975), pp. 371–381; R.J. Bullock and Mark E. Tubbs, "The Case Meta-Analysis Method for OD," *Research in Organizational Change and Development,* vol. 1 (1987), pp. 171–228; Rikard A. Larsson, "Case Survey Methodology: Quantitative Analysis of Patterns across Case Studies," *Academy of Management Journal,* vol. 36, no. 6 (1993), pp. 1515–1546. Following guidance from these sources, one of three researchers, or pairs of researchers, coded each case. To ensure consistent coding among researchers, we used an intercoder reliability testing and training process. This involved pairs of researchers reading and coding the same subset of case studies independently and then comparing codes. The standard required coders to consistently achieve two-thirds agreement, a level of reliability regarded as satisfactory in the literature. About 10% of cases underwent this intercoder reliability process.

15. Concerning public participation specifically is Jeffrey K. Berry, Kent Portney, and Ken Thompson, "Public Involvement in Administration: The Structural Determinants of Effective Citizen Participation," *Journal of Voluntary Action Research,* vol. 13 (1984), pp. 7–23. A study of the related topic of nongovernmental administration of public programs is Yin and Heald, "Using the Case Survey Method To Analyze Policy Studies." Examples of case survey methodology in the business and management literature include Lawrence Jauch, Richard Osborn, and Thomas Martin, "Structured Content Analysis of Cases: A Complementary Method for Organizational Research," *Academy of Management Review,* vol. 5, no. 4 (1980), pp. 517–525; and Henry Mintzberg, Duru Raisinghani, and Andre Theoret, "The Structure of 'Unstructured' Decision Processes," *Administrative Science Quarterly,* vol. 21 (1976), pp. 246–275.

16. A complete discussion of possible bias is contained in Beierle and Cayford, *Democracy in Practice.*

17. Not every case was coded for every goal. For example, 86 cases were coded for building trust, and 195 cases were coded for incorporating public values into decisions, by counting only those cases with moderate- to high-quality data. About 95% of the cases received scores for at least one social goal, and almost 70% received scores for three or more.

18. All of the differences between the dispute resolution cases and the cases that used other forms of participation are statistically significant using standard statistical tests. Statistical significance was determined using a chi-squared test. The significance of the differences between sets of cases was less than 0.05 for all but one of the goals; it was 0.054 for the building trust goal.

19. The differences between the dispute resolution cases and those using other forms of public participation were statistically significant ($p < 0.05$) for "socioeconomic representativeness" and "consultation with wider public" but not for "educational outreach to wider public." The difference between negotiation and mediation cases and all other forms of public participation on "educational outreach to wider public" was moderately statistically significant ($p < 0.10$).

20. Edward P. Weber, "Negotiating Effective Regulation," *Environment,* vol. 40, no. 9 (1998), p. 15.

21. Edward P. Weber and Anne M. Khademian, "From Agitation to Collaboration: Clearing the Air through Negotiation," *Public Administration Review,* vol. 57, no. 5 (1997), p. 407.

22. Rebecca J. Stout and Barbara A. Knuth, *Evaluation of a Citizen Task Force Approach To Resolve Suburban Deer Management Issues,* HDRU [Human Dimensions Research Unit] Series No. 94–3 (Cornell University, August 1994), p. 62.

4

Is Satisfaction Success?
Evaluating Public Participation in Regulatory Policymaking

CARY COGLIANESE

Dispute resolution seeks to find satisfactory solutions to conflicts, and researchers who evaluate dispute resolution procedures understandably want to consider whether disputants using these procedures are eventually satisfied with the resulting outcomes. A similar emphasis on satisfaction pervades the literature on techniques for resolving disputes and involving the public in regulatory policymaking. These techniques include the broad range of procedures and methods available to government for allowing input, feedback, and dialogue on regulatory policymaking, including comment solicitation, public hearings, workshops, dialogue groups, advisory committees, and negotiated rulemaking processes.

Researchers evaluating these various techniques have often used participant satisfaction as a key evaluative criterion. This criterion may seem suitable for evaluating private dispute resolution techniques, but those who disagree in policymaking processes are not disputants in the same sense that landlords and tenants, creditors and debtors, or tortfeasors and victims are disputants in private life. Disputes in regulatory policymaking arise over *public* policy, not over *private* grievances.[1] Yet in recent years, various public participation techniques, such as regulatory negotiation, have been advocated specifically as means of reducing conflict and increasing the acceptance of regulatory policy.[2] Moreover, policies that satisfy participants in regulatory policymaking are often assumed to be better policies. Numerous studies purport to evaluate the success of public participation by asking individuals what they think of the regulatory processes and outcomes in which they were involved.

Data on participant satisfaction can potentially provide useful feedback to those who facilitate various kinds of policy deliberations, but is satisfaction an appropriate basis for evaluating the overall public value of different participation methods? Even though data can be collected relatively easily by distributing surveys to participants or by interviewing them, researchers need to rely on data that can provide a meaningful basis for evaluating public participation in regulatory policymaking. What may be appropriate for evaluating dispute resolution techniques used to address private grievances is not necessarily appropriate for assessing public participation and dispute resolution in regulatory policymaking, which not only affects those individuals who happen to participate in regulatory proceedings but also affects the broader public.

This chapter raises caution about using participant satisfaction, or other measures based on participants' attitudes and opinions, in evaluating dispute resolution and public participation in regulatory policymaking. I first elaborate on the use of satisfaction as a policy evaluation criterion and illustrate the role it has come to play in research evaluating public participation in regulatory policymaking, most particularly in the area of environmental and natural resources policy. I then draw out two conceptual limitations on the use of participant satisfaction, namely that (a) satisfaction does not necessarily equate with good public policy and (b) participant satisfaction is at best an incomplete measure because it excludes those who do not participate. Finally, I detail a series of problems in applying, measuring, and interpreting participant satisfaction that make it a problematic metric for evaluating public participation in regulatory processes. In light of the conceptual and measurement problems with relying on satisfaction, evaluation researchers should resist relying on participant surveys to evaluate public participation techniques and instead should focus attention directly on the effectiveness, efficiency, and equity of the decisions that result from different forms of public participation.[3]

Satisfaction as an Evaluative Criterion

Many [evaluations of consensus building] have used surveys or in-depth interviews to assess participants' levels of satisfaction and their perceptions about what was achieved.[4]

It is not surprising that researchers have relied on measures of satisfaction to evaluate different techniques of public participation. Satisfaction is not only relatively easy to measure, but in various guises it is central to policy evaluation and analysis. Welfare economics, for example, focuses on how policies affect individual preferences, and the Pareto optimality criterion explicitly emphasizes that people should be made better off.[5] Negotiation theory takes a similar approach, seeking win–win opportunities that extend the Pareto frontier and satisfy preferences to the greatest extent possible without making anyone worse off.[6] According to one common conception of democratic theory, public decisionmaking is all about the aggregation of—and ultimately the satisfaction of—public preferences.[7] Public opinion polling has long sought explicitly to measure the general public's satisfaction with a variety of issues. More recently, public managers and public administration scholars have appropriated private-sector customer satisfaction models and applied them as the standard for assessing government-run services.[8] Attention to satisfaction and concepts related to satisfaction would therefore appear to have a firm basis in a number of social science and evaluation fields. Indeed, satisfaction appeals to common sense, too. After all, is it not a good thing if a policy satisfies people, rather than causing them to complain?

When evaluating different approaches to public participation, researchers have apparently thought the answer to this question is "yes." Evaluations of public participation and dispute resolution strategies increasingly rely explicitly on satisfaction, as well as on surveys that measure participant opinions or attitudes. Several recent applications of this evaluation approach illustrate how researchers and government agencies are using satisfaction to study participatory procedures.

Perhaps the starkest example is a recent study that uses a "participant satisfaction scorecard," a survey instrument that asks participants in participatory processes to rate their satisfaction along a number of dimensions.[9] Researchers at the Montana Consensus Council (MCC), a state agency that supports dispute resolution efforts, developed the participant satisfaction scorecard. MCC researchers surveyed a nonrandom sample of 280 participants in environmental impact assessment proceedings (including government officials) and received responses from about 93 participants (a response rate of about 33%). The overwhelming majority of respondents (88%) viewed the public participation processes in which they were involved favorably, and 74% reported that public participation improved the government project overall. The MCC researchers suggested that the participant satisfaction scorecard results showed that the participatory processes it studied worked well.[10]

Another recent study relied on interviews with participants in eight U.S. Environmental Protection Agency (EPA) negotiated rulemakings, as well as with individuals who submitted to the agency written comments in six conventional, nonnegotiated rulemakings.[11] The authors of the study compiled interview data from 101 participants (including government officials) in negotiated rulemakings and 51 individuals (none of whom were government officials) who filed comments about conventional rulemakings. They found that "participants in reg-negs [regulatory negotiations] are more satisfied with the overall process than are participants in conventional rulemaking."[12]

In addition, participants in negotiated rulemaking reportedly gave negotiated rules better ratings along a number of dimensions, including economic efficiency, the quality of scientific analysis, and use of appropriate technologies.

In yet another study, political scientist Mark Lubell analyzed results from a survey of participants and other interested parties active in 20 estuaries included in the EPA's National Estuary Program (NEP), a collaborative program for managing coastal ecosystems where freshwater merges with saltwater.[13] The NEP brought together public and private actors at local, state, and federal levels for collaborative management of estuaries. Lubell compared the results of a survey of NEP participants with those of a comparable set of actors involved in 10 estuaries that were not part of the NEP. Lubell's research showed that, all things being equal, the more collaborative NEP approach to public participation resulted in higher levels of attitudinal support, or "general satisfaction."

Of course, some research shows that participants have been dissatisfied with regulatory processes, even after extensive opportunities for collaboration.[14] For example, in the mid-1990s, the National Marine Fisheries Service (NMFS) convened several negotiation processes to develop plans for reducing the accidental capture of protected marine mammals in fishing nets. Participants in the negotiations included fishermen, conservation groups, researchers, and state and federal regulators. After the final meeting, facilitators surveyed participants to find out how satisfied they were with the negotiation processes and their outcomes.[15] Strikingly, about 60% of the participants reported that they were dissatisfied with the results of the process.[16] The study's authors surmised that the dissatisfaction arose in part because participants thought that they would have much more control over the decisions NMFS made than was realistically—and legally—possible.

Although many researchers have relied on participant satisfaction, others have recognized that there are other ways to conceptualize the success of public participation besides relying solely on participant satisfaction. Thomas Beierle, for example, has recently articulated six social values served by various forms of public participation: (a) educating the public, (b) incorporating public values into policymaking, (c) improving the substantive quality of public policy, (d) increasing public trust, (e) reducing conflict, and (f) achieving cost-effective public policy.[17] Nevertheless, even when success has been conceived more broadly, researchers have still surveyed participants to measure these additional dimensions of success. As already noted, the Langbein and Kerwin study asked participants to rate policy outcomes along several dimensions. In another study, researchers identified six attributes of success for evaluating public participation at nine environmental remediation sites in the United States and then used survey responses from the participants in public outreach and dialogue sessions to measure perceptions about these different attributes.[18]

Even when researchers invoke other attributes of success, they have sometimes directly equated participant satisfaction with these other attributes. Participant satisfaction, for example, is often considered an indicator of or proxy for the quality of regulatory policy. In claiming that consensus building produces high-quality decisions, Judith Innes has written that "[t]hese agreements are typically of higher quality than decisions made through majority-rule voting or litigation, *because they satisfy the interests and concerns of all parties* and are based on the knowledge and expertise of those parties."[19] Similarly, Philip Harter has argued that negotiated rulemaking has resulted in better environmental regulations, citing the Langbein and Kerwin study showing that participants in negotiated rulemakings rated regulations at a higher level of satisfaction.[20] Harter has argued that this evidence shows that negotiated rulemaking leads to "the best, most effective, or most efficient way of solving a regulatory controversy."[21]

These examples illustrate how researchers have used participant satisfaction both as a direct measure of the effect of public participation techniques and as a proxy for other evaluative measures. Researchers perceive satisfaction as something that is either conceptually appealing or simply easily measurable. For those who evaluate methods of involving the public in the regulatory process, the key question is whether satisfaction is an appropriate standard for evaluating techniques of public participation and dispute resolution in the regulatory process.

Is Satisfaction an Important Goal of Public Participation?

[I]t is unreasonable to imagine that regulatory and enforcement agencies find their justification in the satisfactions of those whom they compel to contribute to public purposes.[22]

Based on the existing literature, it might well seem that satisfaction is an appropriate, even important, criterion to use in evaluating public participation in the regulatory process. But is it? In this part of the chapter, I suggest that participant satisfaction is not an important or meaningful goal for two related reasons. First, satisfaction does not equate to high-quality public policy. The mere fact that participants in a regulatory proceeding are satisfied with a policy decision does not mean that the decision is a good one.[23] Second, the participants in any given regulatory proceeding are not the only people whom the regulatory policy will affect, so their satisfaction is at best a partial representation of overall social welfare. Satisfaction could be a meaningful measure of success if everyone affected by a decision were able to make a well-informed judgement about it, but this is simply not possible with regulatory policymaking, especially in the area of environmental policy.

Participant Satisfaction Does Not Necessarily Mean
That a Policy Is Better

A policy decision may satisfy those who participate in making it, but this does not mean that it is a good policy decision. Regulatory policies that satisfy everyone involved in the decisionmaking process may simply be based on the lowest common denominator of the policymaking group, or they may result from the predominance of a "group think" mentality that leads participants to believe that the policy is a good one.[24] The reality is that any policy may prove to be effective or ineffective, efficient or inefficient, and just or unjust. Policies that greatly satisfy participants can still be prone to error. Consider two examples of policies that were satisfactory to participants when they were created but later turned out to be significant policy failures.

The first example is from the early 1990s, when the EPA convened a negotiated rulemaking to develop standards for new, reformulated blends of gasoline that would reduce urban smog. The agency selected representatives from the automobile and petroleum industries, as well as from the environmental community, in an effort to agree on standards for the new fuels. According to one commentator, "the entire process met with considerable success" because "[s]atisfaction with the negotiated outcome led all participants to sign a consensus agreement."[25] Despite the participants' presumed satisfaction, the resulting negotiated regulation has turned out to be one of the most problematic rules in the history of the EPA. Following implementation of the rule, the key chemical additive used in reformulated gasoline—MTBE (methyl tertiary butyl ether)—was discovered to be leaking into groundwater across the country, causing substantial public alarm about drinking water contamination and resulting in legislative and regulatory efforts to ban the use of this fuel additive.[26] A few years ago, *Discover* magazine included the reformulated gasoline decision in its list of "Twenty of the Greatest Blunders in Science in the Last Twenty Years."[27]

The second example is another recent, enormous policy blunder that was remarkably satisfactory to participants at the time the policy was adopted: the restructuring of California's electricity markets. State legislators adopted electricity restructuring in California after extensive public participation, structuring the new market in a way that effectively achieved a compromise among the major players.[28] Because wholesale rates were low, this compromise, which among other things capped retail rates but allowed wholesale rates to float with the market, was generally satisfactory to the participants when the legislation passed. Indeed, at the outset of the restructuring, there were surprisingly few critics of the plan in California. Only later, under a taxing set of circumstances, did California's "satisfactory" compromise reveal itself to be seriously flawed, wreaking havoc on Californians and utility companies as wholesale rates and demand for power increased.[29]

In both of these examples, it should be obvious that the resulting policies failed, no matter what the original participants may have thought about the policies when they were adopted. Undoubtedly, many of the participants in these policymaking processes would probably agree with such an assessment now and are probably no longer satisfied with the policies they helped create. These examples starkly demonstrate that satisfaction at the time a policy is created does not necessarily equate with good public policy. When evaluating the effectiveness of different forms of public participation, researchers must consider the underlying outcomes of policy decisions that result from different participatory processes rather than relying on measures of participant satisfaction.

Of course, participant satisfaction could very well be a relevant consideration to take into account if policies that garnered satisfaction were also—at the margin or on average—higher quality policies. Yet, how likely is this? Regulation is needed to solve problems of externalities, information asymmetries, and monopoly, problems that arise in situations where private interests are not well aligned with overall social interest.[30] The goal of regulation is to change the behavior of firms and individuals to support improved social welfare. As such, the targets of regulation presumably should not find it satisfying to be regulated.

As the principal target of regulation, business firms are usually well-organized and participate regularly and intensively, and in greater numbers, in public policymaking.[31] If these targets of regulation come away satisfied with a policymaking process, it may well be that the resulting policy decision has not been as effective as it needs to be. For this reason, participant satisfaction in many cases could actually indicate that a policy is of a lesser quality, not greater.

To Be Meaningful, Satisfaction Measures Require Full Representation

Satisfaction could be a meaningful measure of policy success, at least conceptually, if all the people affected by a policy were able to make well-informed judgements about how the policy affected them. Yet studies that measure participants' satisfaction only survey a fraction of the entire population that a regulatory policy affects.[32] Even when those involved in a process of public participation are satisfied, the broader public—especially underrepresented segments of the broader public—may not be well served by a policy.

In some cases, as noted in the previous section, participant satisfaction could even indicate that the policy badly serves the public. Participants in policymaking at regulatory agencies that have been "captured" by a segment of the market or society are, by definition, quite satisfied.[33] Yet situations of regulatory capture are properly deplored because the satisfaction of a select group of interests in the regulatory context typically comes at the expense of the interests of the broader public. As Judith Innes has noted, when "a group produces outcomes that harm the larger community, this would not be a positive result even if the stakeholders at the table are satisfied."[34]

Innes has described a regional mass transit planning process in California that did not include representation of poor, inner-city residents, and as a result, cut some of the bus service to these communities.[35] Even if this planning process had been one that participants rated highly, it was clearly not a good policy for those who were not represented in the process. To be sure, lack of adequate representation is generally considered a problem of pluralist policymaking,[36] but it is also something that is particularly acute in processes that demand an intensive level of public time and resources, such as collaborative decisionmaking processes. A recent study by researchers at Resources for the Future, for example, examined 30 case studies of public environmental planning committees in the Great Lakes region and found that "[f]or the most part, participants did not appear to be representative of the wider public."[37] These findings are not uncommon. As Daniel J. Fiorino has observed about negotiated rulemaking, "negotiation is biased toward organized, influential interests in society. The most well-balanced committee is not competent institutionally to represent unorganized or uninfluential groups or broad conceptions of the public interest."[38]

It may be that intensive collaborative enterprises can help participants find creative solutions to problems that expand their Pareto frontier, making everyone around the table better off than they would have been had the agency pursued less intensive techniques of policymaking and public participation. Yet even in the rare cases where this might be so, this does not necessarily mean that the broader public will be better off. From society's standpoint, it is far from clear that increased satisfaction from *represented* interests should be the aim of public policymaking. It might very well be best to secure a level of satisfaction that is simply "good enough" among these participants, or even in many cases to create outright dissatisfaction on the part of some or most of those who participate in a regulatory process. Regulatory officials may well be correct to believe that if they displease both sides of a policy dispute, then "we must have done the right thing."[39]

A full social welfare analysis must also consider whether any relative gains a collaborative or participatory process might achieve would outweigh the costs of using an intensive public participation process, including the opportunity costs associated with the additional time demanded of everyone.[40] Does attempting to increase the satisfaction of those represented in a policy process justify spending the additional investment in time and resources demanded of intensive collaborative processes that seek to satisfy those who participate in the policy process? When public officials devote more of their time to trying to expand the pie on certain policy issues, they have less time available to devote to other issues.

If we could be sure that the issues for which public officials invest the additional time demanded of intensive public collaboration are indeed the ones over which there is the most to be gained from this additional investment, then perhaps we need not worry as much. However, collaboration on the most important issues seldom occurs. Agency experience with negotiated rulemaking indicates that agencies have tended to address "second tier" kinds of issues that probably yield fewer social gains from the greater resource and time demands necessitated by negotiation.[41] Indeed, if policymakers aim first and foremost to satisfy those involved in the policy process, they can easily begin to focus on the most tractable policy issues—those that are more likely to please everyone—rather than on the issues that are most important to society overall.[42]

In the end, even if everyone "at the table" could be made better off by certain forms of public participation, this does not mean that society overall would be better off. Moreover, if the intensity of a participatory process means that participants are not representative of the public at large, then collaborative processes aimed at satisfying participants could actually serve the broader public rather poorly, even if they succeed in satisfying participants greatly. Finally, the kinds of regulatory actions that are necessary to improve overall social welfare will sometimes make participants in the regulatory process decidedly *dis*satisfied. Imposing costly new controls on an industry is not likely to satisfy the regulated firms, even though in some cases, this is exactly what is needed to improve the overall welfare of society.

Problems with Satisfaction in Public Participation Evaluation

> [U]sing subjective perceptions like happiness and satisfaction as an index of the efficiency of negotiated outcomes is problematic.[43]

I have argued that using participant satisfaction as an evaluative measure is misguided as a conceptual matter because such a measure does not necessarily equate to good public policy and tends to exclude consideration of the broader public, who did not participate in the policy process. In addition, reliance on individuals' reported satisfaction or opinions raises a number of measurement problems. In this section of the chapter, I highlight four methodological or measurement concerns associated with using satisfaction in evaluation research. Even if researchers were to conclude that participant satisfaction could be an appropriate performance measure for dispute resolution and public participation, they would nevertheless face significant challenges in interpreting their results.

Surveys Tend To Truncate Extreme Satisfaction and Dissatisfaction

A policy that moderately satisfies most of those who participate in a regulatory proceeding but causes extreme dissatisfaction to one (or a few) of the participants can still appear to fare well in terms of an average level of participant satisfaction, as reflected in survey research. This is a general problem of survey research methods, but it is something that especially needs to be borne in mind whenever participant surveys form the basis of public participation evaluation.

In poorly designed surveys, it can be virtually impossible to assess the relative level of participants' satisfaction or dissatisfaction. For example, in the study of the National Marine Fisheries Service processes noted earlier, participants were simply asked whether they were "satisfied" or "dissatisfied" with the policy outcome.[44] It is difficult to interpret the fact that 60% of the participants indicated that they were "dissatisfied," because this survey instrument does not reveal whether they were mildly or extremely dissatisfied and whether the remaining 40% were mildly or extremely satisfied.

Survey instruments can be designed to capture at least some indication of the intensity of participants' satisfaction. Many researchers ask participants to array their preferences along a scale, from extremely dissatisfied to extremely satisfied. The Langbein and Kerwin study, for example, used an 11-point scale (from −5 to +5) to assess participants' levels of satisfaction and their perceptions of the policy outcomes.[45] Whereas such an approach is clearly superior to the binary measures used in the NMFS study, it too can be rather limited. Those participants who experience extreme levels of net benefits or costs from a rule are unlikely to have the fullness of their satisfaction or dissatisfaction reflected in a survey. One need only consider a policy decision that imposes highly disproportionate burdens on a minority group, especially one that was ill represented in a policy process. For example, even if a regional mass transit planning process like the one described by Innes had included a representative from an inner-city community, the impact of any resulting decision to cut bus services to this community—such as devastating residents' ability to commute to work—might not be well reflected on a Likert scale. If many more people involved in the planning process were not affected by these cuts and were overall moderately satisfied with the process, the process could look much more successful than it would if the overall social costs and benefits were analyzed. In this way, surveys that ask participants to rank satisfaction on a fixed scale tend to truncate the views of outliers.

Perceptions Are Often Erroneous

Asking participants to assess what was accomplished in a policy process is an imperfect measure of what was really accomplished. Such survey results are, at best, evidence of participants' *perceptions*, and not evidence of the underlying qualities of the public policy. Participants' perceptions often do not match reality.

To illustrate the unreliability of participant perceptions, consider Langbein and Kerwin's findings on how well negotiated rulemaking prevented litigation. Langbein and Kerwin asked their respondents to rate the likelihood that the rulemakings in which they were involved would resist legal challenge. The average rating for negotiated rules (3.3) was significantly higher (that is, more resistant) than the average rating for conventional rules (1.9).[46] Does this mean that negotiated rules are really more resistant to legal challenge? Actually, the record is quite the opposite. Data collected from court filings show that negotiated rules are challenged at a higher rate than conventional rules.[47] Along other dimensions, such as the average number of petitions filed and the rate of settlement, negotiated rulemaking exhibits no greater degree of resistance to litigation.[48] It is precisely this type of data, not data on participants' perceptions that is needed to make judgements about the actual resistance of negotiated rules to legal challenge. Similarly, to assess other qualities of policy decisions, it is better to seek direct evidence of those qualities rather than to rely on participants' perceptions of them.

Satisfaction Can Be Affected by Irrelevant Factors

Evaluations of participant satisfaction can be difficult to interpret because satisfaction and perception can be affected by numerous factors that are not relevant to evaluators. Perceptions and feelings of satisfaction can be highly contextual. As Leigh Thompson and Richard Gonzalez have noted, "[t]he same objective outcome may be viewed quite differently, as a function of arbitrary aspects of the context. For example, people report greater life satisfaction on days when the weather is sunny than when the weather is cloudy."[49] For this reason, evaluation researchers should be extremely careful in interpreting survey results. As Judith Innes has cautioned, participant surveys "will probably not provide a meaningful assessment [because] participants responding to a survey could say that they were satisfied with a process when they were actually manipulated and misled, or they could say they were dissatisfied when they actually accomplished a great deal but had unrealistic expectations."[50]

From what we understand about cognitive dissonance, we might expect respondents to give higher ratings to forms of public participation that are more time- and resource-intensive. For many years, social psychologists have told us that individuals adjust their views to avoid dissonance because the existence of incompatible or dissonant cognitions is psychologically uncomfortable.[51] One paradigmatic case of cognitive dissonance occurs when individuals respond to the effects of effort, the so-called "effort-justification paradigm."[52] The more effort an individual must expend at some task, and the more unpleasant that effort, the more dissonance that individual experiences. Individuals who find themselves in such situations can reduce dissonance "by exaggerating the desirability of the outcome."[53] In the classic study demonstrating this effect, women were asked to undertake either a severe or a mild rite of "initiation" to join a discussion group.[54] In either case, the discussion group turned out to be an equally boring one, but the women who were assigned to undertake the more intensive initiation evaluated the group more favorably than did the women who went through the mild initiation.[55] As Eliot Aronson has explained, "going through hell and high water to gain admission to a boring discussion group was dissonant with one's self-concept as a smart and reasonable person, who makes smart and reasonable decisions."[56]

Collaborative forms of public participation, such as dialogue groups and negotiated rulemaking, tend to be quite effort-intensive. When participants expend more, we can expect that they will reduce their dissonance by viewing the outcome of these intensive processes more favorably. This appears to explain well the findings in the study of negotiated rulemaking by Langbein and Kerwin, where participants in negotiated rulemakings rated the resulting policy decisions more favorably than did participants in ordinary rulemaking proceedings. Despite the significantly higher costs of participating in negotiated rulemaking, Langbein and Kerwin's respondents from both negotiated and conventional rulemakings reported no statistically significant differences in terms of their perceived net benefits from participating in the rulemaking process.[57] The overwhelming majority of respondents in both samples said that the benefits they realized from their participation equaled or exceeded the costs.[58] Because the costs of participating in negotiated rulemaking were so much higher, individuals could be expected to exaggerate the desirable qualities of the rulemaking process's outcome, thus holding their net satisfaction level constant and avoiding cognitive dissonance. The higher levels of satisfaction reported by participants in negotiated rulemakings may simply be an artifact of cognitive dissonance, an internal psychological adjustment participants make to justify for themselves their heavier commitment of time and energy to these rulemakings.

Modes of Public Participation Can Select on the Dependent Variable

Participants in policy processes are not randomly selected from the overall population. As a result, the sample of participants in specific policy processes is likely to be biased. This can become a problem when different modes of public participation attract different kinds of participants, and as a result, create biases that are correlated with the dependent variable, such as satisfaction.[59]

For example, suppose researchers compared two methods by which a company's managers could involve workers in management efforts to design a safer, more productive work environment. One method would be to solicit complaints, such as by installing a suggestion box and instructing workers to submit their ideas to management by using this box. Another method would be for management to identify a select group of workers from throughout the firm who could serve on a worker advisory committee that would engage in ongoing dialogue with management on workplace conditions. Participation is not randomly determined in either process. Who would likely submit comments in the suggestion box? Presumably those workers who are most dissatisfied and most inclined to complain would disproportionately represent themselves in the suggestion box process. Who would likely participate on the worker advisory committee? It is likely that management would avoid selecting the most unreasonable, pessimistic, or rebellious workers to participate in a workplace committee, but instead would disproportionately select cooperative workers and those tending to be more satisfied at the outset.

Of course, the bias does not need to occur in both groups for any resulting comparison of participants' satisfaction to be biased. Even if those invited to serve on the committee were selected randomly, for example, the likelihood that a suggestion box would attract disproportionately more disgruntled employees would continue to bias the comparison. Nevertheless, it seems unlikely that a firm would simply pick workers at random to participate on an internal advisory committee. As a result, the two worker–management processes would likely attract participants with different preexisting propensities for satisfaction.

An evaluator could try to control for these differences by administering a pretest designed to measure participants' prior attitudes about the workplace or their attitudes more generally.[60] In the absence of such a control, however, any resulting comparison of participants' satisfaction with the process would be affected not only by whatever effects the process had, but also by the different kinds of people who participated in the two processes, the selection of which is also an effect of the process. It would be impossible to determine whether any resulting differences in the two groups' reported satisfaction arose from the differences in the processes used or from the ways in which the two processes selected different participants.

This example is not entirely hypothetical. Langbein and Kerwin's study of negotiated rulemaking encounters precisely this challenge.[61] Recall that the researchers compared the satisfaction of people who filed comments in the normal rulemaking process with the satisfaction of people who sat on agency-created negotiated rulemaking committees.[62] We could reasonably expect that the individuals who file comments in administrative rulemakings are more likely to be motivated by perceived problems with proposed regulations and are individuals who, at the outset, would tend to be less satisfied with the agency overall. The individuals the agency selects to participate in negotiated rulemaking are those whom the agency presumably believes will be willing to negotiate in good faith and for whom the agency believes there is a likelihood of reaching a consensus.[63] To minimize the problem of procedure-induced selection bias in a study of this kind, it would be better either to compare the satisfaction of commentators in both rules or those with whom the agency held extensive discussions in both rulemaking. For the latter approach, researchers would probably get a less biased measure if they compared the satisfaction of those on negotiated rulemaking committees with those with whom the agency held repeated ex parte discussions during a conventional rulemaking process. The latter are usually noted in the administrative records or could be identified through interviews with agency rule drafters.

Conclusions

For both conceptual and methodological reasons, researchers who evaluate methods of public participation in regulatory policymaking should not be satisfied with using participant satisfaction as an evaluative criterion. Participant satisfaction is at best a relatively unimportant criterion because it does not equate with the quality or effectiveness of regulatory policy and because focusing on participants' opinions can easily lead one to overlook the broader public's interests. Satisfaction is also a problematic measure because participants' perceptions can be wrong or influenced by irrelevant factors and because certain kinds of procedures can sort out individuals according to preexisting propensities to being satisfied.

The chief lesson to be learned from the foregoing examination of participant satisfaction is that other, more reliable measures of policy success should be employed when evaluating public participation techniques. In the absence of good methods of evaluating public participation based on satisfaction, researchers should employ alternative strategies for evaluating policy outcomes, such as impact analysis, cost-effectiveness analysis, or benefit–cost analysis. Evaluators should seek to compare systematically the policies that result from different participatory processes. They should test whether processes yield the kind of results they are intended to achieve. For those procedures intended to prevent litigation, researchers should seek to learn whether litigation was in fact reduced. For those procedures designed to provide agencies with more information to make better policy decisions, researchers should seek to identify whether decisions really are improved.

The evaluation community needs to ask not whether participants are satisfied, but whether the public overall would likely be better served with the results of a given procedure than if the procedure had not been performed or if alternative procedures had been implemented instead. Although researchers will undoubtedly find this task much more challenging than simply surveying the specified set of individuals who participated in a policy process, researchers will only be able to conclude with any confidence that particular processes yield improvements over their alternatives if they compare the different processes' results.

Notes

1. There are also broader public implications to the handling of private disputes. For a discussion of policy considerations in the context of private dispute resolution, see Owen Fiss, "Against Settlement," *Yale Law Journal*, vol. 93 (1984), p. 1073.

2. See, e.g., Philip Harter, "Negotiating Regulations: A Cure for Malaise," *Georgetown Law Journal*, vol. 71 (1982), p. 1; Lawrence Susskind and Gerard McMahon, "The Theory and Practice of Negotiated Rulemaking," *Yale Journal of Regulation*, vol. 3 (1985), p. 133.

3. For a related claim that designers of democratic processes should evaluate empirically how well different decisionmaking procedures reach correct results, see Frederick Schauer, "Talking as Decision Procedure," *The Good Society*, vol. 7 (1997), pp. 47–52.

4. Judith E. Innes, "Evaluating Consensus Building," in Lawrence Susskind, Sarah McKearnan, and Jennifer Thomas-Larmer, eds., *The Consensus Building Handbook: A Comprehensive Guide to Reaching Agreement* (Thousand Oaks, CA: Sage Publications, 1999), p. 639.

5. For a recent discussion of the fundamentals of welfare economics, see Louis Kaplow and Steven Shavell, "Fairness versus Welfare," *Harvard Law Review*, vol. 114 (2001), pp. 961, 977–998.

6. See, e.g., Howard Raiffa, *The Art and Science of Negotiation* (Cambridge, MA: Belknap Press, Harvard University Press, 1982); Thomas Schelling, *The Strategy of Conflict* (New York: Oxford University Press, 1960).

7. For a concise description of the social choice theory of democracy and its emphasis on the aggregation of preferences, see Jon Elster, "The Market and the Forum: Three Varieties of Political Theory," in Jon Elster and Aanund Hylland, eds., *Foundations of Social Choice Theory* (New York: Cambridge University Press, 1989), pp. 103–112.

8. For an overview of this recent movement toward customer service, see Donald F. Kettl, *Reinventing Government? Appraising the National Performance Review* (Brookings Institution, 1994).

9. Matthew McKinney and Will Harmon, "Public Participation in Environmental Decision Making: Is it Working?" (manuscript).

10. See Will Harmon, "Montana Group Tries Scorecard Approach," *Consensus* (January 1999).

11. Laura Langbein and Cornelius Kerwin, "Regulatory Negotiation versus Conventional Rule Making: Claims, Counterclaims, and Empirical Evidence," *Journal of Public Administration Research and Theory*, vol. 10 (2000), pp. 599–632.

12. Ibid, p. 622.

13. Mark Lubell, "Attitudinal Support for Environmental Governance: Do Institutions Matter?" paper presented at the annual meeting of the American Political Science Association (Washington, DC, August 31, 2000).

14. For example, the evidence on negotiated rulemaking shows that using this procedure has not prevented subsequent litigation over regulations. See Cary Coglianese, "Assessing Consensus: The Promise and Performance of Negotiated Rulemaking," *Duke Law Journal*, vol. 46 (1997), p. 1255.

15. *The National Marine Fisheries Service Take Reduction Team Negotiation Process Evaluation* (Washington, DC: RESOLVE, Inc., July 19, 1999).

16. Ibid., p. 11.

17. Thomas C. Beierle, "Public Participation in Environmental Decisions: An Evaluation Framework Using Social Goals," Discussion Paper 99–06 (Washington, DC: Resources for the Future, November 1998).

18. Sam A. Carnes, Martin Schweitzer, Elizabeth B. Peele, and J.F. Munro, "Performance Measures for Evaluating Public Participation Activities in DOE's Office of Environmental Management" (Oak Ridge National Laboratory, August 1996).

19. Judith E. Innes, "Evaluating Consensus Building," p. 634 (emphasis added).

20. Philip Harter, "Assessing the Assessors: The Actual Performance of Negotiated Rulemaking," *New York University Environmental Law Journal*, vol. 9 (2000), p. 32.

21. Ibid., p. 38. In addition, Charles Fox of EPA has claimed, apparently on the basis of the Kerwin and Langbein study, that negotiated rules at EPA have been "more practical and cost efficient, contained more innovative solutions, were more technically and scientifically current, and had greater legitimacy." J. Charles Fox, "A Real Public Role," *Environmental Forum*, vol. 15, no. 19 (1998), p. 24.

22. Mark Moore, *Creating Public Value: Strategic Management in Government* (Harvard University Press, 1995), p. 37.

23. Correspondingly, dissatisfaction does not necessarily mean that a decision is a bad one. Derek Bok, "Measuring the Performance of Government," in Joseph S. Nye, Jr., Philip D. Zelikow, and David C. King, eds., *Why People Don't Trust Government* (Harvard University Press, 1997), p. 56.

24. Cary Coglianese, "Is Consensus an Appropriate Basis for Regulatory Policy?" in Eric Orts and Kurt Deketelaere, eds., *Environmental Contracts: Comparative Approaches to Regulatory Innovation in the United States* (Kluwer Academic Publishers, 2001).

25. Edward P. Weber, "Successful Collaboration: Negotiating Effective Regulations," *Environment* (November 1998), p. 15.

26. See Marla Cone, "EPA To Ban Gas Additive Nationwide," *Los Angeles Times*, March 21, 2000, p. A3; Marla Cone, "Elimination of Additive from Gas Is Ordered," *Los Angeles Times*, Dec. 10, 1999, p. A2.

27. Judith Newman, "Twenty of the Greatest Blunders in Science in the Last Twenty Years," *Discover*, vol. 21, no. 10 (October 2000), pp. 78–83.

28. Paul Joskow has written, "The ultimate design of the wholesale market institutions [in California] represented a series of compromises made by design committees including interest group representatives. … Getting it done fast and in a way that pandered to the many interests involved became more important than getting it right." Paul L. Joskow, "California's Electricity Crisis," Working Paper W8442 (Cambridge, MA: National Bureau of Economic Research, August 2001).

29. William W. Hogan, "The California Meltdown," *Harvard Magazine* (September–October 2001).

30. For an analysis of the rationales for regulation, see W. Kip Viscusi, John M. Vernon, and Joseph E. Harrington, *Economics of Regulation and Antitrust* (MIT Press, 2000).

31. It is generally accepted that business interests are well represented in the policy process, even though they are not always victorious. In my own research, I have found that business groups vastly outnumber environmental groups in the rulemaking process at the EPA. Cary Coglianese, "Litigating within Relationships: Disputes and Disturbance in the Regulatory Process," *Law & Society Review,* vol. 30 (1996), pp. 735–765.

32. Moreover, some studies even include government officials' satisfaction in their participant surveys. See Langbein and Kerwin, "Regulatory Negotiation versus Conventional Rule Making"; Lubell, "Attitudinal Support for Environmental Governance." It is far from conceptually clear why government agency officials' satisfaction should count alongside satisfaction of those who represent public constituencies and will be affected by the regulation the agency adopts.

33. Regulatory capture arises when an interest group comes to dominate the outcomes of an agency so as to yield private benefits to the group. For an economic analysis of interest group rent-seeking behavior, see George J. Stigler, "The Theory of Economic Regulation," *Bell Journal of Economic and Management Science,* vol. 2 (1971), p. 3.

34. Innes, "Evaluating Consensus Building," p. 653.

35. Ibid., p. 641.

36. E.E. Shattschneider, *The Semi-Sovereign People: A Realist's View of Democracy in America* (Holt, Rinehart, and Winston, 1960).

37. Thomas C. Beierle and David M. Konisky, "Public Participation in Environmental Planning in the Great Lakes Region," Discussion Paper 99–50 (Washington, DC: Resources for the Future, September 1999), pp. 23–24. For further discussion of the "representation problem" in dispute resolution and public participation processes, see J. Clarence Davies, "Environmental ADR and Public Participation," *Valparaiso University Law Review,* vol. 34 (2000), pp. 389, 396–397.

38. Dan Fiorino, "Regulatory Negotiation as a Policy Process," *Public Administration Review,* vol. 48 (1988), p. 764.

39. Coglianese, "Litigating within Relationships," p. 747.

40. It is well understood that collaborative processes tend to take more time and demand more resources. For a discussion of the literature on this point, see Cary Coglianese, "Assessing the Advocacy of Negotiated Rulemaking," *New York University Environmental Law Journal,* vol. 9 (2001), pp. 386, 415.

41. Coglianese, "Assessing Consensus," pp. 1317–1321.

42. See Coglianese, "Is Consensus an Appropriate Basis for Regulatory Policy?"

43. Leigh L. Thompson and Richard Gonzalez, "Environmental Disputes: Competition for Scarce Resources and Clashing of Values," in Max Bazerman, David M. Messick, Ann E. Tenbrunsel, and Kimberly A. Wade-Benzoni, eds., *Environment, Ethics and Behavior* (San Francisco: New Lexington Press, 1997).

44. *The National Marine Fisheries Service Take Reduction Team Negotiation Process Evaluation.*

45. Langbein and Kerwin, "Regulatory Negotiation versus Conventional Rule Making."

46. Langbein and Kerwin, "Regulatory Negotiation versus Conventional Rule Making," p. 604 (Exhibit 1). These ratings are on an 11-point scale, with a "5" indicating that the respondent believed the rule had the most possible resistance to legal challenge, and a "–5" indicating a belief that the rule had the least possible resistance.

47. Coglianese, "Assessing Consensus."

48. Coglianese, "Assessing the Advocacy of Negotiated Rulemaking."

49. Thompson and Gonzalez, "Environmental Disputes," pp. 83–84.

50. Innes, "Evaluating Consensus Building," p. 642.

51. Leon Festinger, *A Theory of Cognitive Dissonance* (Stanford University Press, 1957).

52. Eddie Harmon-Jones and Judson Mills, "An Introduction to Cognitive Dissonance Theory and an Overview of Current Perspectives on the Theory," in Eddie Harmon-Jones and Judson Mills, eds., *Cognitive Dissonance: Progress on a Pivotal Theory in Social Psychology* (American Psychological Association, 1999), pp. 3, 7–8.

53. Ibid., p. 7.

54. Eliot Aronson and J. Mills, "The Effect of Severity of Initiation on Liking for a Group," *Journal of Abnormal & Social Psychology,* vol. 59 (1959), p. 177.

55. Ibid.

56. Eliot Aronson, "Dissonance, Hypocrisy, and the Self-Concept," in Eddie Harmon-Jones and Judson Mills, eds., *Cognitive Dissonance: Progress on a Pivotal Theory in Social Psychology* (American Psychological Association, 1999), pp. 103, 112.

57. Cornelius Kerwin and Laura Langbein, "An Evaluation of Negotiated Rulemaking at the Environmental Protection Agency: Phase II, A Comparison of Conventional and Negotiated Rulemaking" (draft report to the U.S. Environmental Protection Agency) (August 1997), p. 26.

58. Ibid.

59. See Gary King, Robert Keohane, and Sidney Verba, *Designing Social Inquiry: Scientific Inference in Qualitative Research* (Princeton University Press, 1994).

60. For a discussion of the use of pretests, see Lawrence Mohr, *Impact Analysis for Program Evaluation, 2nd ed.* (Thousand Oaks, CA: Sage Publications, 1995).

61. Langbein and Kerwin, "Regulatory Negotiation versus Conventional Rule Making."

62. The Langbein and Kerwin study is further confounded due to the fact that their two samples had substantially different percentages of government officials. About 36% of the participants in their negotiated rulemaking sample came from federal and state agencies, whereas only about 6% of their conventional rulemaking sample came from government. (About 11% of the "participants" in the negotiated rulemaking sample were EPA officials, whereas none of those in their conventional rulemaking sample were.) It should be hardly surprising that a sample so disproportionately composed of government officials would tend to rate government regulations more favorably.

63. Indeed, the Negotiated Rulemaking Act specifically directs the EPA and other agencies to use the procedure only when there is a "reasonable likelihood that a committee will reach a consensus" and a willingness by the participants "to negotiate in good faith." 5 U.S.C. § 563(a).

PART III

Midstream Environmental Conflict Resolution

T he chapters in this section push our thinking about how to evaluate the "black box" of conflict resolution practices in the environmental arena. The first three chapters examine aspects of transformative mediation. The premise of transformative mediation is that "the mediation process contains within it a unique potential for transforming people—engendering moral growth—by helping them wrestle with difficult circumstances and bridge human differences, in the very midst of conflict."[1] Its potential as a means of transformation lies in its power to give people control over resolving their own conflicts. The goals of transformative mediation are empowerment and recognition. Empowerment is achieved when parties "grow calmer, clearer, more confident, more organized, and more decisive—and thereby establish or regain a sense of strength and take control of their situation."[2] Recognition is achieved when parties "voluntarily choose to become more open, attentive, sympathetic, and responsive to the situation of the other party, thereby expanding their perspective to include an appreciation for another's situation."[3]

Much of the literature concludes that through its capacity to generate empowerment and recognition for the parties, mediation can help participants learn how to better address future conflict. Empowerment and recognition often result in an agreement. However, this is only a secondary effect. The theory is that experiencing empowerment and recognition will improve each party's ability to approach and resolve both current and future problems. The two major proponents of transformative mediation, Robert A. Baruch Bush and Joseph P. Folger, argue that it can create opportunities to improve individuals, workplaces, and society.[4] However, it is important to distinguish the transformative model from therapeutic mediation. In the transformative model, the mediators do not provide therapy for the parties. Rather, they help create opportunities for the parties to take control of their own decisionmaking.[5]

Marcia Caton Campbell applies the theory of tranformative mediation to intractable environmental conflicts in Chapter 5. Campbell argues that empirical research on intractability in environmental disputes is missing. After reviewing the multiple definitions of intractability, Campbell analyzes the characteristics of intractability. She maintains that a transformative approach to environmental conflict resolution (ECR) is appropriate in intractable conflicts. Campbell highlights lessons learned from her analysis both for research and for practice.

In Chapter 6, Tamra Pearson d'Estrée focuses on one criterion for measuring success in ECR processes: a positive change in the relationship of the parties. After examining past efforts at measuring relationship change (including her own efforts in collaboration with Bonnie G. Colby), d'Estrée takes her analysis one step further by presenting new thoughts on the assessment of relationship change. Urging a focus on communication measure, relational schemas, and interdependence, d'Estrée moves to a discussion of how increased coordination among parties can lead to a building of social capital. The author closes with a suggestion of work for the future, including better documentation of ECR processes and outcomes, a common language with which to understand measures and constructs, and the challenges inherent in doing interdisciplinary ECR evaluative research.

Sanda Kaufman and Barbara Gray present a new evaluation approach in Chapter 7. Their approach uses frame elicitation, an analytic technique based primarily on interviews with the participating stakeholders that reveal how the participants interpret critical aspects of the disputes. Evaluators then provide feedback to participants about their frames and how they may be influencing their views about the issues and the process, as well as how their frames might unnecessarily reduce the realm of possible agreements. The goal is to assist parties in reflecting on the ECR process while it is still underway, rather than waiting for an assessment after the fact. Environmental examples demonstrate applications of the Kaufman–Gray approach.

Concluding this section are William Leach and Paul Sabatier, who take a different approach to examining the "black box" of conflict resolution practices. They investigate the qualities of effective facilitators and coordinators that promote conflict resolution within multistakeholder watershed partnerships in California and Washington state. Contrary to conventional wisdom, as well as much of the current ADR literature, they find that disinterested, professional facilitators are not necessarily more effective than stakeholders who serve as facilitators. Their conclusion, which is sure to rock boats in the conventional ADR ocean, is that watershed partnerships should attempt to recruit facilitators and coordinators from within the partnership, if possible, rather than automatically investing in a professional, third-party facilitator.

Notes

1. R.A. Baruch Bush and J.P. Folger, *The Promise of Mediation: Responding to Conflict through Empowerment and Recognition* (San Francisco: Jossey-Bass, 1994).

2. Bush and Folger, *The Promise of Mediation*, p. 2.

3. Bush and Folger, *The Promise of Mediation*, p. 89.

4. Bush and Folger, *The Promise of Mediation*, pp. 81–95.

5. Bush and Folger, *The Promise of Mediation*, pp. 95–99.

5

Intractable Conflict

MARCIA CATON CAMPBELL

C onflict and its management are central to the resolution of disputes in the
public sector, the urban and regional planning process, and planning
practice. Given that conflict and conditions of uncertainty are endemic to
planning, planners may have to adopt many different roles in their practice to
be effective.[1] Although the nature of the planner's role has been debated in the
planning literature,[2] one area in which it is argued that planners need greater
facility and skill is the practice of dispute resolution.[3] Dealing with land-use,
environmental, and other public policy disputes has become an important
component of the planner's work.[4] Because of their nature, these kinds of disputes can appear particularly resistant to resolution, or "intractable." This
chapter maps the literature on intractable environmental and public policy
disputes, discusses their characteristics and potential for transformation, and
draws lessons for dispute resolution research and planning practice.

As a function of the kind of work they do, planners have special training
and abilities that make them good mediators of land-use, environmental, and
public policy disputes.[5] Planners are typically taught to identify, generate,
and assess a range of options and alternatives, to accommodate a range of
competing interests, and to focus on the need for implementation of agreements or plans.[6] Mediators also need and use all of these skills.[7] Susskind and
others describe an emergent "mediation model of land use planning," which

---

This chapter is adapted from Marcia Caton Campbell, "Intractability in Environmental Disputes: Exploring a Complex Construct," *Journal of Planning Literature,* vol.
17, no. 3 (February 2003), pp. 360–371.

they claim is coming to replace both the technocratic, physical planning of the early part of the twentieth century and the advocacy planning model that developed in response during the 1960s.[8] The consensus-building and negotiation techniques central to this model, they argue, can help planners involve the public in crafting creative and enduring solutions to land-use, environmental, and other planning problems.

Environmental and public policy disputes occur in many different and complex forms. They can stem from disagreements about facility siting issues, from technical debates about environmental policy choices, or from uncertainties about the level of an environmental risk facing the public.[9] Disputes may involve multiple stakeholders, pronounced imbalances of power between disputing parties, lack of agreement on the issues at stake, the potential for irreversible environmental effects, and difficulty calculating costs to the environment and to human health.[10] Environmental disputes often arise from conflicts over the disputing parties' fundamental beliefs or conflicts over rights.[11] They can also arise from clashes of values or world views between the disputing parties.[12]

Disputes based on fundamental values and beliefs, rights, conflicting world views, and other significant issues such as power are more likely to resist resolution than disputes in which the parties can ultimately find common interests or goals.[13] Thus, these types of disputes may require a different type of management than one that would be more appropriate for interest-based disputes or disputes resulting from miscommunication between parties.[14] These resolution-resistant disputes are often termed "intractable," but what exactly does that term mean?

Although the situation is beginning to change, particularly with respect to dispute resolution programs and case evaluation,[15] very little empirical research has been done on environmental disputes.[16] Researchers have studied procedural issues such as conflict management skills, consensus building, citizen involvement, and participant satisfaction as measures of successful mediation outcomes. The sizeable case study literature on environmental disputes documents actual mediation efforts and their successes and failures, but few long-term or cross-sectional empirical analyses of environmental disputes have been made.[17] Because they may lack specialized dispute resolution or facilitation training, planners do not necessarily know which dispute resolution tactics or strategies might help make a seemingly intractable dispute more readily resolvable. Nor can they look to the literature on environmental and public policy disputes for explicit guidance, because confirmable research in the field is relatively scarce.

An area of growing research interest in conflict resolution is the handling of intractable disputes, including intractable environmental and public policy disputes. A January 1998 conference held by the William and Flora Hewlett Foundation, one of the major national funders of conflict resolution research, was entirely devoted to the subject of intractable conflict.[18] The conference identified intractability, particularly the exploration of its meaning and the dispute characteristics that are indicators of or contributors to it, as one important direction for research.

Because empirical research on intractability and environmental disputes is lacking, we must know more about the characteristics that contribute to intractability in environmental conflict. This chapter discusses existing definitions of intractability and identifies from the literature a set of characteristics that are not only promising for further research, but can also be useful in its preliminary form to mediators, planners, and others who work to resolve environmental and public policy disputes. Identifying and discussing characteristics of intractability found in the broader conflict resolution literature helps deepen our understanding of the problematic aspects of some environmental and public policy disputes.

Defining Intractability: A Difficult Task

In the realm of conflict resolution, intractability is a problematic construct. Like Justice Potter Stewart's definition of obscenity, intractability is something researchers and mediators know when they see it, but they are hard pressed to define precisely. At a recent theory-building conference on intractable disputes, conference participants spent several hours over two days discussing the meaning of intractability. Unable to arrive at a precise definition, they agreed that they would substitute "resolution resistant" for "intractable" until the field developed a better understanding of the term.[19] Nevertheless, "intractable" is a term that has been, and will likely continue to be, widely used in the literature.

How is intractability described? Burgess and Burgess denote intractable conflicts as "conflicts which defy resolution." Although individual dispute episodes can be resolved, the longstanding underlying conflict persists over time.[20] Coleman characterizes intractable conflict as "recalcitrant, intense, deadlocked, and extremely difficult to resolve."[21] Gray defines an intractable dispute *not* as one that cannot be resolved, but as one that must be fundamentally transformed to reach agreement (e.g., by making changes in the stakeholder parties, in the context of the dispute, or in how the dispute is framed).[22] In this chapter, "intractable" is understood to mean "resistant to resolution" rather than "unresolvable" (cf. the discussion of intractability below).

Susskind and Cruikshank distinguish distributional disputes from constitutional disputes.[23] Distributional disputes are those that revolve around participants' interests and the allocation of resources, and in which gains and losses are tangible. These types of disputes are considered relatively easy to resolve. Constitutional disputes are those in which constitutional issues, basic rights, and basic values are at stake. These disputes are the least likely to respond to mediation or negotiation.[24] Schön and Rein liken this distinction between distributional and constitutional disputes to their own differentiation of policy disagreements, which can be subjected to reasoned discourse about the facts of the situation, from policy controversies, which they describe as "stubbornly resistant to resolution through the exercise of reason" and "immune to resolution by appeal to the facts."[25] In their estimation, policy controversies are the types of disputes that are most likely to become intractable.

Some researchers argue that intractable conflicts should not be defined in binary terms (i.e., intractable/tractable), but rather should be thought of as occurring along a continuum or varying by degree, and as being deeply embedded in a historical and social context that must be understood.[26] Northrup defines an intractable conflict as "a prolonged conflictual psychosocial process between (or among) parties that has three primary characteristics: (1) it is resistant to being resolved, (2) it has some conflict-intensifying features not related to the initial issues in contention, (3) it involves attempts (and/or successes) to harm the other party, by at least one of the parties."[27] Tractable conflicts, on the other hand, are "part of a normal process of relationship between individuals or parties who perceive that they have incompatible goals" and are characterized by open communication, flexibility, and goodwill among the parties.[28] Northrup contends that there may be a continuum ranging from high tractability on one end to high intractability on the other, and that tractability levels may vary over the course of a conflict and by issue at stake in the conflict. Greenhalgh's conflict diagnostic model illustrates this continuum.[29]

Hunter argues that to fully understand a conflict and its degree of tractability, we must understand the level at which it occurs.[30] She theorizes that everyone has an ontology (a view of "what is" about the world) that structures their existence and constrains their values, goals, and actions. This ontology frames individuals' interactions with others and the range of problem-solving strategies to which they are open. Hunter maintains that the values held as a function of this ontology are deeply rooted and not amenable to debate, mediation, or negotiation.[31] Thus, the most intractable environmental and public policy conflicts come about because of differences in the stakeholder groups' ontologies.[32]

These descriptions of intractability mesh well with the contention made by a number of environmental dispute resolution researchers that not all environmental disputes (particularly those based on values, rights, or conflicting world views) are susceptible to mediation.[33] Forester concedes that public skepticism may be justified regarding the efficacy of applying mediation or consensus-building techniques to disputes involving deeply felt values or moral disagreements.[34] As counterpoint, though, he argues that it is possible to achieve solutions to these deep disagreements not by altering or compromising any group's fundamental world view, but by working toward more immediate, practical responses to problems that disagreeing parties can mutually support despite their fundamental differences. Similarly, Dukes describes a process in which parties simultaneously transcend their differences while remaining faithful to their own world views.[35] However, using dispute resolution or consensus-building techniques under these circumstances requires a more detailed and sophisticated understanding of the nature and dynamics of deeply value-laden, resolution-resistant disputes than we now have. In seeking that understanding, a reasonable point of departure is the growing theoretical literature on intractable conflict.

Characteristics of Intractability

Taking "resistance to resolution" as a nutshell definition of intractability, what are some of the defining characteristics that combine to bring about, indicate, or predict that resistance? Box 5-1 lists characteristics of intractability discussed in the theoretical literature on environmental conflict and international conflict. One of the primary characteristics of intractability presented in these works is variously described as fundamental or deep-rooted moral conflict;[36] fundamental value differences that cause parties to disagree about the nature of the conflict;[37] or fundamental conflict rooted in different world views, values, principles, or societal structures.[38]

A second frequently offered characteristic of intractability is the length of time a conflict has persisted,[39] although length alone may not be a causal determinant.[40] Some conflicts have become virtually institutionalized after numerous failed attempts at resolution, such as the jobs versus the environmental conflict embodied in the spotted owl–logging industry dispute in the Pacific Northwest[41] or, outside the environmental realm, ethnic or religious conflicts such as the clashes between Protestants and Catholics in Northern Ireland.

Several conflict theorists identify power imbalances between dispute participants as a contributor to intractability.[42] Hunter notes that actual or perceived power differentials affect choices of dispute resolution strategies.[43] The potential for the conflict to involve coercion of one or more parties, or for it to escalate to the level of violence, is an important hallmark of intractability in the international conflict resolution literature[44] that may also be germane to environmental conflict in some instances.

Intractable disputes are often marked by a high level of hostility and parties assuming rigid positions.[45] Strongly held beliefs or positions can also influence resistance to resolution.[46] This has been termed "irreconcilable interests," particularly in the event that neither party has sufficient power or authority to force the other to comply.[47] The condition of irreconcilable interests should be distinguished from disputes based on clashes of interests, which many theorists agree are more easily mediated.[48]

Box 5-1. Characteristics of Intractability

— Fundamental or deep-rooted moral conflict
— Fundamental value differences between parties
— Fundamental conflict about world views, values, principles, or societal structures
— Length of time conflict has persisted
— Severe power imbalances between disputing parties
— Potential for coercion or escalation to violence
— Rigidity of positions, high level of hostility
— Strongly held beliefs or positions
— Complexity of issues, interlocking issues
— High-stakes distributional questions
— Pecking order conflicts or one-upmanship
— Threats to parties' individual or collective identities
— Frames held by disputants

Complexity of issues, interlocking issues, or sheer numbers of issues involved may also contribute to dispute intractability, as in the case of conflict overlay issues versus core issues described in the Burgesses' constructive confrontation technique[49] or in the case of factual uncertainties created by research biases.[50] Again, issue complexity alone may not be a causal determinant.[51] In their work on constructive confrontation, Burgess and Burgess also identify high-stakes distributional questions ("who gets what") as another indicator of a conflict's resistance to resolution.[52] Examples of this type of issue include social program reform and federal budget setting.[53] High-level environmental policymaking, such as questions about whether to rely on nuclear power or another energy resource or how to set acceptable levels of risk to public health, would be another example. The participants can elevate environmental policymaking involving concrete and irrevocable change, such as species extinction or loss of open land, to intractable levels.[54] Burgess and Burgess also point out that "pecking order conflicts," in which disputants jockey for position in a social order, can contribute to intractability.[55]

Perceived threats to disputing parties' individual or collective identities and culture or lifestyle often characterize intractable disputes in general and environmental disputes in particular.[56] Northrup describes identity as a core construct that is central to the individual (and sometimes to a group) and helps stabilize the relationship between the individual (or group) and the world.[57] Thus, when identity is threatened, the individual's (or group's) sense of safety and stability is threatened, and hostility and severe conflict may result. In Albuquerque, New Mexico, for example, a conflict arose in the late

1990s between the city government and the Pueblo, Zuni, and Hopi tribes about the siting of a six-lane highway extension. With the city landlocked on three sides and a 50% increase in population projected over the next 25 years, planners proposed that one-quarter mile of the highway extension be routed through the Petroglyph National Monument, necessitating the loss of 8.5 acres of parkland and the relocation of about a dozen ancient carvings. Central to the tribal histories, however, was the creation belief that their ancestors had emerged from the earth; thus, to the tribes, routing any part of the highway through the monument constituted a significant threat to their identity as indigenous peoples.[58] Northrup contends that "one of the major implications is that intractable conflicts involving threatened identities are not likely to be readily changed from within."[59]

The framing of disputes and issues within disputes has recently emerged as an area of research interest in the domain of environmental conflict[60] and in research on social conflict.[61] Gray argues that the "risk frames" disputants hold can also affect a dispute's resistance to resolution, as determined by five factors:

1. whether the participants expect gains or losses;
2. whether participants see the dispute as involving rights, interests, or values;
3. whether there are discrepancies between parties' value frameworks; or
4. whether there are threats to identity (individual or collective); and
5. whether participants hold differing perceptions of how fairly the risk is distributed among them.

If the disputants in a conflict have compatible risk frames, the potential for resolving conflict is enhanced; however, if risk frames are incompatible, then the conflict is more likely to become intractable.[62] Dietz adds another dimension: whether the potential risks are voluntarily assumed or involuntarily imposed—if the latter, they can create a sense of moral outrage that fuels conflict.[63]

Hunter uses frame analysis to explore the level and intensity of environmental conflict. She reports the results of three mail questionnaires used in a case study of deep-rooted environmental conflict in the Lake Tahoe region.[64] The questionnaires surveyed government officials, environmentalists, and local property owners to learn whether these groups framed the Lake Tahoe conflict (and the world) differently. Hunter finds that frame analysis can help identify possible sources of agreement that may then yield stable settlements.[65] How disputes are framed may not just influence the way a problem is defined and which facts are considered relevant,[66] but may also favor "production sciences" (e.g., engineering, economics, and chemistry) over "impact sciences" (e.g., ecology and sociology) in the debate over environmental policy, thereby privileging private-sector interests over public interest groups.[67]

Research on the characteristics of intractable environmental and public policy disputes is in a nascent state.[68] Thus, little is known about the relative frequency of these characteristics or any relative hierarchy of their importance to dispute intractability. Research on the framing of intractable environmental and public policy disputes is underway, but has not yet been published.[69] Both of these areas are likely to be fruitful avenues for research to enhance our understanding of intractable disputes and explore the possibility of transforming them into more tractable ones.

Transforming Intractability

An emergent theme in conflict resolution theory concerns the transformative approach to conflict resolution. Bush and Folger compare this approach, which seeks growth and improvements in the disputing parties themselves, to the more dominant problem-solving approach to mediation, which seeks improvements in the parties' situations.[70] They argue that the transformative approach should become the norm for all areas of conflict resolution. In transformative mediation, disputes and conflicts are not necessarily seen as destructive events, but rather as opportunities for growth and transformation in the disputing parties. Two important results of transformative dispute resolution processes are (1) the empowerment of the parties in mediation, which involves clarifying goals and objectives, identifying options, improving conflict resolution skills, creating awareness of resources, and enhancing decisionmaking capabilities and (2) achieving recognition in the mediation process, in which disputing parties come to understand and acknowledge each other's points of view. These results are separate objectives from that of resolving the problems at issue in a dispute.[71]

Conflict theorists have applied Bush and Folger's notion of transformative mediation to the realm of environmental and public policy conflict. Burgess and Burgess have developed an approach to intractable public policy conflicts that they call "constructive confrontation."[72] Their method entails three steps:

1. diagnosis of the dispute and preparation of a conflict "map" showing disputants, potential third-party intermediaries, interests and positions, and the long-term conflict context in which the dispute has arisen;
2. identification of "conflict overlays" (although often important, these are surface issues in a dispute that obscure deeper underlying issues) and their separation from core aspects of the dispute; and
3. reduction of the conflict overlay problems so that the core issues of the dispute can be addressed.

Burgess and Burgess acknowledge that several of the goals of constructive confrontation are similar to those of Bush and Folger's transformative approach. These goals include helping disputants reach an understanding of each other's perspectives and helping them separate overlay issues from the core issues in the dispute.[73] They go on, however, to encourage disputants to create strategies for constructive confrontation that keep their own interests and objectives at the forefront. Burgess and Burgess state that the major difference between constructive confrontation and transformative mediation is that the former has been developed specifically for disputing parties to use on their own in intractable public policy conflicts, whereas the latter is intended primarily for interpersonal disputes mediated by a third party.

Maser contends that Bush and Folger's transformative mediation is the "only ... facilitative approach that can begin to accomplish [the] resolution" of destructive environmental conflicts.[74] On the other hand, Dukes disagrees with Bush and Folger's position that the individual moral growth of disputing parties should be the primary goal of mediation.[75] He argues instead that the transformative approach can be used in the public policy realm to create a "healthy democracy" by "embracing conflict ... as central to the political process."[76] Dukes' method for realizing the transformative agenda in society involves fostering public discourse, making government more responsive to the public, and improving societal capacity for problem solving, while emphasizing that parties need not alter their fundamental world views to achieve workable solutions.[77] Forester argues that "the wisdom of mediation involves knowing what does not need to change while enabling participants to climb down from the abstract peaks of value systems, paradigms, and world views to get their hands dirty in a collaborative discussion of actual options they may endorse together."[78] He maintains that planners and policymakers can potentially shape this policy discourse by influencing the public's understanding of the dynamics of public disputes.[79]

Because it encourages disputants to work to understand each other's viewpoints and confront issues that are central to the dispute, transformative mediation has potential as a method for managing environmental disputes that arise out of clashes of fundamental values, rights, and world views. However, its application to these disputes should also be informed by the literature on intractable disputes, which offers additional insights into the sources, nature, and dimensions of environmental conflict. This literature suggests that there exists a continuum of dispute intractability because many of these characteristics are also found in disputes that are not considered intractable.[80] A critical aspect of managing and resolving intractable disputes, then, is ascertaining how far along the continuum of intractability a dispute has gone—to what degree a dispute is intractable and what makes it so—and how to work with stakeholders on aspects of the dispute to help pull it back along the continuum into the range of less resolution-resistant disputes. Empirical research on both successful and failed mediations of intractable environmental disputes is called for.

Lessons Learned for Research

The practice of environmental dispute resolution is well established.[81] However, little empirical research has been done on any aspect of environmental dispute resolution, placing the empirical foundations of the field on less solid ground.[82] As recently as the early 1990s, the bulk of the research (empirical or otherwise) had been done by practitioners and proponents of environmental dispute resolution, thereby raising the specter of a conflict of interest concerning their assessments of its usefulness.[83] O'Leary has commented:

> How do we know what we know? We know what we know primarily through atheoretical case studies, "wisdom" derived from the experiences of environmental mediators, and conceptual thinking. There is little rigorous empirical evidence. ... We need survey research and additional comparative case studies that examine not just the spectacular cases but also the less-spectacular, less-successful cases of environmental mediation. Moreover, we need additional studies of what interventions are more likely to be successful than others under what conditions, as well as long-range longitudinal studies of the outcomes of environmental mediation efforts. ... Most important, we need an adequate theoretical base from which researchers can predict effects, test them, and ascertain, in a more systematic and rigorous fashion, the impact of environmental mediation.[84]

Researchers are beginning to address the lack of systematic empirical research. For example, Sipe's five-year study of mediation activities at the Florida Department of Environmental Protection tested two important claims about mediation: (1) mediation results in higher settlement rates than more traditional forms of dispute resolution, such as adjudication and legislation, and (2) rates of compliance with and implementation of mediated settlements are higher than those for settlements resulting from traditional processes.[85] The results of Sipe's logistic regression analysis support the first claim but not the second.

O'Leary and others[86] have also evaluated the status of state environmental dispute resolution programs, grading the programs on their management abilities, institutional capacity, and commitment to environmental protection, and looking for evidence of leadership innovation. D'Estrée and Colby have developed a guidebook for the systematic assessment of environmental conflict resolution outcomes based on a set of six categories of criteria, which they have used to evaluate eight southwestern U.S. water conflicts.[87,88] This guidebook, which can be used to analyze negotiated, mediated, or adjudicated cases, has significant potential for developing a uniform body of detailed, comparative case analyses.

Concerning local and regional disputes, two recently published reports offer findings from a Consensus Building Institute (CBI) study of 100 multiparty land-use disputes around the United States, all of which were mediated by professional third-party neutrals. Of the 100 cases, two-thirds were considered settled and one-third remained unsettled. CBI researchers conducted person-to-person interviews with at least three key participants from each case.[89] Respondents identified three groups of obstacles to mediated settlements: tension among stakeholders, procedural obstacles, and substantive obstacles. Of these, stakeholder tensions, including distrust among parties, entrenched positions, conflicting values, and personality issues, presented the greatest overall challenge to mediated settlements and were reported as obstacles 52% of the time. These kinds of tensions closely parallel the characteristics of intractability presented in Box 5-1 and are worthy of greater research attention.

Although some empirical assessments of intractability are underway, few have been published.[90] Interesting research questions emerge concerning which characteristics contribute most strongly to dispute intractability, as well as how they combine and interact to render disputes intractable. What are the relative contributions to intractability of the different characteristics listed in Box 5-1? Does any one characteristic "trump" the others in rendering disputes intractable? How do these different aspects of intractability interact in the context of a dispute? What influence does issue framing have on intractability? At what point are mediators aware that they have an intractable dispute on their hands, and what do they do to manage it?

In an exploratory study, I describe the results from a survey of professional environmental and public policy mediators' views on a set of 11 characteristics of intractability.[91] Mediators ranked these characteristics for their relative contributions to dispute intractability, first within the context of four specific dispute scenarios, and then in an ordinal ranking independent of the dispute scenarios. In both instances, "moral differences between the parties" was judged to be the leading contributor to intractability. However, in the independent ordinal ranking, "differences in issue framing" emerged as a significant contributor to intractability, despite mediators having ranked this characteristic extremely low as a contributor when considered in the context of the dispute scenarios.

Although this study did not develop a definitive sorting of the characteristics of intractability, the findings could have some bearing on future research. For example, because the mediators' judgements could well have been bound to the four scenarios presented in the survey, their varying assessments of issue framing's importance might indicate a cognitive bias on their part about what they think drives intractable disputes.[92] This bias could lead mediators to make very different judgements about dispute intractability in the abstract than they would make about intractability in context-specific situations.

Comprehensive evaluation studies of mediation and consensus-building efforts are essential to learning more about what works and what does not. Improved knowledge of the dimensions of intractability could also be used in program and case study evaluation to identify more productive ways of moving a dispute away from impasse and toward settlement. The studies reported above are important developments in the direction that O'Leary recommends researchers take,[93] but much research remains to be done.

Conclusions

Unlike other disputes that might revolve around a central issue (or issues) and have a small number of disputing parties, environmental and public policy conflicts typically involve multiple parties, have potential stakeholders who are difficult to represent at the table (future generations or the public interest, for example), and are riddled with complex and highly technical issues that sometimes have significant spillover effects.[94] These kinds of disputes can be difficult to resolve, and, as several commentators have suggested,[95] they may bring greater responsibilities upon the mediator than those who attend the mediation of less complicated disputes.

The planner's role in environmental and public policy dispute resolution is a growing one. Planners are accustomed to working with tools for data forecasting and analysis, policy analysis, and site design, but they are also knowledgeable practitioners of the strategizing, facilitating, and communicating that are required of a third-party intermediary.[96] Planners may represent one of the stakeholder parties at the table (such as a local or regional agency, a local government, or a neighborhood community) or they may be asked to take on the role of a trusted third-party intermediary, for example, between the local community and a developer.[97]

As employees of city, state, regional, or federal governments and agencies, planners cannot really claim to be neutral third parties.[98] Susskind and Ozawa argue that a nonneutral or activist stance is an important aspect of the planner's role as mediator because it enables planners to ensure that critical minority interests are represented at the table.[99] Glavovic, Dukes, and Lynott go further by arguing that mediators who practice transformative environmental mediation must be advocates for the environment and be "proactive and persistent in seeking to ensure that the negotiation process promotes ... inclusivity ... empowerment ... mutual learning ... and shared responsibility."[100] Few public-sector professionals, however, have the combination of qualities and experience the authors use to describe the "consummate environmental mediator."[101] Nor do they have the skills or problem-solving ability required to practice the art of frame-reflective policy design that Schön and Rein advocate as a method for handling intractable policy controversies.[102]

Despite the specialized training that planners undergo to become professionals in their field, few of them are sufficiently trained in dispute resolution to handle intractable environmental or public policy disputes. To do so effectively, planners would need specific and extensive mediation and consensus-building skills training and practice, a critical component of which involves learning how to conduct conflict assessments. These assessments are used to probe the dimensions of a dispute and assess the feasibility of engaging in consensus building, mediation, or other dispute resolution processes. Mediators conducting conflict assessments identify the range of stakeholders (primary and secondary); tease out the critical issues at stake (including which ones are central to the dispute and which are side issues); explore how stakeholders frame these issues and how strongly they hold their views; examine the historical, social, and cultural contexts of the dispute; and design a preliminary dispute resolution process.[103]

Research on intractability has ramifications for planners attempting to resolve disputes. A more nuanced understanding of intractability (and mediators' responses to it) could inform planners' initial conflict assessments and lead to more effective management of resolution-resistant disputes. This would apply both to individual episodes of disputes and to evolving or re-emergent conflicts. In some cases, greater knowledge of the problems inherent in managing intractable disputes might reasonably make planners wary of tackling them on their own. Planners could then encourage decision-makers to seek the services of more highly skilled dispute resolution professionals.

Dispute resolution systems design could also incorporate this greater awareness of a dispute's complexity and potential for resistance to resolution. The New Community Meeting Model, developed by the Conflict Resolution Center in Eugene, Oregon, is an example of this type of attention to process design. This model was used to build high levels of trust and effective working relationships among community leaders who had become alienated from one another—in one case, over religious conflict, and in the other, over a land-use planning process to control regional growth. The 12-month consensus-building process involved building participants' listening skills, identifying core values and operating principles, and structured dialogue and facilitated consensus building.[104]

Until professional requirements for conflict resolution training become common or environmental and public policy dispute resolution courses are more widely offered in planning schools' curricula,[105,106] it is important for planners to be able to identify environmental and public policy disputes that have the potential to become intractable, so that they can determine how best to respond to them. Planners may be able to make use of assessment tools, strategies, and tactics commonly employed by dispute resolution professionals to prevent a dispute from spiraling into unmanageable conflict.[107] In other instances, they may be able to help disputing parties reframe portions of the dispute to enable the parties to work more productively toward a settlement. In still others, planners may be able to help parties achieve workable, practical solutions to seemingly intractable environmental and public policy problems.

Notes

1. Karen S. Christensen, "Coping with Uncertainty in Planning," *Journal of the American Planning Association,* vol. 51, no. 1 (1985), pp. 63–71.

2. Compare Norman Beckman, "The Planner as Bureaucrat," reprinted in Andreas Faludi, ed., *A Reader in Planning Theory* (Pergamon, 1973); Anthony J. Cantanese, *The Politics of Planning and Development* (Beverly Hills, CA: Sage Publications, 1984); Elizabeth Howe, "Professional Roles and the Public Interest in Planning," *Journal of Planning Literature,* vol. 6, no. 3 (1992), pp. 230–248; Martin Meyerson and Edward Banfield, eds., *Politics, Planning and the Public Interest* (New York: Free Press, 1955); Lawrence Susskind, Mieke van der Wansem, and Armand Ciccarelli, *Mediating Land Use Disputes: Pros and Cons* (Cambridge, MA: Lincoln Institute of Land Policy, 2000).

3. A. Bruce Dotson, David Godschalk, and Jerome Kaufman, *The Planner as Dispute Resolver: Concepts and Teaching Materials* (Washington DC: National Institute for Dispute Resolution, 1989); John Friedmann and Carol Kuester, "Planning Education for the Late Twentieth Century: An Initial Inquiry," *Journal of Planning Education and Research,* vol. 14, no. 1 (1994), pp. 55–64.

4. Dotson, Godschalk, and Kaufman, *The Planner as Dispute Resolver;* John Forester and David Stitzel, "Beyond Neutrality: The Possibilities of Activist Mediation in Public Sector Conflicts," *Negotiation Journal,* vol. 5, no. 3 (1989), pp. 251–264; Francine F. Rabinovitz, "The Role of Negotiation in Planning, Management, and Policy Analysis," *Journal of Planning Education and Research,* vol. 8, no. 2 (1989), pp. 87–95; Lawrence Susskind and Connie Ozawa, "Mediated Negotiation in the Public Sector: The Planner as Mediator," *Journal of Planning Education and Research,* vol. 4, no. 1 (1984), pp. 5–15; see also Noah Dorius, "Land Use Negotiation: Reducing Conflict and Creating Wanted Land Uses," *Journal of the American Planning Association,* vol. 59, no. 1 (1993), pp. 101–106; David R. Godschalk, "Negotiating Intergovernmental Policy Conflicts: Practice-Based Guidelines," *Journal of the American Planning Association,* vol. 58, no. 3 (1992), pp. 368–378; Daphne Spain, "Been-Heres Versus Come-Heres: Negotiating Conflicting Community Identities," *Journal of the American Planning Association,* vol. 59, no. 2 (1993), pp. 156–171.

5. Dotson, Godschalk, and Kaufman, *The Planner as Dispute Resolver;* Susskind and Ozawa, "Mediated Negotiation in the Public Sector: The Planner as Mediator."

6. John Forester, "Planning in the Face of Conflict: Negotiation and Mediation Strategies in Local Land Use Regulations," *Journal of the American Planning Association,* vol. 53, no. 3 (1987), pp. 303–314; Susskind and Ozawa, "Mediated Negotiation in the Public Sector: The Planner as Mediator."

7. John Forester, "Envisioning the Politics of Public Sector Dispute Resolution," in S.S. Silbey and A. Sarat, eds., *Studies in Law, Politics, and Society* (Greenwich, CT: JAI Press, 1992); Susskind and Ozawa, "Mediated Negotiation in the Public Sector: The Planner as Mediator."

8. Susskind, van der Wansem, and Ciccarelli, *Mediating Land Use Disputes.*

9. Lawrence J. MacDonnell, "Natural Resources Dispute Resolution: An Overview," *Natural Resources Journal,* vol. 28, no. 1 (1988), pp. 5–19; Connie P. Ozawa, *Recasting Science: Consensual Procedures in Public Policy Making* (Boulder, CO: Westview Press, 1991).

10. Thomas Dietz, "Thinking about Environmental Conflicts," in Lisa M. Kadous, ed., *Celebrating Scholarship* (Fairfax, VA: College of Arts and Sciences, George Mason University, 2001); Barbara Gray, "Framing and Reframing of Intractable Environmental Disputes," in Roy J. Lewicki, Robert J. Bies, and Blair H. Sheppard, eds., *Research on Negotiation in Organizations* (Greenwich, CT: JAI Press, 1997); Michael S. Hamilton, "Environmental Mediation: Requirements for Successful Institutionalization," in Miriam K. Mills, ed., *Alternative Dispute Resolution in the Public Sector* (Chicago: Nelson-Hall, 1991).

11. Gray, "Framing and Reframing of Intractable Environmental Disputes"; Susan Hunter, "The Roots of Environmental Conflict in the Tahoe Basin," in Louis Kriesberg, Terrell A. Northrup, and Stuart J. Thorson, eds., *Intractable Conflicts and Their Transformation* (Syracuse University Press, 1989).

12. Marcia Caton Campbell and Donald W. Floyd, "Thinking Critically about Environmental Mediation," *Journal of Planning Literature,* vol. 10, no. 3 (1996), pp. 235–247; see also John Forester, "Dealing with Deep Value Differences," in Lawrence Susskind, Sarah McKearnan, and Jennifer Thomas-Larmer, eds., *The Consensus Building Handbook: A Comprehensive Guide to Reaching Agreement* (Thousand Oaks, CA: Sage Publications, 1999); Robert H. Socolow, "Failures of Discourse: Obstacles to the Integration of Environmental Values into Natural Resource Policy," in Laurence H. Tribe, Corinne S. Schelling, and John Voss, eds., *When Values Conflict: Essays on Environmental Analysis, Discourse, and Decision* (Cambridge, MA: Ballinger, 1976); Laurence H. Tribe, Corinne S. Schelling, and John Voss, eds., *When Values Conflict: Essays on Environmental Analysis, Discourse, and Decision* (Cambridge, MA: Ballinger, 1976).

13. Guy Burgess and Heidi Burgess, "Beyond the Limits: Dispute Resolution of Intractable Environmental Conflicts," in J. Walton Blackburn and Willa Marie Bruce, eds., *Mediating Environmental Conflicts: Theory and Practice* (Westport, CT: Quorum Books, 1995); William L. Ury, Jeanne M. Brett, and Stephen B. Goldberg, *Getting Disputes Resolved: Designing Systems To Cut the Costs of Conflict* (San Francisco: Jossey-Bass, 1988).

14. Douglas J. Amy, *The Politics of Environmental Mediation* (Columbia University Press, 1987); Terrell A. Northrup, "The Dynamic of Identity in Personal and Social Conflict," in Louis Kriesberg, Terrell A. Northrup, and Stuart J. Thorson, eds., *Intractable Conflicts and Their Transformation* (Syracuse University Press, 1989); Lawrence Susskind and Patrick Field, *Dealing with an Angry Public: The Mutual Gains Approach to Resolving Disputes* (New York: The Free Press, 1996).

15. Tamra Pearson d'Estrée and Bonnie G. Colby, *Guidebook for Analyzing Success in Environmental Conflict Resolution Cases* (Fairfax, VA: The Institute for Conflict Analysis and Resolution, George Mason University, 2000); Judith E. Innes, "Evaluating Consensus Building," in Lawrence Susskind, Sarah McKearnan, and Jennifer Thomas-Larmer, eds., *The Consensus Building Handbook: A Comprehensive Guide to Reaching Agreement* (Thousand Oaks, CA: Sage Publications, 1999); Judith E. Innes

and David E. Booher, "Consensus Building and Complex Adaptive Systems: A Framework for Evaluating Collaborative Planning," *Journal of the American Planning Association,* vol. 65, no. 4 (1999), pp. 412–423; Rosemary O'Leary and Tracy Yandle, "The State of the States in Environmental Dispute Resolution: Implications for Public Management in the New Millenium," *Journal of Public Administration Research and Theory,* vol. 10, no. 1 (2000), pp. 137–156.

16. Rosemary O'Leary, "Environmental Mediation: What Do We Know and How Do We Know It?" in J. Walton Blackburn and Willa Marie Bruce, eds., *Mediating Environmental Conflicts: Theory and Practice* (Westport, CT: Quorum Books, 1995); Barry G. Rabe, "Impediments to Environmental Dispute Resolution in the American Political Context," in Miriam K. Mills, ed., *Alternative Dispute Resolution in the Public Sector* (Chicago: Nelson-Hall, 1991).

17. Caton Campbell and Floyd, "Thinking Critically about Environmental Mediation"; for one example, see Neil G. Sipe, "An Empirical Analysis of Environmental Mediation," *Journal of the American Planning Association,* vol. 64, no. 3 (1998), pp. 275–285.

18. Guy Burgess and Heidi Burgess, "Some Thoughts on Intractable Conflict: Conference Framing Essay," paper presented at meeting of the Hewlett-Funded Centers for the Study of Conflict Resolution and Negotiation, January 9–11, 1998, at Palo Alto, CA.

19. This determination was made by participants in "The Meaning of Intractability: Sessions I and II," at the 1998 meeting of the Hewlett-Funded Centers for the Study of Conflict Resolution and Negotiation, January 9–11 in Palo Alto, CA. The overall conference theme was intractable conflict. Conference participants were leading researchers and practitioners in the dispute resolution field.

20. Burgess and Burgess, "Some Thoughts on Intractable Conflict."

21. Peter T. Coleman, "Intractable Conflict," in Morton Deutsch and Peter T. Coleman, eds., *The Handbook of Conflict Resolution* (San Francisco: Jossey-Bass, 2000), p. 429.

22. Gray, "Framing and Reframing of Intractable Environmental Disputes."

23. Lawrence Susskind and Jeffrey Cruikshank, *Breaking the Impasse: Consensual Approaches to Resolving Public Disputes* (New York: Basic Books, 1987).

24. Donald A. Schön and Martin Rein, *Frame Reflection: Toward the Resolution of Intractable Policy Controversies* (New York: Basic Books, 1994); Susskind and Cruikshank, *Breaking the Impasse.*

25. Schön and Rein, *Frame Reflection.*

26. Coleman, "Intractable Conflict"; Robert A. Rubinstein, "Intractable Conflicts and Possibilities for Resolution: Conference Framing Essay," paper presented at meeting of the Hewlett-Funded Centers for the Study of Conflict Resolution and Negotiation, January 9–11, 1998, Palo Alto, CA; Stuart J. Thorson, "Introduction: Conceptual Issues," in Louis Kriesberg, Terrell A. Northrup, and Stuart J. Thorson, eds., *Intractable Conflicts and Their Transformation* (Syracuse University Press, 1989).

27. Northrup, "The Dynamic of Identity in Personal and Social Conflict," p. 62.

28. Northrup's assertion to the contrary, all is not rosy in apparently tractable conflicts. They are still capable of yielding upset or angry disputing parties who have difficulty arriving at a solution to their dispute, and if one accepts that intractability occurs along a continuum, tractable conflicts still have the potential to become

intractable if something goes awry (see Leonard Greenhalgh, "Managing Conflict," *Sloan Management Review* (Summer 1986), pp. 45–51).

29. Greenhalgh, "Managing Conflict."

30. Hunter, "The Roots of Environmental Conflict in the Tahoe Basin."

31. Ibid., p. 26.

32. A clear example of an ontology is the set of views held by deep ecologists, who view humans and nonhuman species both as having equal value, and value in and of themselves. Deep ecologists believe that humans do not have the right to engage in direct or indirect actions that would adversely affect nonhuman species or reduce biodiversity anywhere on earth. Examples of impermissible direct actions include agriculture, mining, forestry, and technology. Impermissible indirect actions include population growth or any social or economic policies that harm humans or nonhuman species. This ontology leads deep ecologists to call for drastic reductions in the population, for example, or to spike trees in forests so that they cannot be cut down (Peter Borrelli, "The Ecophilosophers," in Peter Borrelli, ed., *Crossroads: Environmental Priorities for the Future* (Washington, DC: Island Press, 1988).

33. Amy, *The Politics of Environmental Mediation*; Douglas J. Amy, "Environmental Dispute Resolution: The Promise and the Pitfalls," in Norman J. Vig and Michael E. Kraft, eds., *Environmental Policy in the 1990s: Toward a New Agenda* (Washington, DC: CQ Press, 1990); Caton Campbell and Floyd, "Thinking Critically about Environmental Mediation"; Donald W. Floyd, "Managing Rangeland Resource Conflicts," *Rangelands,* vol. 15, no. 1 (1993), pp. 27–30; Harvey M. Jacobs and Richard G. RuBino, *Predicting the Utility of Environmental Mediation: Natural Resource and Conflict Typologies as a Guide to Environmental Conflict Assessment* (Institute for Legal Studies, University of Wisconsin—Madison Law School, 1988); MacDonnell, "Natural Resources Dispute Resolution"; O'Leary, "Environmental Mediation"; An Painter, "The Future of Environmental Dispute Resolution," *Natural Resources Journal,* vol. 28, no. 1 (1988), pp. 145–170; Rabe, "Impediments to Environmental Dispute Resolution in the American Political Context."

34. Forester, "Dealing with Deep Value Differences."

35. E. Franklin Dukes, *Resolving Public Conflict: Transforming Community and Governance* (Manchester University Press, 1996), p. 175; see also Bryan G. Norton, *Toward Unity among Environmentalists* (New York: Oxford University Press, 1991).

36. Burgess and Burgess, "Beyond the Limits"; Heidi Burgess and Guy Burgess, "Constructive Confrontation: A Transformative Approach to Intractable Conflicts," *Mediation Quarterly,* vol. 13, no. 4 (1996), pp. 305–322.

37. Dietz, "Thinking about Environmental Conflicts"; Louis Kriesberg, "Intractable Conflicts," *Peace Review,* vol. 5, no. 4 (1993), pp. 417–421; Painter, "The Future of Environmental Dispute Resolution."

38. Amy, *The Politics of Environmental Mediation*; Hunter, "The Roots of Environmental Conflict in the Tahoe Basin."

39. Kriesberg, "Intractable Conflicts"; Rubinstein, "Intractable Conflicts and Possibilities for Resolution."

40. Northrup, "The Dynamic of Identity in Personal and Social Conflict."

41. The spotted owl–logging dispute has apparently reached a point of resolution in the Quincy Library agreement, though there is serious concern about the long-range implications of the agreement (see Julia M. Wondolleck and Steven L. Yaffee,

Making Collaboration Work: Lessons from Innovation in Natural Resource Management (Washington, DC: Island Press, 2000)). However, protracted, underlying conflict pervades the Pacific Northwest over stakeholder identity, way of life, and environmental management and protection. In its most recent incarnation, a seemingly intractable dispute has arisen in Oregon's Klamath Basin between farmers and environmentalists over water use (irrigation) rights and endangered species (fish and bald eagles).

42. Gray, "Framing and Reframing of Intractable Environmental Disputes"; Northrup, "The Dynamic of Identity in Personal and Social Conflict"; Rubinstein, "Intractable Conflicts and Possibilities for Resolution."

43. Hunter, "The Roots of Environmental Conflict in the Tahoe Basin."

44. See, e.g., Burgess and Burgess, "Constructive Confrontation"; Dean G. Pruitt and Paul V. Olczak, "Beyond Hope: Approaches to Resolving Seemingly Intractable Conflict," in Barbara B. Bunker and Jeffrey Z. Rubin, eds., *Conflict, Cooperation, and Justice: Essays Inspired by the Work of Morton Deutsch* (San Francisco: Jossey-Bass, 1995); Rubinstein, "Intractable Conflicts and Possibilities for Resolution"; I. William Zartman and Johannes Aurik, "Power Strategies in De-Escalation," in Louis Kriesberg and Stuart J. Thorson, eds., *Timing and the De-Escalation of International Conflicts* (Syracuse University Press, 1991).

45. Northrup, "The Dynamic of Identity in Personal and Social Conflict."

46. Hunter, "The Roots of Environmental Conflict in the Tahoe Basin."

47. Kriesberg, "Intractable Conflicts."

48. Amy, *The Politics of Environmental Mediation*; Jeanne M. Brett, Debra L. Shapiro, and Anne L. Lytle, "Refocusing Rights- and Power-Oriented Negotiators toward Integrative Negotiations: Process and Outcome Effects" (paper presented at Academy of Management Proceedings, 1996); Gray, "Framing and Reframing of Intractable Environmental Disputes."

49. Burgess and Burgess, "Beyond the Limits"; "Some Thoughts on Intractable Conflict"; "Constructive Confrontation."

50. Dietz, "Thinking about Environmental Conflicts."

51. Northrup, "The Dynamic of Identity in Personal and Social Conflict."

52. Burgess and Burgess, "Beyond the Limits"; "Constructive Confrontation." Burgess and Burgess's definition of high-stakes distributional issues and the ease with which they can be resolved runs counter to the definition of distributional disputes given by Susskind and Cruikshank, *Breaking the Impasse*, discussed earlier in this chapter. One interpretation of the difference might be that the Burgesses' definition bears a closer resemblance to the policy controversies described by Schön and Rein, *Frame Reflection*; another interpretation is that the two sets of researchers simply disagree.

53. Burgess and Burgess, "Some Thoughts on Intractable Conflict"; "Constructive Confrontation."

54. Dietz, "Thinking about Environmental Conflicts."

55. Burgess and Burgess, "Constructive Confrontation."

56. Burgess and Burgess, "Some Thoughts on Intractable Conflict"; Dietz, "Thinking about Environmental Conflicts"; Louis Kriesberg, "Conclusion: Research and Policy Implications," in Louis Kriesberg, Terrell A. Northrup, and Stuart J. Thorson, eds., *Intractable Conflicts and Their Transformation* (Syracuse University Press, 1989); Northrup, "The Dynamic of Identity in Personal and Social Conflict";

Leo F. Smyth, "Intractable Conflict and the Role of Identity," *Negotiation Journal,* vol. 10, no. 4 (1994), pp. 311–321.

57. Northrup, "The Dynamic of Identity in Personal and Social Conflict."

58. A subsequent proposal to tunnel under the monument for the highway extension was similarly rejected by the tribes. The Paseo del Norte conflict remains unresolved at the time of this writing.

59. Northrup, "The Dynamic of Identity in Personal and Social Conflict," p. 76.

60. Gray, "Framing and Reframing of Intractable Environmental Disputes"; Roy J. Lewicki and Craig Davis, "Environmental Conflict Frames in Wetland Conversion Disputes," paper presented at the Center for Research in Conflict and Negotiation, Pennsylvania State University, April 1996; Gi-Chul Yi, "An Analysis of Disputants' Conflict Frames Relating to Ohio Wetland Conversion Disputes," Ph.D. dissertation, School of Natural Resources, The Ohio State University, 1992.

61. Barbara Gray, Julie Younglove-Webb, and Jill M. Purdy, "Frame Repertoires, Conflict Styles and Negotiation Outcomes," paper presented at International Association of Conflict Management, Bonn, Germany, 1997; Robin L. Pinkley, "Dimensions of Conflict Frames: Disputant Interpretations of Conflict," *Journal of Applied Psychology,* vol. 75, no. 2 (1990), pp. 117–126; Robin L. Pinkley and Gregory B. Northcraft, "Conflict Frames of Reference: Implications for Dispute Processes and Outcomes," *Academy of Management Journal,* vol. 37, no. 1 (1994), pp. 193–205.

62. Gray, "Framing and Reframing of Intractable Environmental Disputes."

63. Dietz, "Thinking about Environmental Conflicts."

64. Hunter, "The Roots of Environmental Conflict in the Tahoe Basin."

65. But compare Burgess and Burgess, "Some Thoughts on Intractable Conflict"; see framing as an overlay problem.

66. Schön and Rein, *Frame Reflection.*

67. Dietz, "Thinking about Environmental Conflicts."

68. See Marcia Caton Campbell, "Exploring the Characteristics of Intractable Environmental Disputes," paper presented at the 40th Annual Conference of the Association of Collegiate Schools of Planning, November 4–8, 1998, Pasadena, CA.

69. A consortium of researchers from The Ohio State University, Pennsylvania State University, Cleveland State University, and the Georgia Institute of Technology are now preparing for publication the results of a Hewlett Foundation–funded study of framing in intractable environmental disputes.

70. Robert A. Baruch Bush and Joseph P. Folger, *The Promise of Mediation: Responding to Conflict through Empowerment and Recognition* (San Francisco: Jossey-Bass, 1994).

71. Ibid.

72. Burgess and Burgess, "Beyond the Limits"; "Constructive Confrontation."

73. Burgess and Burgess, "Constructive Confrontation."

74. Chris Maser, *Resolving Environmental Conflict: Towards Sustainable Community Development* (Delray Beach, FL: St. Lucie Press, 1996), p. xv.

75. Dukes, *Resolving Public Conflict.*

76. Ibid., p. 164.

77. Dukes, *Resolving Public Conflict.*

78. Forester, "Dealing with Deep Value Differences," p. 485.

79. Forester, "Envisioning the Politics of Public Sector Dispute Resolution." See also the discussion in Schön and Rein, *Frame Reflection*, pp. 50–51.

80. See Greenhalgh, "Managing Conflict."

81. See Gail Bingham, *Resolving Environmental Disputes: A Decade of Experience* (Washington, DC: Conservation Foundation, 1986); Caton Campbell and Floyd, "Thinking Critically about Environmental Mediation."

82. Caton Campbell and Floyd, "Thinking Critically about Environmental Mediation"; Morton Deutsch, "A Framework for Thinking about Research on Conflict Resolution Training," in Morton Deutsch and Peter T. Coleman, eds., *The Handbook of Conflict Resolution* (San Francisco: Jossey-Bass, 2000); O'Leary, "Environmental Mediation"; Sipe, "An Empirical Analysis of Environmental Mediation."

83. Rabe, "Impediments to Environmental Dispute Resolution in the American Political Context."

84. O'Leary, "Environmental Mediation," p. 32.

85. Sipe, "An Empirical Analysis of Environmental Mediation."

86. Rosemary O'Leary, Tracy Yandle, and Tamilyn Moore, "The State of the States in Environmental Dispute Resolution: Implications for Public Management in the New Millenium," *Ohio State Journal on Dispute Resolution,* vol. 14, no. 2 (1999), pp. 515–613.

87. D'Estrée and Colby, *Guidebook for Analyzing Success in Environmental Conflict Resolution Cases.* The categories are outcome reached, process quality, outcome quality, relationship of parties to outcome, relationship between parties, and social capital. Within each category, d'Estrée and Colby have identified many more detailed evaluation criteria.

88. I am also using the guidebook to evaluate outcomes of cases from Wisconsin's Waste Facility Siting Program, which has been in existence for 12 years and handles the siting of solid waste landfills and hazardous waste facilities around the state.

89. Lawrence Susskind and the Consensus Building Institute, *Using Assisted Negotiation To Settle Land Use Disputes: A Guidebook for Public Officials* (Cambridge, MA: Lincoln Institute of Land Policy, 1999), p. 3; Susskind, van der Wansem, and Ciccarelli, *Mediating Land Use Disputes*, p. 14.

90. Northrup, "The Dynamic of Identity in Personal and Social Conflict."

91. Caton Campbell, "Exploring the Characteristics of Intractable Environmental Disputes."

92. See Blair H. Sheppard, "Third Party Conflict Intervention: A Procedural Framework," in B.M. Staw and L.L. Cummings, eds., *Research in Organizational Behavior* (Greenwich, CT: JAI Press, 1984).

93. O'Leary, "Environmental Mediation," p. 32.

94. Lawrence Susskind and Connie Ozawa, "Mediated Negotiation in the Public Sector: Mediator Accountability and the Public Interest Problem," *American Behavioral Scientist,* vol. 27, no. 2 (1983), pp. 255–279.

95. Dukes, *Resolving Public Conflict;* Forester, "Envisioning the Politics of Public Sector Dispute Resolution"; Forester and Stitzel, "Beyond Neutrality"; Bruce C. Glavovic, E. Franklin Dukes, and Jana M. Lynott, "Training and Educating Environmental Mediators: Lessons from Experience in the United States," *Mediation Quarterly,* vol. 14, no. 4 (1997), pp. 269–292; Susskind and Ozawa, "Mediated

Negotiation in the Public Sector: Mediator Accountability and the Public Interest Problem."

96. Dotson, Godschalk, and Kaufman, *The Planner as Dispute Resolver;* Forester, "Planning in the Face of Conflict"; Rabinovitz, "The Role of Negotiation in Planning, Management, and Policy Analysis"; Susskind and Ozawa, "Mediated Negotiation in the Public Sector: The Planner as Mediator."

97. Dukes, *Resolving Public Conflict;* Forester, "Planning in the Face of Conflict."

98. Forester and Stitzel, "Beyond Neutrality."

99. Susskind and Ozawa, "Mediated Negotiation in the Public Sector: Mediator Accountability and the Public Interest Problem"; "Mediated Negotiation in the Public Sector: The Planner as Mediator."

100. Glavovic, Dukes, and Lynott, "Training and Educating Environmental Mediators," p. 262.

101. Ibid., p. 279. The authors say, "The consummate environmental mediator, one who is both ethical and effective, needs to embody at least the following six qualities: 1. Advocacy for sustainable development, 2. Environmental literacy, that is, familiarity with the language and substance of environmental science and public policy, 3. Significant life experience, 4. Commitment, integrity, and trustworthiness, 5. The ability to adopt different dispute resolution styles and behaviors, and 6. Superb planning and organizational capacity."

102. Schön and Rein, *Frame Reflection.*

103. See Susan L. Carpenter and W.J.D. Kennedy, *Managing Public Disputes: A Practical Guide for Government, Business, and Citizens' Groups,* 2nd ed. (San Francisco: Jossey-Bass, 2001); Roy J. Lewicki, David M. Saunders, and John W. Minton, *Essentials of Negotiation,* 2nd ed. (McGraw-Hill/Irwin, 2001); Lawrence Susskind and Jennifer Thomas-Larmer, "Conducting a Conflict Assessment," in Lawrence Susskind, Sarah McKearnan, and Jennifer Thomas-Larmer, eds., *The Consensus Building Handbook: A Comprehensive Guide to Reaching Agreement* (Thousand Oaks, CA: Sage Publications, 1999), for excellent examples of conflict assessments.

104. Lynne Fessenden, *Evaluating the Success of the New Community Meeting on Issues of Growth and Sustainable Development in Central Lane County, Oregon (1997–1998)* (Eugene, OR: The Conflict Resolution Center, 1999).

105. Dotson, Godschalk, and Kaufman, *The Planner as Dispute Resolver;* Friedmann and Kuester, "Planning Education for the Late Twentieth Century."

106. A 1987 study by the National Institute for Dispute Resolution reported that 25% of planning schools offered at least one course on disputes and dispute resolution; however, only 3% of the members of the Association of Collegiate Schools of Planning (20 planning educators overall) taught dispute resolution courses at that time (see Dotson, Godschalk, and Kaufman, *The Planner as Dispute Resolver*). Respondents to a more recent survey (Friedmann and Kuester, "Planning Education for the Late Twentieth Century"), which was based on a nonscientific sample, ranked the mediator–negotiator role as the third most important role a planner could assume (see also Caton Campbell and Floyd, "Thinking Critically about Environmental Mediation," for discussion of the planner-as-mediator role).

107. See, e.g., Carpenter and Kennedy, *Managing Public Disputes.*

6

Achievement of Relationship Change

TAMRA PEARSON D'ESTRÉE

In d'Estrée, Beck, and Colby[1] and d'Estrée and Colby,[2] we attempted to address the question, what is success in environmental conflict resolution? Much of the research in the field often reduces the answer to three or four basic criteria for success, such as settlement reached, level of party satisfaction, and substantive outcome quality.[3] However, a comprehensive survey of the field[4] suggests a much longer menu of criteria for success that reflects the range of possible goals and emphases for conflict resolution processes. We organize these many criteria into six general categories that reflect how practitioners, parties, and communities conceive success in environmental conflict resolution: agreement or outcome reached, process quality, outcome quality, relationship of parties to outcome, relationship between parties or relationship quality, and social capital (see Box 6-1). These criteria can be applied not only to extralegal processes and outcomes such as negotiation, mediation, and consensus building, but can also be applied to more traditional, institutionalized dispute resolution such as litigation, administrative rulemaking, and legislation.[5]

Asking the question, "what is success?" is the key to understanding the practice of environmental conflict resolution (ECR); it is the Rosetta stone for unlocking the meaning within a whole self-contained and self-referential system. The question centers around identifying basic goals and underlying assumptions about correct and desirable practice. It is a form of epistemological inquiry for the domain of praxis (practice) rather than for knowledge. Instead of asking, "how do we know what is real or true?" it asks, "how do we know what we are trying to accomplish and how to do it?" It is a type of

Box 6-1. Effective Environmental Conflict Resolution Criteria Categories

I.   Outcome reached
     Unanimity or consensus
     Verifiable terms
     Public acknowledgement of outcome
     Ratification

II. Process quality
     Procedurally just
     Procedurally accessible and inclusive
     Reasonable process costs

III. Outcome quality
     Cost-effective implementation
     Perceived economic efficiency
     Financial feasibility and sustainability
     Cultural sustainability and community self-determination
     Environmental sustainability
     Clarity of outcome
     Feasibility and realism (legal, political, and scientific)
     Public acceptability

IV. Relationship of parties to outcome
     Outcome satisfaction and fairness as assessed by parties
     Compliance with outcome over time
     Flexibility
     Stability and durability

V.   Relationship between parties (relationship quality)
     Reduction in conflict and hostility
     Improved relations
     Cognitive and affective shift
     Ability to resolve subsequent disputes
     Transformation

VI. Social capital
     Enhanced citizen capacity to draw on collective potential resources
     Increased community capacity for environmental and policy
         decisionmaking
     Social system transformation

Source: Tamra Pearson d'Estrée and Bonnie G. Colby, *Guidebook for Analyzing Success in Environmental Conflict Resolution,* ICAR report 3 (Fairfax, VA: Institute for Conflict Analysis and Resolution, 2000).

inquiry with a long tradition in early philosophy, such as Plato's exploration of "how do we set up the ideal state?" It has enjoyed a more recent resurgence in the work on reflective practice.[6]

In our efforts to define success, both in the program of work described above and in parallel work in international conflict resolution,[7] we have used two approaches when asking practitioners about their success. The first approach is to ask directly, "What are your goals?" The second approach is to ask a question that elicits a practitioner's intuitive knowledge about the "right" solution: "How do you know when you are succeeding?" (or, more colloquially, "How do you know when you've 'made it'?").

Answers to the first question are often statements like "an agreement all parties can live with" or "a durable, stable, and efficient agreement." However, answers to the second question are interestingly different. They are comments such as "the parties are willing to *talk* with each other," "parties start using 'we' rather than 'they'," and "they seem willing to go the long haul together." In these cases, reaching an agreement on paper seems like more of a by-product, or at least an outgrowth, of something else equally or more important—changes in relationships.

I do not intend to replay something akin to Bush and Folger's argument that transformation of individuals, relationships, and society should supplant "reaching agreement" as the primary goal of conflict resolution practice.[8] However, I do want to recognize that throughout the history of this field, both in practitioners' talk and in theory, we have stressed the critical goal of relationship change.

This chapter reviews this emphasis and suggests new directions for documenting, researching, and evaluating criteria in the category of "relationship between the parties" or, in other words, the achievement of relationship change.

Theoretical Sources

Mediation Theory

In Bush and Folger's controversial book, *The Promise of Mediation*, they argue that the most important goal of mediation should be the transformation of individuals, to change their ability to act as agents in the world ("empowerment") and to relate to others with empathy ("recognition"). The authors define success as follows: "... transformative mediation is successful when the parties experience growth in both dimensions of moral development mentioned ... developing the capacity for strength of self and the capacity for relating to others."[9] They explicitly contrast these goals of mediation with those of what they refer to as the more prevalent "satisfaction" approach to mediation. For Bush and Folger, an emphasis on settlement neglects attention to the more important goals of relationship development.

By contrast, Gulliver's[10] earlier seminal review of the anthropological evidence on dispute settlement, and more specifically on negotiation (mediated or unmediated), suggests that the goal of reaching a settlement is entwined with the process of developing or mending the relationship between the parties. Whether the conflict resolution process used is adjudicatory or negotiating, parties must make intermediate decisions on such issues as forum, agenda, and procedure. When an adjudication process is used, the authoritative third party makes these decisions. However, when the process is mediated or unmediated negotiation, the parties make these intermediate decisions themselves. Gulliver stresses the importance of these intermediate exercises in interpersonal learning and coordination for building the foundation for achieving and implementing the final settlement. In fact, he suggests that "each joint decision by the negotiators contributes to the development of a coordination between them that is essential to the ultimate end of an agreed outcome."[11]

For Gulliver, the relational achievements in turn allow for higher quality substantive outcomes. Parties must develop coordination and trust to be able to share the information necessary to construct integrative agreements. Such integrative agreements represent an alternative to compromise[12] that emerges from information sharing and mutual learning in interaction. The result is new, joint decisions based on new information, adjusted expectations, and a gradually increasing willingness to coordinate.

Moore suggests that the goals of a conflict resolution process should be tied to an assessment of the conflict's causes, which often include relational elements.[13] He suggests five likely categories of causes for (or sources of) conflict: structural, value, relationship, data, and interest conflicts. Most conflicts have multiple sources, and it is the mediator's job, in consultation with the parties, to diagnose the sources and develop strategies for remedying them. Relationship sources may include strong emotions, miscommunication, stereotyping and misperception, and repetitive negative behaviors. Parts of the intervention should be consciously focused to address these relational elements. Assuming that these are causes of conflict will necessitate drawing, either explicitly or implicitly, on theories that explain conflict this way and suggest remedies along the same lines. Moore suggests that intervention is actually a form of repeated hypothesis testing. According to Moore, however, reaching an agreement is the ultimate test of usefulness and relevance.[14]

Conceiving "success" to mean changes in relationships may be tough for many people because conflict resolution processes are typically not focused on relational change as the primary goal. It is achieved as a by-product of a task focus, the exchange of information, and the development of a coordinated response. However, many authors acknowledge that achieving quality substantive outcomes is difficult or impossible without relationship change.

For others, relationship change is a goal not only for the improved outcomes it produces, but also because of the increased likelihood for long-term success that such change reinforces. Improved relationships make it more likely that agreements will endure, new difficulties can be addressed, and agreements will be implemented.

Environmental and Public Policy Disputes

Environmental and public policy disputes often have an enduring nature. They are seldom "settled" in the long term, but rather return as new versions of an ongoing debate between conflicting values and a pattern of conflict. This character necessitates producing agreements that also improve relationships to increase the likelihood that agreements endure and will be implemented.

Susskind and Ozawa include relationship change in one of the earliest discussions of evaluation criteria particular to this domain.[15] In addition to the standard settlement criteria of satisfaction, fairness, substantive quality, and reasonable cost, they propose that the process should improve the relationship among the disputants.

Dukes argues for reconceptualizing the goals of "public conflict resolution" to include transforming governance and transforming the public's perception of and participation in self-governance.[16] Fundamental to this idea is the building and restoration of relationships, considered as the basis for community life. Though individual dispute settlement may be desirable, the much larger agenda of public conflict resolution is breathing new life into a disintegrated system. According to Dukes, it is only through addressing this broader vision that the purported postmodern "crisis of governance" can be dispelled. Refocusing on these larger goals clarifies the importance of expanding the tools of public conflict resolution beyond mediation to include policy dialogues, consensus building, visioning, and facilitated dialogues, processes that are just as oriented toward forging relationships as toward reaching any specific agreements.

Past Assessment of Relationship Change

Although many writers in ECR identify relationship change as a desirable goal and a criterion of success, few articulate how such changes are observed and verified. Practitioners often rely on intuitive assessments, such as changes in the participants' tone with each other. We review some of the ways that relationship change has been researched and measured in the past, both within conflict resolution contexts and in related contexts, such as psychotherapy. This review draws heavily on that done for Criteria Category V: Relationship Change in d'Estrée, Beck, and Colby.[17]

Measures for relationship change, though sparse in conflict resolution literature, are more common in the context of marital and divorce mediation and marital therapy. They include direct measurements of relationship quality, ability to resolve future disputes, reduction in conflict and hostility, cognitive–affective shift, and transformation. These criteria are usually assessed over time, and in the long term.

Pruitt and others assessed relationship change in terms of improvement in long-term quality of the relationship.[18] Participants were asked to rate two things: current relationship with the other party (very unpleasant to very pleasant) and whether the relationship worsened, remained the same, or improved. Pruitt and colleagues also coded for any development of new problems.

Another relationship-quality criterion some researchers use is the ability of parties to resolve future disputes. Davis and Roberts coded the parties' ability to negotiate together in the future.[19] In their study of relitigation, Keilitz, Daley, and Hanson allude to the importance of parties' ability to maintain amicable and cooperative relationships, an ability they say a limited time in mediation is unlikely to permanently improve.[20] However, they do not actually assess this ability.

Johnston, Campbell, and Tall considered the reduction in conflict and hostility as an indicator of success in their studies of marital conflict.[21] The researchers measured conflict using the Straus Conflict Tactics Scale, which is composed of 18 behavioral questions on how disagreements have been managed during the previous year.[22] Other mediation studies measured reductions in hostility using scales such as the O'Leary–Porter Scale, on which parents self-report the frequency of hostility displayed in front of a child. Although these measures have been developed primarily for use in marital and family conflict, analog measures could be developed for other domains. In fact, in studies of international conflict, one of the primary indicators of successful "resolution" is an end or reduction in overt hostilities.[23] Note, however, that the measurements of hostility reduction in marital conflict were self-reported, while in international relations, hostilities are coded based on identifiable changes in behavior, such as injury rates.

In d'Estrée and Colby,[24] interviews with both local and regional ECR practitioners produced repeated references to a "shifting" in the way parties saw each other as a result of an effective process, in essence a *"cognitive shift."*[25] Although commonly noted by practitioners, both within and outside the environmental domain, little research has been done directly on cognitive shift within conflict resolution research. Other areas of research, however, have sought to capture or measure such changes or shifts in relationships. Primary among these is psychotherapy research.

Therapy literature authors who discuss "cognitive shifts" fall into three basic theoretical camps: behavioral, cognitive, and family systems. In the behavioral school of therapy, shifts are noted as the focus of the party changes from the faults of the other to the increasing recognition of the independence of the self. As the party becomes more independent, the faults and imperfections of the other party become less critical. Shifts are assessed through interviews, where changes in feelings are noted.

In the cognitive–behavioral school, shifts come as information-processing errors that cause conflicts are corrected. Focus is placed on modifying selective attention, attributions, expectancies, assumptions, and standards.[26] Shifts are assessed through self-report questionnaires, interviews, and scoring of observational data taken during conflict resolution sessions.

Finally, in the family systems school, shifts can occur in the shared narrative stories that define "reality" within families, organizations, and communities. Whereas one narrative reality can exclude certain other interpretations of reality, changes in specific stories or the relationship between stories can change the parties' experience of the existing reality. Cognitive shifts are assessed through noting changes on the dimensions of the story told such as *time, space, causality, interactions, values,* and *telling style.*[27] It may be more accurate to conceive of these shifts as both affective and cognitive, that is, not only do parties change their stored representation of the others' characteristics and behaviors, but they also change their basic evaluation or emotion associated with the other from more negative to more positive. As such, changes should be notable not only in how the other is conceived, but also in how the other is evaluated.

Fundamental changes in how one thinks and feels about another person become permanent. Once a new explanation for another's behavior is understood, the old ways of explaining and perceiving become useless and are discarded.

Such changes have the feel of changes in physical states, where matter takes on a new appearance and form. A criterion for success that is currently broadly discussed throughout the conflict resolution field is the notion of "transformation."[28] As noted earlier, some argue that conflict presents an opportunity for individual and collective moral growth.[29] Transformation may appear as the parties' renewed sense of their own capacity to handle challenges, as empathy for and acknowledgement of the others' circumstances,[30] and as other major shifts in perception besides that of the other, such as perception of relationship context, paradigm, social and political context, or tools and solutions.

Assessing Relationship Change in Environmental Conflict Resolution

In d'Estrée and Colby, we build and expand upon the review above and propose several ways to measure relationship change.[31]

Reduction in Conflict and Hostility

A common measure of improvement in conflictual relationships is a reduction in hostility. This criterion captures a sense of whether the conflict is de-escalating or escalating in actions, rhetoric, or tone of communication. Various factors from the literature on conflict escalation, such as the presence or absence of various possibilities for nonalignment (indicates the level of polarization), are also included as indicators.[32]

Improved Relations

Theorists have sought to conceptualize "peace" or "good relations" as something beyond the lack of hostilities.[33,34] What represents "good relations" in terms of the presence, rather than the absence, of something? This criterion seeks to capture changes in the way parties see and relate to one another that may reflect the essence of successful resolution. To note change, one must first note the nature of the original relationship as a baseline for comparison. Indicators to explore for change include discussions of the relationship itself, as well as the tone of communication among the parties (e.g., hostile, conciliatory), the effort parties expend to protect themselves, and their sense of trust as indicated by the necessity or lack of enforcement clauses or other formalities.

Cognitive and Affective Shift

This criterion is designed to provide evidence of the phenomenon that many practitioners (and even parties) note a shift in parties' framing of the conflict, the relationship, or both. Indicators include noting the ways parties refer to one another and the way they describe or explain the other parties' behavior (pre- and postagreement). Building on literature from family systems theories, it also includes a bit of narrative analysis of the way that stories are told about the conflict—whether narratives change (pre- to postagreement) in their description of causality, interactions, values, and so on.

Ability To Resolve Subsequent Disputes

This criterion addresses how well parties handle subsequent related conflict, such as problems with implementing the agreement outcome. Indicators include evidence that problems are handled constructively; that an ongoing relationship has emerged, making it possible to address future concerns; and possibly that an ongoing forum for conflict management has emerged also. This considers the parties' subsequent joint track record in terms of actions, rather than simply their perceptions.

Transformation

Bush and Folger's argument that conflict presents an opportunity for individual and collective moral growth can be extended beyond interpersonal conflict resolution. Again, their view is that this moral growth is toward a social vision that integrates individual freedom and social conscience and integrates concerns over justice and rights with concerns about care and relationships.[35] This moral growth can occur if ECR processes help people to change their old ways of operating and to achieve new understanding and new relationships. Following Bush and Folger's definition, our indicators include evidence of empowerment (i.e., the parties' renewed sense of their own capacity to handle challenges) and evidence of recognition (i.e., empathy for and acknowledgement of others' circumstances). However, we expand transformation to include evidence of other major shifts in perception (e.g., relationship context, paradigm, social and political context, and tools and solutions).

New Thoughts on the Assessment of Relationship Change

Whereas these ways of operationalizing relationship change are a good start to tracking this more elusive goal of conflict resolution, a more diverse set of measures should be developed and tested. At least three difficulties with these existing measures can be identified.

One of the difficulties in measuring a variable like relationship change leads directly to the primary debate within social science epistemology: does one assess truth externally and objectively or through examining the meanings derived and imputed by the actors studied? Do we look for evidence that relationships have changed by developing externally verifiable indicators that can be "objectively" observed by outsiders to that relationship (to avoid "bias"), or do we look for evidence that relationships have changed by asking the parties themselves (to avoid "outsider misunderstanding")?

If one chooses to focus on "objective" outside assessments of relationship change, a second difficulty is encountered. One common source for "objective" information is news media, especially in archival research. The media constitute one of the few sources where information about most criteria can be found or inferred. In d'Estrée and Colby, we suggest this as a useful source for assessments of changes in such criteria as rhetoric, framing the conflict, and attitudes toward other parties.[36] However, news media are not neutral or objective sources. For example, in our research on the Mono Lake case, we found that although the parties seemed to interact professionally according to varied sources, the media consistently portrayed the case as a "battle," with sides pitted against one another. Perhaps conflict and confrontation sell more newspapers. Therefore, one must consider the limitations of objectivity when using information from media sources.

Finally, perhaps the ideal evidence for relational change is behavioral change—change that can be verified independently from different observations. However, what types of behaviors would represent positive relationship change?

In this section, I propose a few areas of research with potentially fruitful answers to these three challenges.

Communication Measures

One area of research that seeks to bridge the two epistemologies of explaining objectively versus understanding from within the area of relationships is communication research. Talk is both a reflection of parties' internal states and understandings and an external manifestation of a relationship that is directly observable by outsiders. Research on relationships using a communication paradigm may suggest new measures for tracking relationship change in conflict resolution. Though these are primarily developed in the context of interpersonal relationships, analogies to intergroup relations can be fruitfully gleaned.

One of the most important indicators that a relationship is growing is increase in self-disclosure. In social penetration theory, Altman and Taylor propose self-disclosure as the primary mechanism through which intimacy is created.[37] It is particularly important at early stages of a relationship's development, when parties look for similarity and compatibility. Parties need mutual trustworthiness before moving to more intimate stages, and this is produced through matching breadth and depth of self-disclosure. Parties continue to use self-disclosure to regulate intimacy and to signal the desire to move to a different level. More recent work on this topic has called for further examination of how self-disclosure is used strategically to increase or to decrease intimacy.[38]

Thus, one new measure that might be used to substantiate relationship change in conflict resolution is evidence of "reciprocal self-disclosure." In this context, self-disclosure does not refer to "intimate" or personal information but rather to disclosure of information an outside observer considers sensitive or "insider." One could also interview parties about whether they share information with other parties and what type of information they share (i.e., sensitive, strategic, or about priorities). Did the level of information they felt comfortable sharing with other parties change?

Other communication-oriented definitions of relationship growth may be exploited to develop measures of relationship change. I summarize several of these from Duck and Pittman,[39] and for each, describe a measurement that might be suggested:

1. *A relationship is the process of construction of one mental creation from two different mental creations. It is the symbolic or direct negotiation of different views of the world into "a shared system of understanding."* As parties negotiate, they construct a shared symbol system and a shared meaning. Even if tensions remain, evidence that parties have begun to use a common vocabulary would signal the existence of a relationship. As parties interact, important words with symbolic attachment to conflict issues are clarified from each perspective, and parties develop a common understanding of these loaded words.[40] Noting how these words shift in the parties' vocabularies could provide evidence of a move to "a shared system of understanding."

2. *An essential part of a relationship involves rhetorical activity where partners present their own views to each other in hopes of persuading the other to adopt them.* Implied is that the lack of "presentation" to the other would imply little or no engagement. Therefore, public and semipublic statements could be explored to determine to what degree parties even consider each other as 'interlocutors.' Moving from a state where no communication is addressed toward persuading the other, to a state where persuasion toward the other is clearly imbedded in the language used would be a signal that a relationship has begun. In fact, to be persuasive toward another, one must learn about the other's desires and motivators.

3. *Partners in relationship reveal constructions of the relationship in casual talk.* This talk contains symbolic terminology (shared by the partners) and also has celebratory functions, such as playfulness.[41] Though difficult to capture in traditional archival sources such as the media, casual talk of the parties could reveal relationship development in the form of shared symbolic terminology (similar to the first definition), playfulness, and shared humor. Brown and Levinson discuss similar language markers of shared in-group identification in their politeness theory.[42]

4. *The mental creation of a relationship involves recognizing similarity.* Duck and Pittman state that sharing has two components: first, the parties recognize that similarity exists between them, and then they acknowledge it in their talk.[43] This realization of sharing "is itself an important message about the stability, nature, and futurity of the relationship."[44] Parties' acknowledgement in public statements that they are "similar" to other parties with whom they are in dispute, that they share certain things, is an early step toward recognizing that they have a relationship. Implied as well is a time dimension—the relationship is ongoing, and the parties have a future together that frames their interaction.

5. *Hays found that partners developing successful close friendships had a greater breadth of interaction as well as depth of intimacy.[45]* Anecdotes of successful policy dialogues and consensus-building processes describe how parties began to share and interact around topics and activities beyond those related to the conflict itself, such as sharing meals together or supporting

common community events or activities. Though difficult to assess from archival sources, this may be recorded in secondary reports on the conflict process and could be assessed through interviews.

Relational Schemas

It may also be useful to explore the cognitive representation of intergroup relationships and how these may change. Mark Baldwin has suggested that *relational schemas* guide our processing of, and behavior in, interpersonal relationships.[46] These relational schemas include cognitive representations of three components: the self, the "other," and the script for interaction. Implied is that as one of these linked representations changes, the other pieces will also change. For example, changes in one's perception of the other will also lead to changes in one's self-perceptions.

Changes in conflicted relationships may be noted through relational schema characteristics. My research has shown that the more intimate and personal the type of relationship, the less likely that there was a clear or uniform way to respond to conflict (larger schema variance). As clear and rigid scripts for responding to conflicts break down, this could both signal that the relationship was deepening and also that the relational schema was changing. Whereas these notions have mainly been applied to interpersonal relationships, I have also been engaged in a program of research to extend them to intergroup relationships.

Ability To Recognize Interdependence

D'Estrée and Walch found that a better predictor for cooperative behavior than altruistic or egoistic motivation is a negotiator's cognitive orientation.[47] In other words, did the negotiator see all parties as independent, with self as the focus, or did the negotiator recognize the interdependence of parties within a system (a system focus)? Those who had this latter "sociocentric" orientation were literally more able to conceive of joint gains and to coordinate with others to achieve the payoffs of cooperative behavior.

Parties who have this orientation are more likely to share information, to conceive of coordinated options, and to recognize the necessity of working together. Parties may actually note their interdependence. Increases in shared information and coordinated responses would likely indicate increased recognition of interdependence, as well as of the growth of the relationship.

Interaction with the Building of Social Capital

Increased coordination among parties leads to structures that can endure beyond the particular conflict at hand. In our work,[48] we group such changes under a separate but related criteria category: social capital (see Box 6-1). Social capital (V in Box 6-1) is the capacity of individuals to command resources that comes from having relationships with others. Social capital does not reside in the individuals themselves but is a characteristic or possession of relationships and communities. In our criteria framework, we identify three types of criteria that represent social capital goals of conflict resolution processes.[49]

The first, *enhanced citizen capacity to draw on collective potential resources*, builds on group members' ability to rely on potential assistance from others in their network. This increased growth of collective potential resources stems from the new relationships and networks woven among people through ECR processes. The increased potential for assistance and greater collective resources allows for greater risk-taking and creativity, yielding further increases in capacity. Fukuyama argued that such trust, even among strangers, allows for people to spontaneously work together for common purposes.[50] Our framework notes new partnerships and projects for the parties to the former conflict and also notes evidence of reciprocity and mutual assistance.

Innes argued that the most important result of a consensus-building process may not be an agreement per se, but the increased ability for a community to handle future challenges.[51] Two more of our criteria attempt to capture this new ability to respond. First is the *increased community capacity for environmental and public policy decisionmaking*. Many practitioners have argued that the true test of success is whether or not parties can translate their new way of resolving conflict to new issues, new relationships, and new domains. This is to be distinguished from handling subsequent conflicts within the same relationship (see "Ability To Resolve Subsequent Disputes"). New skills learned and new patterns of behavior in formerly conflicting parties provide an increased capacity for cooperation in the community. As a result, activities are better coordinated, and redundancies and inefficiencies are reduced. Information is shared rather than hoarded, and in fact, information gathering becomes a joint activity. As a system, the community is more integrated and more able to act proactively, rather than reactively, to new challenges. Attempts to aggregate resources, share information, divide and coordinate tasks, set up joint decisionmaking structures, share costs, and respond jointly to new crises and challenges may indicate increased capacity for environmental and public policy decisionmaking.

Finally, successful conflict resolution produces a community that has truly evolved into an integrated, adaptive, learning system—one that has undergone *social system transformation*. The community's (or system's) integration results in coordinated responses to new crises and challenges. The system is more resilient and has increased capacity for continued learning and improved action.[52] The community engages in "double-loop learning,"[53] where assumptions and ways of problem framing are themselves reexamined and creative new responses are considered. Responsibility is framed to encompass the entire community system. Indicators include assistance and support provided to the general community, unified and coordinated response to crises, increased civic discourse, evidence of collaboration among groups to examine assumptions and joint needs, and evidence of proactivity and prevention.

As one assesses increases in social capital, it is important also to assess the presence of two types of conditions that foster its formation and use: networks and perceptions. Whereas networks usually are thought to include such "horizontal associations" as volunteer groups, community associations, and social clubs, other types of networks that are vital to social capital are communication channels between diverse groups and community residential stability. Perceptual conditions fostering social capital include the perception that one has relationships based on mutual reciprocity and assistance, that one is interdependent with others in the community, and that one can trust the community's members and institutions. Whereas these networks and perceptions form a foundation for the development and strengthening of social capital, they, in turn, are further strengthened as social capital is used; networks and positive perceptions are reinforced as those relationships are drawn upon for mutual assistance, engendering still more social capital. Means for assessing networks are voluminous and will not be discussed here beyond noting their usefulness. General assessments of levels of trust and distrust in a community system can be done through media sources, interviews, or both.

To maintain conceptual clarity, we distinguish in our framework between changes in relationships between the parties themselves and changes in community capacities and resources stemming from those improved relationships, which we label social capital. However, increases or improvements in relationships are often used to indicate increased social capital. These concepts are clearly intimately linked. This is discussed further below.

Conclusions

Many fruitful avenues exist for new and creative ways to assess relationship change. Communication sources suggest additional focus on joint meaning systems, the way language is used and changed to mark relational shifts, and evidence that similarity is recognized by the parties. Research on trust, relational schemas and how they change, and parties' abilities to recognize interdependence may provide further means for capturing evidence that ECR processes are producing changes in relationships.

Additional difficulties in assessing relationship change remain to be addressed. These difficulties are common to the assessment of other potential changes resulting from ECR and represent methodological challenges requiring further collaborative work among researchers of ECR evaluation. First, clear documentation is lacking. We share common difficulties with other researchers in the lack of data reporting, highlighting a need for clear documentation protocols. We need to develop clear and manageable procedures for practitioners and researchers alike to record and report information.

Second, a common language is necessary to understand measures and constructs. Researchers appear to interpret ambiguous measures as they see fit, reducing our ability for comparisons or accumulation of wisdom. In a related matter, different researchers may consider the same behaviors to indicate different criteria (this can sometimes happen within the *same* evaluation framework, also). For example, during a recent conference exchange, it was learned that one research group's framework coded the indicator of parties "joking together; knowing each others' views" as "social and political capital," while our framework would have coded it as evidence of "improved relations" and "transformation."

Finally, developing theory and ways to measure change in this domain of ECR evaluation is necessarily an interdisciplinary endeavor. The complex and multifaceted nature of the changes in the interests of the parties requires drawing on the expertise, theories, and measures developed in varying research traditions. Rather than working independently and reinventing many wheels, we need to work collaboratively to pool our relative expertise to develop joint conceptual frameworks, measures, and protocols. As we increase our ability to measure relationship changes, we will increase our capacity for discriminating those processes and practices that best create the results considered to be successful ECR.

Notes

1. Tamra Pearson d'Estrée, Connie A. Beck, and Bonnie G. Colby, *Criteria for Evaluating Successful Environmental Conflict Resolution* (unpublished manuscript, 1999).

2. Tamra Pearson d'Estrée and Bonnie G. Colby, *Guidebook for Analyzing Success in Environmental Conflict Resolution*, ICAR report 3 (Fairfax, VA: Institute for Conflict Analysis and Resolution, 2000); Tamra Pearson d'Estrée and Bonnie G. Colby, *Braving the Currents: Evaluating Conflict Resolution in the River Basins of the American West* (New York: Kluwer, 2003).

3. See Dean G. Pruitt, Robert S. Pierce, Neil B. McGillicuddy, Gary L. Welton, and Lynn M. Castrianno, "Long-Term Success in Mediation," *Law and Human Behavior*, vol. 17, no. 3 (1993), pp. 313–330; Jessica Pearson and Nancy Thuennes, "Mediating and Litigating Custody Disputes: A Longitudinal Evaluation," *Family Law Quarterly*, vol. 17 (1984), pp. 497–524.

4. d'Estrée and Colby, *Braving the Currents.*

5. In d'Estrée and Colby, *Braving the Currents,* we make tentative comparisons among these processes and their ability to address the various success criteria in cases of southwestern U.S. water disputes.

6. Donald A. Schön, *The Reflective Practitioner* (New York: Basic Books, 1983); Chris R. Argyris, Robert Putnam, and Diana M. Smith, *Action Science: Concepts, Methods, and Skills for Research and Intervention* (San Francisco: Jossey-Bass, 1985); Michael D. Lang and Alison Taylor, *The Making of a Mediator: Developing Artistry in Practice* (San Francisco: Jossey-Bass, 2000).

7. Tamra P. d'Estrée, Joshua Weiss, Monica Jakobsen, Larissa Fast, and Sandra N. Funk, *A Framework for Evaluating Intergroup Interactive Conflict Resolution* (unpublished manuscript, 2000); Tamra P. d'Estrée, Larissa A. Fast, Joshua Weiss, and Monica S. Jakobsen, "Changing the Debate about 'Success' in Conflict Resolution Efforts," *Negotiation Journal,* vol. 17, no. 2 (2001), pp. 101–113.

8. Robert A. Baruch Bush and Joseph P. Folger, *The Promise of Mediation* (San Francisco: Jossey-Bass, 1994).

9. Ibid., p. 84.

10. P.H. Gulliver, *Disputes and Negotiations* (Academic Press, 1979).

11. Ibid., p. 7.

12. See Richard E. Walton and Robert B. McKersie, *A Behavioral Theory of Labor Negotiations: An Analysis of a Social Interactions System* (McGraw-Hill, 1965).

13. Christopher Moore, *The Mediation Process* (San Francisco: Jossey-Bass, 1996).

14. Ibid., p. 62. "Each intervention is a test of the hypothesis that part of the dispute is caused by [e.g.,] communication problems and that if these difficulties can be lessened or eliminated, the parties will have a better chance of *reaching an agreement.* If the desired effect is not achieved, the intervener may reject the specific approach as ineffective and try another" (emphasis added).

15. Larry Susskind and Connie Ozawa, "Mediated Negotiations in the Public Sector," *American Behavioral Scientist,* vol. 27, no. 2 (1983), pp. 255–279.

16. E. Frank Dukes, *Resolving Public Conflict: Transforming Community and Governance* (Manchester University Press, 1996).

17. d'Estrée, Beck, and Colby, *Criteria for Evaluating Successful Environmental Conflict Resolution.*

18. Pruitt and others, "Long-Term Success in Mediation."

19. Gwynn Davis and Marian Roberts, *Access to Agreement* (Milton Keyes, U.K.: Open University Press, 1988).

20. Susan L. Keilitz, Harry W.K. Daley, and Roger A. Hanson, *Multi-State Assessment of Divorce Mediation and Traditional Court Processing,* project report for the State Justice Institute, Williamsburg, VA, 1992.

21. Joseph R. Johnston, Linda E.G. Campbell, and Mary C. Tall, "Impasses to the Resolution of Custody and Visitation Disputes," *American Journal of Orthopsychiatry,* vol. 55 (1985), pp. 112–119.

22. Murray A. Straus, "Measuring Intrafamily Conflict and Violence: The Conflict Tactics (CT) Scales," *Journal of Marriage and the Family,* vol. 41 (1979), pp. 75–86.

23. Patrick Regan, "Conditions of Successful Third-Party Intervention in Interstate Conflicts," *Journal of Conflict Resolution,* vol. 40 (1996), pp. 336–359; Duane

Bratt, "Assessing the Success of U.N. Peacekeeping Operations," *International Peacekeeping,* vol. 3 (1997), pp. 64–81.

24. d'Estrée and Colby, *Braving the Currents.*

25. One practitioner described this shift as one noticeable in "the way they held their arms" and in the pronouns parties used to referred to others and to themselves ("we" versus "us" and "them").

26. Donald H. Baucom and Norman Epstein, "Will the Real Cognitive–Behavioral Marital Therapy Please Stand Up?" *Journal of Family Psychology,* vol. 4 (1991), pp. 394–401.

27. For more details on these dimensions, consult Carlos E. Sluzki, "Transformations: A Blueprint for Narrative Changes in Therapy," *Family Process,* vol. 31 (1992), pp. 217–230 or our research guidebook, d'Estrée and Colby, *Guidebook for Analyzing Success in Environmental Conflict Resolution,* reprinted in d'Estrée and Colby, *Braving the Currents.*

28. Bush and Folger, *The Promise of Mediation.*

29. Ibid.

30. Ibid.

31. d'Estrée and Colby, *Guidebook for Analyzing Success in Environmental Conflict Resolution;* d'Estrée and Colby, *Braving the Currents.*

32. Jeffrey Z. Rubin, Dean G. Pruitt, and Sung Hee Kim, *Social Conflict: Escalation, Stalemate, and Settlement, 2nd ed.* (Colin McGraw Hill, 1994).

33. Adam Curle, *Making Peace* (London: Tavistock, 1971).

34. Jonathan Galtung, *Peace by Peaceful Means: Peace and Conflict Development and Civilization* (Thousand Oaks, CA: Sage Publications, 1996).

35. Virginia Held, *Justice and Care: Essential Readings in Feminist Ethics* (Boulder, CO: Westview Press, 1995).

36. d'Estrée and Colby, *Braving the Currents.*

37. Irwin Altman and Dalmas A. Taylor, *Social Penetration: The Development of Interpersonal Relationships* (Holt, Rinehart & Winston, 1973).

38. Steve Duck and Garth Pittman, "Social and Personal Relationships," in M.L. Knapp and G.R. Miller, eds., *Handbook of Interpersonal Communication* (Thousand Oaks, CA: Sage Publications, 1994).

39. Ibid.

40. See also Tamra P. d'Estrée, "Learning What To Say To Resolve Conflict: An Ethnography of Intergroup Interaction in a Workshop Setting," presented at the annual meeting of the International Communication Association, Miami, FL, May 1992.

41. Leslie A. Baxter, "Forms and Functions of Intimate Play in Personal Relationships," *Human Communication Research,* vol. 18 (1992), pp. 336–363.

42. Penelope Brown and Stephen Levinson, "Universals in Language Usage: Politeness Phenomena," in E.N. Goody, ed., *Questions and Politeness* (Cambridge University Press, 1978).

43. Duck and Pittman, "Social and Personal Relationships."

44. Ibid., p. 682.

45. R.B. Hays, "The Development and Maintenance of Friendship," *Journal of Social and Personal Relationships,* vol. 1 (1984), pp. 75–98.

46. Mark W. Baldwin, "Relational Schemas and the Processing of Social Information," *Psychological Bulletin,* vol. 112 (1992), pp. 461–484.

47. Tamra P. d'Estrée and Karen S. Walch, *Egocentric vs. Sociocentric Behavior in Multilateral Negotiations,* unpublished manuscript, 2001.

48. d'Estrée and Colby, *Guidebook for Analyzing Success in Environmental Conflict Resolution;* d'Estrée, Beck, and Colby, *Criteria for Evaluating Successful Environmental Conflict Resolution;* d'Estrée and Colby, *Braving the Currents.*

49. d'Estrée and Colby, *Guidebook for Analyzing Success in Environmental Conflict Resolution;* d'Estrée and Colby, *Braving the Currents.*

50. Francis Fukuyama, *Trust: The Social Virtues and the Creation of Prosperity* (New York: Free Press, 1995).

51. Judith E. Innes, "Evaluating Consensus Building," in Larry Susskind, Sarah McKearnan, and Jennifer Thomas-Larmer, eds., *The Consensus Building Handbook: A Comprehensive Guide to Reaching Agreement* (Thousand Oaks, CA: Sage Publications, 1999).

52. Ibid.

53. Argyris, Putnam, and Smith, *Action Science.*

7

Retrospective and Prospective Frame Elicitation

Sanda Kaufman and Barbara Gray

E valuation of intervention in environmental disputes has largely relied on retrospective approaches. These are summative in nature, usually conducted at the conclusion of the intervention effort, and sometimes after one or two years. Retrospective analyses can offer insight into whether or not an intervention process was successful on a number of important dimensions. For example, success can be assessed according to whether or not an agreement was reached and whether it was feasible and sustainable; how the decisionmaking process compared with other alternative processes; whether the agreed-upon actions were taken; whether adversarial relationships changed; to what extent parties were satisfied with process, representation, and outcome; whether parties' relationships improved, or at least did not worsen; and whether or not learning took place.[1] All of these kinds of assessments can be particularly beneficial to the stakeholders should they participate in subsequent intervention or consensus-building efforts. They provide cumulative insights for agencies and alternative dispute resolution (ADR) practitioners, and even for stakeholders who meet again. However, they do not offer participants an opportunity for reflection about a process in which they are currently engaged and which could benefit from the evaluative information.

In this chapter, we propose a more contemporaneous approach to evaluation that interjects opportunities for "on line" sensemaking[2] about the intervention or consensus-building process and about participants' own views of the issues while the conflict interactions are unfolding. The proposed evaluation approach uses frame elicitation, an analytic technique based primarily on interviews with the participating stakeholders, which reveals how the par-

129

ticipants are interpreting critical aspects of the dispute. Evaluators then provide feedback to participants about their frames and how these may be influencing their views about the issues and the process, as well as how their frames might unnecessarily reduce the realm of possible agreements. The goal of providing parties with framing information is to help them and any intervenors reflect on the ADR process while it is still underway, so that they can profit from it, rather than get the information after the process is over. Ideally, this reflection enhances the participants' overall chances of reaching consensus. At the very least, it reveals the implicit choices parties often make through the frames they choose to apply to a dispute and thereby points to avenues for improving current and future dialogue.

We offer two case examples of how frame elicitation techniques have been used for prospective evaluation and intervention in protracted environmental disputes. The first case is that of the 35-year-long controversy about the status and management of Voyageurs National Park. Despite an 18-month mediated effort to resolve the dispute in 1996 and 1997, an agreement was never reached. We examine why mediation was unsuccessful and then describe some subsequent efforts to intervene in the dispute using frame elicitation techniques based on data collected from a broad spectrum of stakeholders. The second case has been unfolding over the past 25 years: a lack of concerted mitigating actions has allowed Doan Brook, an urban watershed, to deteriorate, despite the seeming agreement among parties that such actions are sorely needed. Parties in a recent decisionmaking process (1999–2000) were interviewed, and information on elicited frames was analyzed and offered back to the parties, who found it helpful in crafting joint action to manage the watershed.

The chapter is organized as follows. After a brief review of the status of evaluation research within the ADR field, we describe work that has been done on the framing of environmental disputes and introduce frame elicitation techniques. We then present the two cases and show how frame elicitation was used in each one for evaluation purposes. Finally, we sum up the lessons learned from the two cases and consider the potential of frame elicitation as a prospective evaluation tool for consensus building, particularly its complementarity to the convening process.

Retrospective Evaluation of Environmental Consensus-Building Efforts

The difficulties of evaluating ADR in environmental cases have been amply documented.[3] They include the complexity of issues, the lack of similarities among cases, obstacles to observation and documentation because of confidentiality concerns and the length of time involved, and the fact that several

ADR and legal processes often occur concurrently. In addition, the scientific uncertainty surrounding environmental consequences of various decisions makes it difficult to assess the extent and quality of the outcomes for the environment and to hold them independent of the stakeholders' preferences. However, an even bigger challenge is defining success, and consequently some measures that would signal it or the lack of it, given the fact that the environment is a public good.[4] Rather than being a disagreement among observers, the definition of success goes to the core purpose of ADR, broadly defined here to include facilitation, mediation, and consensus-building processes.

When ADR initiatives were first undertaken, an important concern of those engaged in them—parties, intervenors, and sponsors—was establishing their utility as alternatives to the traditional decisionmaking processes they replaced (e.g., litigation, agency policymaking). During these early years, evaluation sought to demonstrate that ADR processes required less time and money and produced superior outcomes (with respect to both quality of the agreements and stakeholders' commitment to them).[5] This type of evaluation required counterfactual argumentation[6] because it was next to impossible to find cases to compare that had sufficient similarities besides the method they used to resolve the dispute. Another approach to demonstrating the legitimacy and even superiority of ADR processes was the production of descriptive case studies that enabled readers to step into the cases and make vicarious judgements about their effectiveness. In the environmental arena, several volumes of such cases,[7] as well as many accounts of individual cases,[8] helped to establish this early credibility for the field.

As environmental ADR processes have gained acceptance, evaluation efforts have been challenged to capture more of their inherent complexity.[9] Recently, cross-case comparisons have begun to emerge.[10] Still, when evaluation efforts have been undertaken, they have largely been limited to summative findings that relied on retrospective analysis.[11] This type of evaluation was conducted at or shortly after the conclusion of the ADR process, and sometimes even one or two years beyond it, and generally focused on three issues: the results achieved, the process used, and the relationships built.[12] As Birkhoff and Bingham[13] and Innes[14] have pointed out, however, evaluators can use many other criteria to assess each of these issues. For example, retrospective evaluation of outcomes can include the following:

— Was an agreement reached?
— What was the quality of the agreement (as defined by the stakeholders, or other measures)?
— How does the agreement compare with decisions reached by other methods?
— Is the agreement feasible and sustainable? Were agreed-upon actions taken?
— Did relationships among parties change for the better?
— To what extent did learning occur?

— How just is the agreement?

— What second-order effects were triggered?

Criteria for judging process success have also been suggested.[15] These include the following:

— How representative of the diversity in the problem domain are the people seated at the table?

— How much process control do they have?

— To what extent is the discourse characterized by respect and civility?

— Are constituencies included in the process?

— Does it encourage participants to challenge assumptions?

— Does it conform to existing laws and procedures?

— Are the costs (time and money) reasonable?

— Were constructive processes for dealing with differences used?

— Did it avoid setting precedents for nonparticipants?

All of these kinds of retrospective assessments can be particularly beneficial to the participants, should they participate in subsequent ADR or consensus-building efforts, which is often likely when environmental issues are at stake. They also provide cumulative insights for agencies and ADR practitioners. Nonetheless, they do not afford participants an opportunity for in situ reflection about the process in which they are currently engaged or for making any midcourse corrections that might improve their chances of reaching a lasting consensus. Although Innes has suggested the utility of midcourse evaluations, her suggestions for these only focus on "identifying and remedying process problems."[16]

In the next section we explore what we refer to as *prospective evaluation*. It offers an ongoing approach to examining both the content and process issues that may be derailing or blocking intervention and consensus-building activities. This approach relies on the frame elicitation techniques that are also discussed below.

Frame Elicitation as a Prospective Evaluation Technique

Frames and Frame Elicitation

In general, frames refer to the ways people interpret issues and problems.[17] They represent the gestalt that individuals (or groups or organizations) use to make sense of their experience. Bateson used the term to refer to how two people engaged in a playful wrestle can change their interpretation of their interaction to "a fight" if one of them begins to use more aggressive tactics that fall outside the scope of "play."[18] We generate frames to help us make sense of circumstances. These frames have been found to have important influences on the processes and outcomes of disputes and negotiations. Parties develop specific interpretations about the issues in a dispute and about other stakeholders.[19] How participants frame their concerns has profound effects on how they label the dispute[20] and how they evaluate outcomes.[21] Frames also affect parties' preferred approaches to settling a dispute.[22]

The role of framing has been examined specifically in environmental disputes. For example, Hilgartner found that industry and labor representatives differed in whether they were willing to link evaluations of economic and health risks in the same equation; the latter considered this as unacceptable framing.[23] In conflicts over nuclear power and hazardous waste cleanup, important differences have been discovered in the way that lay people and technical experts frame the risks associated with the hazards.[24] Lay stakeholders tend to focus on catastrophic extremes regardless of their small likelihood, whereas technical experts trust probabilistic estimates of risk, with catastrophic but unlikely hazards looming much smaller. Unless the parties are aware of these differences and their causes, and can somehow reconcile them, success in consensus-building efforts is hampered.

In a study of eight intractable environmental disputes, Lewicki and others have identified a typology of 14 different types of frames parties use to make sense of contested issues and their experiences with each other regarding these issues.[25] Differences in how disputants frame various aspects of these disputes can lead to misunderstandings, stalemates, and escalation. If disputants have different views about how social control should be exercised, for example, they may find it difficult to agree on the level of government that should have authority for making natural resource decisions or administering existing policies.

When some disputants frame their conflict as a threat to their identity,[26] it often becomes more intractable, particularly if others discount or denigrate what are core concerns for these stakeholders.[27] On the other hand, if others acknowledge identity concerns (for instance, during a consensus-building process) and regard them as interests that need to be addressed throughout the process, they can alleviate and even overcome the negative effects of these frame differences.[28]

Disputants can hold frames about each other, called characterization frames. Typically, at the outset of an ADR process, disputants' frames about each other are negative and often encapsulated in stereotypes such as "tree huggers," "polluters," "mercurial politicians," or "square scientists." Mischaracterizing others, even with positive intent, leads to misinterpreting motives and interests and to difficulties in crafting mutually satisfactory agreements.

Environmental disputants have also been found to develop frames relating to their beliefs about how disputes should be handled and by whom, who is trustworthy; and what kinds and sources of information they should believe. If a community frames as untrustworthy the government agency with the most access and information about a specific environmental problem, it will discount valuable information from this source and will try to obtain the same information elsewhere, wasting resources and time.

The Process of Prospective Evaluation

Historically, evaluation has been conducted after the implementation of processes, policies, and decisions. Almost exclusively, evaluation textbooks[29] offer models and designs that aim to capture ex post facto the effects of implemented decisions and programs.

Such summative evaluation is immensely useful for those deciding whether to fund or engage in ADR processes. Public agencies are often direct or indirect parties to environmental disputes and have a critical need for evaluative data[30] that can inform their next choice of conflict management mode. Intervenors also benefit from such information because of their recurrent involvement in disputes.

Disputants in a specific case must choose the process they will use, and they need information to compare ADR to other alternatives. An evaluation after the fact, however, most likely will be of little use to the parties. Instead, they may find ongoing feedback more useful for timely reality checks and course corrections that can improve the odds of a satisfactory process and outcome. Such prospective evaluation can also be useful to intervenors. To a large extent, performing conflict assessments before intervention accomplishes some of this.[31] We propose here that framing information, not usually part of an assessment process and not typically shared with the disputants, can add an important information dimension. It affords disputants and intervenors the opportunity to understand each other and to correct misunderstandings that lie at the root of some frames or that are entrenched through their repeated, shared use. Prospective frame evaluation can enrich assessment and convening efforts by analyzing the parties' and even the intervenors' patterns of sensemaking, which may stem from past interactions in environmental conflicts lasting several years and even decades.

Prospective evaluation that elicits stakeholders' frames and offers ongoing feedback cannot, by itself, solve a dispute. In some cases, it may be the very information that leads parties and intervenors to opt out of the ADR process. On the other hand, such information can contribute to a dose of introspection and enhance the mutual understanding of parties, their needs, and role stringencies. It can foster dialogue, which is the most significant achievement in some extremely protracted disputes.[32] Much like Wildavsky's self-evaluating organizations,[33] parties and intervenors can benefit from being made aware of their own and each other's prevailing frames. They can then engage in reframing that contributes to the productive management of dispute episodes, even within the framework of long-term, intractable conflicts.[34]

Evaluators feed back their frame analysis to the parties and intervenors during an ADR initiative that occurs before the parties reach the joint decisionmaking phase. This information can help parties consider whether their own framing of the dispute is useful in light of what they have heard from other stakeholders. It may help stakeholders adopt a constructive collective frame during their deliberations. Next, to illustrate some of these possibilities, we present two cases that used frame elicitation and feedback. We offer suggestions for how to productively couple this evaluation method with retrospective evaluation to enhance the chances of success in multiparty consensus-building processes and to inform sponsoring or participating government agencies on the likelihood of reaching consensus in specific situations.

Two Cases Using Retrospective and Prospective Evaluation

Voyageurs National Park

In 1964, a governor and former governor of Minnesota resurrected proposals originally made by the Minnesota legislature in 1891 for the creation of a U.S. national park in the northern part of the state. The 1964 initiative led to legislation establishing Voyageurs National Park in 1975 but rekindled a firestorm of controversy that continues to this day. The park, on Minnesota's northern border with Canada, is largely a water park, with resources primarily accessible by boat. It consists of parts of several large lakes, a peninsula between two of them, and many small inland lakes. Unlike that of most other national parks, its enabling legislation permitted fishing, snowmobiling, and the use of motorized watercraft. Before becoming a park, the land and water were used for logging operations, private cottages and camps, and hunting and fishing. The federal government's purchase of the land left a bitter taste in the mouths of many, and some challenged the government's appraisals in court and eventually received higher payments for their properties as a result.

By the early 1980s, the park had acquired sufficient lands, but the controversy continued, centering on park officials' decisions to regulate usage to protect certain natural resources, such as gray wolves. For example, the superintendent's decision to close 11 bays to snowmobiling during the winter of 1996 to protect gray wolf habitat infuriated many local residents, who characterized the park's actions as "naïve" and "stupid." In 1995, local resistance culminated in efforts, which eventually failed, to decommission the park. A state-level oversight commission called the Citizens' Council for Voyageurs National Park was at the heart of much of the controversy. The Minnesota legislature in 1975 charged the council with delivering community input to the park's management.

The park's enabling legislation provided that parklands undergo evaluation for wilderness designation—an outcome strongly supported by state and federal environmental groups whose actions also contributed to the conflict's escalation. Motor vehicle users feared loss of all motorized access to the park if this designation were granted for the main peninsula—a prime snowmobiling location. Environmentalists wanted to ensure preservation of the resources for future generations. Through the years, several lawsuits were filed on these issues, and some challenged the park's legal jurisdiction over its water resources. In the early 1990s, a locally brokered compromise proposal and numerous federal court decisions failed to end the conflict. During 1996 and 1997, at the urging of Minnesota's Democratic Senator Paul Wellstone, concerted efforts were made to settle the Voyageurs controversy through mediation. A panel of 18 people representing various interests in the dispute met for almost a year and a half. However, the process concluded without any agreements, despite the assistance of a federal mediator and a proposed compromise for the peninsula—the result of eight different revisions— between wilderness and snowmobiling advocates.

Retrospective evaluation of this ADR effort is based on one of the author's interviews with 17 of the 18 stakeholders who participated in the mediation, one of the mediators, and several others who closely observed the proceedings, as well as summary minutes of the proceedings. The evaluation suggests at least six reasons for the group's failure to reach agreement:

1. Representation issues plagued the makeup of the group at the negotiation table.
2. Participants did not always adhere to or enforce ground rules for civil behavior at the table.
3. One intransigent party stuck to its original position.
4. Class issues appear to have undermined trust building among the parties.
5. Some parties harbored longstanding suspicion about whether promises and agreements had staying power.
6. There were strong political overtones to the conflict controversy that may have helped scuttle the possibility of any agreement.

From 1998 to 2000, the park engaged in a general management planning (GMP) process to establish guidelines for its next 15 to 20 years. In addition to gathering suggestions and reactions from the community at designated points in the GMP process, park management convened a Visitors' Use and Facilities Planning Committee of about 40 stakeholders to provide input on the plan as it was being drafted. During these meetings, many of the previous issues resurfaced.

Following data collection efforts from 1998 to 2000, the researchers, including one of the authors, held two public feedback meetings in which they shared the results of their frame analysis with the attendees. In these meetings, the researchers alternated presenting their analysis conclusions with engaging the participants in brief experiential exercises designed to elicit their frames about the conflict. First, the researchers suggested that the conflict had persisted, in part, because of the different frames that the parties held. To illustrate this, the researchers used "whole story" frames, which encapsulate what parties believe the conflict is about. Participants were asked to exchange their "whole story frames" with someone they did not know. This legitimized each person's frame and allowed participants to hear how others defined and experienced the conflict.

After introducing the notion of negative characterization frames and providing examples from the interviews, the researchers asked people to consider how they acquired information about the conflict and about their opponents, and whether they took it on faith without verification. The purpose here was to raise the possibility that they may wittingly or unwittingly be purveyors of misinformation and rumors that contribute to the perpetuation of their opponents' negative characterizations. The researchers also used direct quotes from the interviews to help the participants understand the sources of their resistance to each other's proposals. The researchers also demonstrated how power and rights frames contributed to the intractability of the conflict. Finally, the researchers presented back to them their espoused views about how to resolve the conflict. Although most of the stakeholders supported joint problem solving, they differed in who should be included at the table. The researchers suggested that these discrepancies also contributed to the conflict's intractability. The presentation ended with recommendations for moving forward.

Although not many people attended the feedback session, some positive outcomes did emerge there. Some people expressed new awareness about why others took the stance they did, and subsequently the local newspaper adopted a more conciliatory tone than it had previously.

Doan Brook

Some environmental disputes are incidents driven by crisis but embedded in a long-term stream of recurrent issues related to some natural resource. Others revolve around the environmental consequences of human activities, and they, too, can extend over many years. Often, environmental disputes are about a mix of resource management and human activity problems. They come to the forefront when some incident occurs and sparks a dispute episode, or when public or private entities make decisions with environmental consequences. Arguably, then, what makes environmental decisions problematic is long spells of inattention to the issues that drive and exacerbate them. For example, during a long-term lack of attention to water quality in a watershed, problems will accumulate and severe deterioration will occur, but most stakeholders will only pay attention to it if an accident occurs or some illness with high public visibility is linked to it. Or, for many years, a local body of water can sustain activities such as boating until there is a public decision to curtail the activity for preservation purposes. Because damage to the environment or to people's health occurs regardless of whether people pay attention to it, dealing with problems when there is no crisis—to the extent possible—would help alleviate some problems before they reach crisis proportions.

The Doan Brook case offers the opportunity to examine some of these issues and to explore the value of frame-based prospective evaluation in enhancing the success of ADR processes.

In a nutshell, in the words of an environmental engineer:

> Doan Brook is a watershed near Lake Erie, an urban stream that is suffering from urbanization, (pollution, erosion …), and we are at the initial stages of putting together a plan to reverse those impacts. We've been collecting information for the past year to understand the issues, and we are going through a public process to solve those problems. It's watershed management in the context of public consensus.

Whereas this is an accurate snapshot of the moment, Doan Brook took many years to become the urban watershed and the public resource it is today. At the beginning of the twentieth century, as open space was rapidly becoming built up, rare foresight and several generous donations of land and maintenance funds made a string of parks out of this brook, which crosses Cleveland and two of its suburbs on its way to Lake Erie. As the buildup continued around it, this piece of nature was subjected to increasing stresses typical for urban watersheds. Today, its water quality is so poor that not much more than looking at the water and at the scant wildlife it still harbors is permitted. In addition, its forced course through channels and culverts is causing disruptive flooding in one of Cleveland's hubs and in several of its neighborhoods.

For the past 25 years, a small but faithful, mostly suburban group of champions led by the Nature Center at Shaker Lakes has been trying to protect and improve Doan Brook. However, although the group counts among its victories the defeat of a highway that threatened to destroy the brook, on the whole it has not been successful in implementing any significant improvement and maintenance decisions outside of sporadic responses to small crises.

This lack of success is rather surprising, given the widespread affection the brook and its parks enjoy. Residents seem to appreciate the brook but are largely unaware of its problems and are therefore reluctant to allocate resources for its improvement, or to modify some of the behaviors that hurt it, such as heavy use of lawn fertilizers and pesticides or cleanup of pet waste.

This situation poses an interesting dilemma for environmental ADR: there is no active dispute, rarely a crisis, but no action either, while this unique resource deteriorates rapidly.

It is worth noting that residents along the brook are not an indifferent lot; past incidents suggest that they react rapidly to oppose actions they deem contrary to their interests. So there is latent conflict around brook issues. For instance, many expect that any move to secure funding through taxing would be met with lively opposition. Past small-scale initiatives for change in the watershed have also been disputed and opposed, especially in the suburban stretch. Recently, for example, some neighbors, claiming aesthetic reasons, secretly planted a couple of willow trees in the middle of the brook, on a sandy high spot favored by nesting birds. This incident made waves far greater than could be expected based on the scale of the deed and generated controversy that went well beyond the willows to include issues of decision-making authority and process.

Over the years, Doan Brook's latent conflict situation has surfaced occasionally in a stream of encounters among stakeholders, with no long-term plan or action and only stop-gap measures in response to crises such as the latest flood or the current "greening" of one of the lakes caused by algae cover fed by fertilizer runoff. Retrospective evaluators examining past interactions regarding the watershed would have to struggle with questions such as the following:

— Is the unit of observation and evaluation a dispute episode or the persistent, latent conflict underlying it?
— What would the evaluators and the stakeholders consider to be a good outcome? For whom? Should the criteria apply to dispute episodes or to the underlying conflict?
— Does framing information help explain the observed dynamics over time?
— If stakeholders appear to be able to live with the status quo, is the continuous environmental deterioration of the watershed a concern? Whose?
— What process would be suitable for making decisions concerning the brook's future, and what would be the alternatives with which such a process would be compared?

The prospective evaluator might be concerned with these issues as well, but he or she would also face the challenge of understanding enough about the situation as a dispute episode is unfolding to be able to make process suggestions to the stakeholders and assist them in understanding the past and present, to benefit the future. In many environmental cases, intervenor and evaluator alike enter at neither beginnings nor ends of incidents, but rather at moments in a long string of interactions involving the same cast of parties (place-, role-, or interest-bound) over many years. By necessity, assessments and evaluations at such points combine retrospective and prospective aspects.

In the case of Doan Brook, after 25 years of attempts to protect and improve the watershed through small-scale initiatives, in 1999, the Northeast Ohio Regional Sewer District (NEORSD), under EPA mandate to attend to water quality, initiated a study of the brook coupled with a participatory process that involved a study group of residents, elected officials, service staff, and NEORSD consulting engineers meeting monthly for two years to understand the problems and explore solutions. The process concluded in the spring of 2001, with the NEORSD reporting on the collected data and on the measures it proposed to undertake: largely building underground tunnels to improve water quality. However, the study group identified many more needs not under NEORSD jurisdiction, which will remain unattended unless the communities surrounding the watershed can come together and decide on management and protection measures. Although study group participants were satisfied with the NEORSD process overall, they were only wiser from it (from the detailed data collection), not further along. Many agreed on needs, but no actions were implemented.

During 1999, 50 hour-long interviews were conducted with participants in this process and other watershed stakeholders. Frame analysis revealed that

— Residents of Cleveland and the suburbs had images of each other (characterization frames) that were not accurate and could account for the past difficulties in involving Cleveland residents in watershed efforts. Suburbanites thought that they understood the problems of the mostly low-income Clevelanders and their supposed lack of interest in environmental matters. The reality is quite different: Cleveland neighborhood organizations have a keen interest in environmental quality in general, and in Doan Brook in particular.

— A few service staff and engineers held characterization frames about environmentalists that could have made dialogue difficult, but many welcomed the dialogue and the opportunity to explain their approach to the lay public. In general, representatives of institutions and agencies offered characterizations of others more frequently than did residents.

— There was widespread resident trust in NEORSD-produced data and in its intricate model of the brook that enabled exploration of various solutions to the flooding and water quality problems.

— Conflict management frames, regarding how decisions should be made and by whom, diverged. Whereas interviewees agreed that someone should do something about Doan Brook, most respondents did not deem themselves to be those who could or should act. In other words, everyone thought that someone else should take charge and solve the brook's problems, with the unsurprising exception of NEORSD engineers, who thought that they should attend strictly to the problem of water quality, over which they had jurisdiction.

One of the authors reported the discovered frame patterns to the inter-viewees, who found the information (particularly the conflict management frames and frames of each other) useful for their next steps. A councilwoman at the February 2001 meeting of the Joint Committee on Doan Brook said the following:

> We were thinking for a year about a structure to go forward. ... We built a foundation and got educated as we considered options. A couple of other resources went into the thinking. One is [one of the author's] project. She studied us and came up with some recommendations about the need for a partnership. They made a significant contribution.

According to the meeting participants, the realization that the long-term lack of action could be related to the widespread perception that Doan Brook was someone else's responsibility contributed to the new-found resolve to form a funded partnership that could make and implement watershed management decisions. The emerging design has focused on ensuring that the new entity will have broad participation from Cleveland and the two suburbs so that they will be in position to engage in implementable projects rather than in friendly discussions that leave the real problems unattended. The informa-tion about the difference between how suburbanites framed Cleveland resi-dents and how the Clevelanders saw themselves is giving impetus to new col-laborative and more inclusive efforts. In the fall of 2001, many of those involved in the NEORSD process joined a Doan Brook Watershed Partner-ship, which is striving to use the framing feedback to increase its efficacy.

The framing analysis revealed information that stakeholders in the Doan Brook found helpful not only in retrospect, but also for their next steps. In this sense, it was prospective. It could not, however, replace other types of evaluation. For instance, as an institution not accustomed to participatory decisionmaking, the NEORSD made some choices that ADR professionals would find inappropriate. Although watershed residents were involved, they did not represent any constituency and did not report to anyone, so aware-ness of this process beyond its participants was extremely low. In addition, the NEORSD structured the process in a way that left relatively little room for input from the participants, who nevertheless went along with it and were quite satisfied, often framing the process as a desirable way of making the kinds of decisions the group faced.

The participants' satisfaction poses an interesting dilemma: should any-one alert happy participants that their participation is less meaningful than it could be? The outcome most participants predicted and thought they would dislike—massive concrete underground tunnels—will be imple-mented. Is the current satisfaction a result of dialogue that worked or of the NEORSD's successful persuasion through use of data that tunnels are the best solution, or is the satisfaction with the outcome a by-product of satisfaction with the consultative process? If the latter is the case, one might expect difficulties at implementation time because (even if it is indeed a good solution or the only feasible one) residents will continue to dislike it and the majority of them were not even involved in the consensus-building effort. Finally, although not all problems other than water quality have been resolved, as the NEORSD has repeatedly reminded everyone throughout the study process, only a small number of the participants are determined to act. They have formed the partnership that could halt deterioration and restore the brook.

A more general evaluation challenge stemming from the Doan Brook case is related to the definition of success in situations that are ongoing, with slow damage over time that does not seem to affect anyone in the short run, until it is too late to act and environmental effects are irreversible or too costly to remedy. In the Doan Brook case, further framing analysis might help decipher the reasons behind the avowed public affection for the brook and seemingly contradictory reluctance to do anything to protect it, whether individually or collectively.

Conclusions

Prospective evaluation, giving feedback to parties while the ADR process is unfolding, is not common practice, but it is gaining currency. It holds considerable promise as an aid to consensus building and intervention. Prospective evaluation can help both the parties and the intervenors by increasing the effectiveness of ADR processes. The goal is to share lessons with those who most need it, rather than wait for the process to be concluded, then studying what went right or wrong. In addition, prospective evaluation may also help intractable conflicts become unblocked. Key obstacles to this practice include lack of awareness of the benefits and techniques, limited funding of ADR processes, and the broader challenges faced by both prospective and retrospective evaluation of ADR discussed earlier.

Both the Voyageurs and the Doan Brook cases took an opportunity to assess parties' frames and then to give the parties feedback that they could use for making progress in the disputes. These cases suggest that there is reason to consider and further explore prospective evaluation activities, including frames elicitation and analysis. The proposed incorporation of framing information in both types of evaluation appeared useful in the two cases presented here. It held some surprises for stakeholders, debunked some myths, and revealed interpretations of the situation and of the other participants that hampered consensus building. The frame analysis was useful in understanding past events and in shaping subsequent interactions among the stakeholders. It seems to be a promising tool to add to the conflict assessment processes that frequently precede intervention.

Like conflict assessment preceding ADR intervention, the frame-based component proposed and illustrated here yields information that helps assess whether intervention is appropriate and has a chance of success, however success is defined in each situation. The framing diagnosis, shared with those who most need it, could become part and parcel of ex ante efforts such as convening, which itself consist of interviewing individuals, as in the two cases described. Whereas it is unlikely to turn a protracted dispute into a series of friendly encounters, it can inform potential sponsors and intervenors about the chances that ADR would succeed and help them understand what kinds of reframing are necessary to get a consensus-building process underway. A framing diagnosis can also help parties consider from the outset of an ADR process whether their own framing of the dispute serves them well in light of what they hear from other stakeholders, and it might help them reframe to improve the dialogue.

Further research is needed to strengthen the usefulness of frames assessment as a prospective evaluation tool. Comparison among cases like Voyageurs and Doan Brook, which extend over many years but differ in their level of protractedness, might suggest which frames could be viewed as "canary in the mine" frames, or indicators of a situation's resistance to resolution. Analysis of environmental disputes that took a long time to resolve might reveal some "inflexion point frames"—frames that changed over time and could be linked to change in the character of the dispute. For example, some characterization frames that changed from negative to positive might be associated with improved relations among stakeholders and even progress toward resolution, whereas changes from positive to negative might be associated with new obstacles to dialogue. Similarly, changes from conflict management frames that favor dialogue to perceptions that a situation is best handled in court might mark a change in strategy or willingness to accommodate others, or even a political change perceived to favor some side in the conflict. On the other hand, because the nature of relations among stakeholders is thought to play an important role and is a typical dimension of retrospective evaluation, finding that such frame changes have no effect on the course of the dispute might help intervenors turn their efforts in different, more productive, directions. Such information might also tell us which frames are susceptible to information and more amenable to change through persuasion or intervention. Lastly, if frames assessment is to be valuable to disputants, it is important to design and test alternative means of frame elicitation and feedback to identify those that are productive in producing frame changes that can help the dispute process.

Can prospective framing information be useful to government agencies involved in or sponsoring ADR processes? Does it hold any retrospective evaluative value? Is it likely that for an agency playing an active role in a consensus-building process, framing information can play the same role as for any other stakeholder? For example, community representatives personalize agencies and often miss their discrete, changing human composition. Instead, agencies tend to be viewed as immutable and having their own "character." Framing information can alert current agency representatives about community views based on interactions with others, especially in disputes unfolding over a number of years. As community representatives react sharply to their own frames of agency bureaucrats, these can better understand and take in stride hostility. Measures can be taken to relieve it through reframing efforts.

More generally, framing information can shed light on some deeply seated concerns and values that are apt to polarize a situation or lead to escalation of conflict. It can also help predict when, in certain situations, consensus building stands little chance of success. Finally, framing information can also be used to make sense retrospectively of the dynamics of an ADR initiative, leading to better understanding for future interactions with the same stakeholders. Although more research is needed to streamline frame elicitation and to make the best of its results during disputes and in their aftermath, it seems that framing information adds a needed dimension to prospective and retrospective evaluation.

Notes

1. Juliana Birkhoff and Gail Bingham, "Defining Success—What Is Success in Mediation and What Does the Field Want To Know about Success?" paper prepared for "Building Bridges between Research and Practice: What Is Success in Public Policy Dispute Resolution?" sponsored by RESOLVE and National Institute for Dispute Resolution, June 1997; Judith E. Innes, "Evaluating Consensus Building," in Lawrence Susskind, Sarah McKearnan, and Jennifer Thomas-Larmer, eds., *Consensus Building Handbook: A Comprehensive Guide to Reaching Agreement* (Thousand Oaks, CA: Sage Publications, 1999), pp. 631–675; Tamra Pearson d'Estrée, Connie A. Beck, and Bonnie G. Colby, "Criteria for Evaluating Successful Environmental Conflict Resolution," unpublished manuscript, 1999.

2. Karl Weick, *The Social Psychology of Organizing, 2nd ed.* (Reading, MA: Addison Wesley, 1979); Karl Weick, *Sensemaking in Organizations* (Thousand Oaks, CA: Sage Publications, 1995).

3. See for example, Gail Bingham, *Resolving Environmental Disputes: A Decade of Experience* (Washington, DC: Conservation Foundation, 1986); D'Estrée, Beck, and Colby, "Criteria for Evaluating Successful Environmental Conflict Resolution."

4. D'Estrée, Beck, and Colby, "Criteria for Evaluating Successful Environmental Conflict Resolution."

5. Bingham, *Resolving Environmental Disputes;* Kenneth Kressel and Dean Pruitt, eds., *Mediation Research* (San Francisco: Jossey-Bass, 1989); Craig A. McE-

wen and Richard J. Maiman, "Mediation and Arbitration: Their Promise and Performance as Alternatives to Courts," in P.L. Dubois, ed., *The Analysis of Judicial Reform* (Lexington, MA: Lexington Books, 1981); Craig A. McEwen and Richard J. Maiman, "Mediation in Small Claims Court: Consensual Processes and Outcomes," in Kenneth Kressel and Dean G. Pruitt, eds., *Mediation Research* (San Francisco: Jossey-Bass, 1989), pp. 53–67; Janice A. Roehl and Royer F. Cook, "Mediation in Interpersonal Disputes: Effectiveness and Limitations," in Kenneth Kressel and Dean G. Pruitt, eds., *Mediation Research* (San Francisco: Jossey-Bass, 1989), pp. 31–52.

6. Patrick D. Larkey, *Evaluating Public Programs: The Impact of General Revenue Sharing on Municipal Government* (Princeton University Press, 1979).

7. Lawrence S. Bacow and Michael Wheeler, *Environmental Dispute Resolution* (Plenum, 1984); Philip J. Harter, "Negotiating Regulations: A Cure for Malaise," *Georgetown Law Journal*, vol. 71, no. 1 (1982), pp. 1–188; Allan R. Talbot, *Six Case Studies in Environmental Mediation* (Washington, DC: Conservation Foundation, 1983); Bingham, *Resolving Environmental Disputes*; James E. Crowfoot and Julia M. Wondolleck, *Environmental Disputes: Community Involvement in Conflict Resolution* (Washington, DC: Island Press, 1990).

8. Francis X. Murray and John C. Curran, *Why They Agreed: A Critique and Analysis of the National Coal Policy Project* (Washington, DC: Georgetown University, Center for Strategic and International Studies, 1982); Lawrence Susskind and Denise Madigan, "New Approaches to Resolving Disputes in the Public Sector," *The Justice System Journal*, vol. 9, no. 2 (1984), pp. 179–203; Barbara Gray and Tina M. Hay, "Political Limits to Interorganizational Consensus and Change," *Journal of Applied Behavioral Science*, vol. 22 , no. 2 (1986), pp. 95–112.

9. Innes, "Evaluating Consensus Building."

10. Judith E. Innes, J. Gruber, M. Neuman, and R. Thompson, "Coordinating Growth and Environmental Management through Consensus Building," in *CPS Report: A Policy Research Report* (California Policy Seminar, University of California—Berkeley, 1994); Julia Wondolleck and Steven Yaffee, *Making Collaboration Work* (Washington, DC: Island Press, 2000); Roy Lewicki, Barbara Gray, and Michael L.P. Elliott, eds., *Making Sense of Intractable Environmental Conflicts: Concepts and Cases* (Washington, DC: Island Press, 2003).

11. Michael L.P. Elliott, "Carpet Policy Dialogue Assessment," final report to the U.S. Environmental Protection Agency, prepared under contract to RESOLVE. Consortium on Negotiation and Conflict Resolution, Georgia Institute of Technology, Atlanta, 1993; Jill M. Purdy and Barbara Gray, "Government Agencies as Mediators in Public Policy Conflicts," *The International Journal of Conflict Management*, vol. 5, no. 2 (1994), pp. 158–180; Cornelius Kerwin and Laura Langbein, *An Evaluation of Negotiated Rulemaking at the Environmental Protection Agency, Phase I* (Washington, DC: Administrative Conference of the United States, 1995).

12. Michael L.P. Elliott, "The Role of Facilitators, Mediators, and Other Consensus Building Practitioners," in Lawrence Susskind, Sarah McKearnan, and Jennifer Thomas-Larmer, eds., *The Consensus Building Handbook: A Comprehensive Guide to Reaching Agreement* (Thousand Oaks, CA: Sage Publications, 1999), pp. 199–238.

13. Birkhoff and Bingham, "Defining Success."

14. Innes, "Evaluating Consensus Building."

15. Barbara Gray, *Collaborating: Finding Common Ground for Multiparty Problems* (San Francisco: Jossey-Bass, 1989); Birkhoff and Bingham, "Defining Success"; Innes, "Evaluating Consensus Building."

16. Innes, "Evaluating Consensus Building."

17. Gregory Bateson, *Steps to an Ecology of Mind* (New York: Ballantine Books, 1972); Deborah Tannen, "What's in a Frame? Surface Evidence of Underlying Expectations," in R. Freedle, ed., *New Dimensions in Discourse Processes* (Norwood, NJ: Ablex, 1979), pp. 137–181.

18. Bateson, *Steps to an Ecology of Mind*.

19. Lewicki, Gray, and Elliott, *Making Sense of Intractable Environmental Conflicts*; Sanda Kaufman and Janet Smith, "Framing and Reframing in Land Use Change Conflicts," *Journal of Architecture Planning and Research*, Special Issue on Managing Conflict in Planning and Design, vol. 16, no. 2 (Summer 1999), pp. 164–180; Barbara Gray, Julie Younglove-Webb, and Jill Purdy, "Frame Repertoires, Conflict Styles and Negotiation Outcomes," working paper, Center for Research in Conflict and Negotiation, Pennsylvania State University, 1998; Barbara Gray, "Framing and Reframing of Intractable Environmental Disputes," in Roy Lewicki, Robert Bies, and Blair Sheppard, eds., *Research on Negotiation in Organizations*, (Greenwich, CT: JAI Press, 1997), pp. 163–188.

20. Lynn Mather and Barbara Yngvesson, "Language, Audience, and the Transformation of Disputes," *Law & Society Review*, vol. 15, no. 3–4 (1980–1981), pp. 775–821; Sally E. Merry and Susan Silbey, "What Do Plaintiffs Want? Reexamining the Concept of Dispute," *Justice System Journal*, vol. 9 (1984), pp. 151–177; Elaine Vaughan and Marianne Seifert, "Variability in the Framing of Risk Issues," *Journal of Social Issues*, vol. 48, no. 4 (1992), pp. 119–135; Gray, "Framing and Reframing of Intractable Environmental Disputes."

21. Amos Tversky and Daniel Kahneman, "The Framing of Decision and the Psychology of Choice," *Science*, vol. 211 (1981), pp. 453–458; Carsten K.W. De Dreu, Peter J.D. Carnevale, Ben J.M. Emans, and Evert Van de Vliert, "Effects of Gain–Loss Frames in Negotiation: Loss Aversion, Mismatching, and Frame Adoption," *Organizational Behavior and Human Decision Processes*, vol. 60 (1994), pp. 90–107.

22. Blair H. Sheppard, Katherine Blumenfeld-Jones, John W. Minton, and Elaine Hyder, "Informal Conflict Intervention: Advice and Dissent," *Employee Rights and Responsibilities Journal*, vol. 7, no. 1 (1994), pp. 53–72; Lewicki, Gray, and Elliott, *Making Sense of Intractable Environmental Conflicts*.

23. Stephen Hilgartner, "The Political Language of Risk: Defining Occupational Health," in D. Nelkin, ed., *The Language of Risk* (Beverly Hills, CA: Sage Publications, 1985), pp. 25–66.

24. Harry J. Otway, Dagmar Maurer, and Kerry Thomas, "Nuclear Power: The Question of Public Acceptance," *Futures*, vol. 10 (1978), pp. 109–118; Elliott, "The Role of Facilitators, Mediators, and Other Consensus Building Practitioners."

25. Lewicki, Gray, and Elliott, *Making Sense of Intractable Environmental Conflicts*.

26. John A. Folk-Williams, "The Use of Negotiated Agreements To Resolve Water Disputes Involving Indian Rights," *Natural Resources Journal*, vol. 28 (1988), pp. 63–103; Jay Rothman, *Resolving Identity-Based Conflict in Nations, Organizations, and Communities* (San Francisco: Jossey-Bass, 1997).

27. Barbara Gray and Ralph Hanke, "Frame Repertoires and Non-Collaborative Behavior in Intractable Environmental Disputes," working paper, Center for Research in Conflict and Negotiation, Pennsylvania State University, 2001.

28. Gray, "Framing and Reframing of Intractable Environmental Disputes"; Lewicki, Gray, and Elliott, *Making Sense of Intractable Environmental Conflicts.*

29. For example, see L. Bernd Mohr, *Impact Analysis for Program Evaluation* (Chicago: The Dorsey Press, 1988); Harry P. Hatry, Richard E. Winnie, and Donald M. Fisk, *Practical Program Evaluation for State and Local Governments* (Washington, DC: The Urban Institute Press, 1981); David Nachmias, *The Practice of Policy Evaluation* (New York: St. Martin Press, 1980); Larkey, *Evaluating Public Programs.*

30. Nachmias, *The Practice of Policy Evaluation.*

31. Lawrence Susskind, Sarah McKearnan, and Jennifer Thomas-Larmer, eds., *Consensus Building Handbook: A Comprehensive Guide to Reaching Agreement* (Thousand Oaks, CA: Sage Publications, 1999).

32. Heidi Burgess and Guy Burgess, "Constructive Confrontation: A Transformative Approach to Intractable Conflicts," *Mediation Quarterly,* vol. 13, no. 4 (1996).

33. Aaron Wildavsky, "The Self-Evaluating Organization," in David Nachmias, ed., *The Practice of Policy Evaluation* (New York: St. Martin Press, 1980).

34. Mather and Yngvesson, "Language, Audience, and the Transformation of Disputes"; Don A. Schön and Martin Rein, *Frame Reflection* (New York: Basic Books, 1994).

8

Facilitators, Coordinators, and Outcomes

WILLIAM LEACH AND PAUL SABATIER

O f the myriad factors that influence environmental conflict and its resolution, few are perceived to be more important than the role of professional facilitators. In a recent review of the empirical research on multistakeholder watershed partnerships, effective facilitation and coordination was second only to financial resources as the most frequently cited factor deemed important for success.[1] Twenty-one of the 37 studies concluded that effective coordination, facilitation, or both promoted success, and no study suggested that skillful coordination and facilitation might impede success. The academic interest in facilitation mirrors the attention that it receives from watershed managers and interest groups, for whom the most active public policy debate since the mid-1990s has been whether state governments should support stakeholder-based planning by providing money to hire dispute resolution professionals in each local watershed. At least four states (Oregon, Washington, Massachusetts, and Ohio) currently provide such funding. Other states, including California, periodically propose legislation to do so (thus far unsuccessfully). At the federal level, the U.S. Environmental Protection Agency funds coordinators and facilitators through the Watershed Assistance Grants program and the Clean Water Act 319h planning grants.

Despite the apparent consensus on the importance of facilitation and coordination, the factors that influence success are not necessarily well understood. Most of the available literature uses research designs with significant limitations. For example, existing case studies generally suffer from subjective methods of data acquisition and analysis, and the lack of replica-

148

tion raises serious concerns about the generality and validity of the results. The surveys of multiple partnerships usually solicit views from only one or two participants per partnership—typically the coordinators or facilitators themselves. This approach is adequate for gathering factual information, but unreliable for examining the politics of high-conflict situations in which perceptions of success are likely to vary across participants. Compared to other participants, coordinators and facilitators typically (a) view their partnership as being more successful, (b) express more trust in their fellow participants, and (c) have stronger pro-environmental values.[2]

This chapter seeks to stimulate debate by empirically addressing the following questions:

1. Do facilitators and coordinators contribute to partnership success?
2. How important is their contribution relative to other factors such as funding, local leadership, trust, and time?
3. Is it better to have a facilitator or coordinator or both?
4. Are trained, professional facilitators more effective than those who, perhaps unexpectedly, find themselves facilitating a partnership in addition to their regular job responsibilities as scientists, farmers, or teachers?
5. Does it matter whether facilitators and coordinators are themselves stakeholders (parties to the conflict) as opposed to disinterested consultants?

To address each question, we present data from 50 watershed partnerships in California and Washington state. Interviews, surveys, and documents are used to characterize the relative success of each partnership, as well as numerous factors thought to be conducive to success, including facilitation and coordination.

Testing the Conventional Wisdom: Hypotheses on Facilitation and Coordination

As outlined above, it is widely accepted that the skill of a facilitator or coordinator can have a significant effect on the success of a partnership.[3] Measuring facilitator skill is not a straightforward task, however. In this chapter, we examine the relationship between partnership success and several tangible traits of coordinators and facilitators (such as formal training in facilitation skills, as described below). We also attempt to measure overall effectiveness, which is probably a function of both tangible traits and intangibles, such as personality, style, and demeanor. To measure overall effectiveness, we simply asked three to five interviewees whether, in their judgement, the facilitator or coordinator was effective.

If the conventional wisdom is correct (i.e., facilitators and coordinators are important)—and if stakeholders can judge the effectiveness of facilitators and coordinators more or less accurately—then these assumptions would lead to the following hypothesis:

Hypothesis 1. Stakeholders' perceptions of the effectiveness of a facilitator or coordinator should be correlated with objective measures of partnership success.

We define a coordinator as someone who handles administrative or secretarial responsibilities for the partnership. Coordinators' duties may include any of the following:[4]

—setting the time and place for meetings and publicizing the meetings;
—soliciting agencies and interest groups to participate, and inviting guest speakers;
—setting the agenda;
—typing and distributing meeting minutes;
—serving as the contact person for all interactions with the public, the media, and nonparticipating government agencies;
—searching for funding opportunities;
—writing grant proposals;
—administering grants received, including any subcontracts;
—submitting periodic accountability reports to funding agencies;
—lobbying nonparticipating state or federal agencies for policy or legislation favorable to the goals of the partnership; and
—drafting policy or planning proposals.

We define a facilitator as the person chiefly responsible for running the meetings and fostering productive discussions and decisionmaking. Facilitators may perform any of the following activities:[5]

—crafting ground rules;
—enforcing ground rules;
—leading exercises (e.g., brainstorming, role playing, focus groups) to identify issues, separate interests from positions, identify goals and objectives, rank priorities for action;
—recording points of agreement and disagreement;
—proposing compromises or solutions; and
—training the participants in listening skills or collaboration skills.

In partnerships that have a facilitator but no coordinator, the facilitator typically assumes many of the essential coordinator responsibilities. In this sense, partnerships with facilitators enjoy some of the benefits of both roles, whereas partnerships with only a coordinator will lack the facilitation role almost entirely. The ideal situation would seem to be to have two individuals, one facilitator and one coordinator, such that each could fully devote themselves to the designated role. This line of reasoning leads to Hypothesis 2:

Hypothesis 2. Having a facilitator alone is better than having a coordinator alone, but having both (two separate individuals) is best.

It is often claimed that facilitators and coordinators ought to be objective and neutral.[6] Neutrality is believed to be critical for gaining the confidence of the stakeholders and for giving the process legitimacy in the eyes of participants as well as outside observers. Neutrality and independence can be achieved by going outside the partnership to hire a professional "third-party" consultant to facilitate or coordinate, rather than asking stakeholders (parties to the dispute) to serve these roles. Wondolleck and Ryan conclude that agency officials can facilitate effectively only if the facilitator is not the same official who is responsible for advocating the agency's interests.[7] However, O'Leary and Raines conclude that such "in-house neutrals would not be acceptable" to the vast majority of industry participants in ADR processes led by the U.S. Environmental Protection Agency.[8]

Hypothesis 3. Facilitators and coordinators are more effective when they are disinterested, rather than parties to the dispute.

Landre and Knuth[9] and the President's Council on Sustainable Development[10] have argued for the importance of professionally trained facilitators. It stands to reason that trained facilitators (as well as those who lack formal training but have ample experience and consider themselves members of the facilitator profession) would have a large repertoire of skills, techniques, and experiences to draw upon.

Hypothesis 4. Professional, trained facilitators are more effective than facilitators who lack training and who do not consider themselves to be career-track facilitators.

Increased funding for facilitators and coordinators was recommended by 11 of the reviewed studies. Availability of funding could potentially enable partnerships to coalesce in watersheds where no single stakeholder is willing to bear the costs of initiating and leading the collaboration. In watersheds where one or more stakeholders are willing to coordinate or facilitate on an in-kind basis, funding could allow them to devote more time and attention to that role, without the distractions of their regular job responsibilities. Funding also gives partnerships the option of hiring a disinterested nonstakeholder to facilitate or coordinate.

Hypothesis 5. Facilitators and coordinators who are explicitly paid to facilitate are more effective than those who volunteer or those who are employed by an agency or interest group and facilitate on an in-kind basis.

Data and Methods

Quantitative case studies of 50 randomly sampled watershed partnerships in California and Washington state were compiled between 1999 and 2001. The field research began with an effort to identify all partnerships in California that were active at any point between 1995 and 2000, including partnerships that are now defunct.[11] To be included in the sampling frame, a partnership needed to meet at least four times per year and needed to focus on managing one or more streams, rivers, or watersheds. To ensure an adequate diversity of stakeholders, each partnership needed to include (a) at least one state or federal official; (b) at least one representative of local government—from a city, county, or special district (such as a water or school district); and (c) at least two opposing interests, such as a resource user and a regulating agency or environmentalist.

Our search revealed a population of 150 partnerships in California, from which 39 were randomly sampled with geographic stratification, such that no more than two partnerships were selected from a single watershed (Box 8-1).[12] In Washington state, we randomly selected 10 watersheds and sampled one or two partnerships from each.[13] Because the selection process was random and the sample size is relatively large, the overall results should be representative of watershed partnerships in the two states.

The sample includes 10 partnerships that had disbanded by the time of our study. Three of these disbanded because they achieved their main objectives. The other 7 disbanded after their negotiations ended in stalemate.

For each selected partnership, we

— interviewed three to six key participants, including the partnership's coordinator or facilitator plus at least one key participant from a pro-environment perspective and at least one participant from a pro-development perspective;
— analyzed relevant documents such as watershed plans and meeting minutes; and
— mailed a survey to all participants sufficiently knowledgeable about the partnership to complete at least part of the questionnaire, plus several knowledgeable nonparticipant observers.

For the survey, the names of the participants and knowledgeable observers were obtained during the interviews. The smallest partnership had 6 survey recipients, and the largest had 76. The resulting dataset includes 182 interviews and 920 surveys (out of 1,421 originally distributed, for a response rate of 65%). Response rates for individual partnerships ranged from 45% to 88%.

Measures of Partnership Success

Drawing upon the survey and interview data, we can construct several measures of success for each partnership. In this chapter we focus on four measures:

Box 8-1. Randomly Sampled Partnerships

California (*n* = 39)
Alameda Creek Watershed
    Management Program Steering
    Committee
American River Watershed Group
Butte Creek Watershed Conservancy
Cache Creek Stakeholders Group
Central Sierra Watershed Committee
Cosumnes River Task Force
Dos Palmas Cooperative
    Management Committee
Dry Creek Cooperative Resource
    Management Program (CRMP)
Eel River Watershed Improvement
    Group
Garcia River Watershed Advisory
    Group
Goose Lake Fisheries Working
    Group
Klamath River Fisheries Task Force
Los Angeles and San Gabriel Rivers
    Watershed Council
Lower Stony Creek Task Force and
    Technical Team
Marin Coastal Watershed
    Enhancement Project
Mokelumne River Watershed Group
Navarro River Watershed Advisory
    Group
Northern Klamath Bioregional
    Group
Oakhurst River Parkway Partnership
Panoche–Silver Creek CRMP
Pescadero–Butano Creek CRMP
Pin Creek CRMP
Russian River Watershed Council
Sacramento River Fisheries and
    Habitat Restoration Program
San Francisquito Creek CRMP

San Joaquin River Management
    Program
San Joaquin Valley Resource
    Conservation. Partnership
San Juan Creek Feasibility Study
    Management Team
Santa Ana River Watershed Group
Scott River Watershed Council
Shasta–Tehama Bioregional Council
Smithneck Creek CRMP
Sonoma–Marin Animal Waste
    Committee
South Fork Dialogue
South Fork Trinity River CRMP
Stanislaus Stakeholders
Tuolumne River Technical Advisory
    Committee
Watsonville Sloughs Water Resources
    Program
Yuba Watershed Council

Washington state (*n* = 11)
Cedar River Council
Clean Water District
Douglas County Watershed Planning
    Association
Elwha–Morse Creek Watershed
    Management Team
Entiat Valley Landowners
    Association
Hood Canal Water Protection
    Council
Jefferson County Water Resources
    Council
Padilla Bay Farm Committee
Skagit IRC
Tolt Fish Habitat Restoration Group
Wenatchee River Watershed
    Management Committee

1. level of agreement reached,
2. restoration projects (i.e., implementation of agreements),
3. perceived effects on human and social capital, and
4. perceived impacts on environmental and social conditions in the watershed.

Level of Agreement Reached. The classic benchmark of success for environmental disputes is consensus on substantive issues. "Level of agreement reached" was measured using interviews and relevant documents to determine whether each partnership had achieved the following levels of agreement, which we treat as an ordinal 5-point scale: 0 = no agreement; 1 = agreement on which issues to discuss and address; 2 = agreement on general goals or principles; 3 = agreement on one or more implementation actions (relatively limited and unintegrated); and 4 = agreement on a relatively comprehensive watershed management plan with specific projects or proposals.

Restoration Projects. A second central measure of success is implementation—the extent to which the members of a partnership have followed through on their commitments. In watershed-based partnerships, stakeholders frequently agree to implement restoration projects designed to improve local environmental or social conditions. We measure this dimension of success by using interviews and partnership documents to evaluate progress on the four main types of restoration projects that watershed partnerships pursue:

— abatement or prevention of point or nonpoint sources of pollution;
— modifications to in-stream flows or water allocation;
— stream channel projects (restoration of vegetation, morphology, or biota); and
— changes in land-use designations (through such actions as purchase, easements, or zoning).

As detailed in Figure 8-1, points are allotted according to the scope and degree of completion of each of the four types of projects attempted, resulting in an index that ranges between 0 and 40.

Perceived Impacts on Human and Social Capital. A third dimension of success is the extent to which a partnership has improved its stakeholders' capacity for achieving future accomplishments. Partnerships have the potential to promote greater knowledge, new interpersonal relationships, and mutual understanding, which are believed to be important ingredients for fruitful collaboration.[14] As conceptualized by Putnam[15] and Coleman,[16] personal networks and relationships are key components of social capital, and pertinent knowledge is a key component of human capital. Both social and human capital can promote agreements and coordinated implementation. This criterion was measured by asking survey respondents to assess, on a 7-

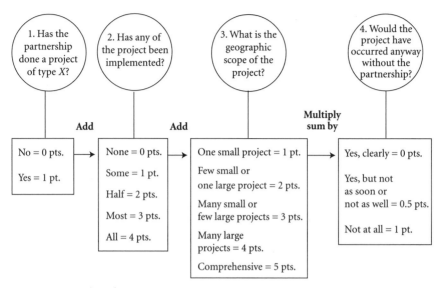

FIGURE 8-1. Index of Restoration Projects

Note: The formula, based on data from interviews and documents, is repeated and summed for four types of restoration projects; see discussion in text.

point Likert scale, whether the partnership has given them (a) "new long-term friendships or professional relationships," (b) "a better understanding of other stakeholders' perspectives," or (c) "a better understanding of the physical or biological processes in the watershed." Each respondent's answers to the three questions were averaged to create a scale. The scale was then averaged across all respondents for a given partnership to create an overall measure of the partnership's perceived impact on social and human capital. The empirical justification for combining these three questions is that they scale reliably (Chronbach's $\alpha = 0.74$), and the factor loadings are quite high ($r = 0.82$, 0.84, and 0.79), respectively.

Perceived Impacts on Environmental and Social Conditions in the Watershed. Finally, the questionnaire asked survey respondents to evaluate their partnerships' impacts on 12 problems ranging from impaired water quality to threats to Native American treaty rights.[17] The scale for each evaluation ranged from -3 (the partnership made the problem much worse) to $+3$ (the partnership made the problem much better), centered at zero (no net impact). Respondents were also asked to assess, on a scale from 0 to 100, the seriousness of each problem in their watershed. To develop an overall score for each respondent, we weighted the 12 impacts by the corresponding seriousness of the problem. (The impact assessment was multiplied by the seriousness assessment, and the product was divided by the sum of all 12 seriousness scores, such that the final score ranges from –3 to +3.) The final index was calculated by averaging across all respondents for each partnership. Weighting each of the 12 assessed impacts by the seriousness of the problem allows each respondent to effectively tailor the uniform list of issues to their own watershed. In this way, the assessments are fairly comparable across partnerships, even if different partnerships face different types of issues.

Measures of Independent Variables

For each partnership, we use information gleaned from the three to five interviews to determine whether there is a coordinator, a facilitator, or both. For each coordinator or facilitator, we characterize them as being either (a) a disinterested consultant versus a stakeholder, (b) hired to facilitate or coordinate versus being "on loan" through an in-kind contribution from an agency, interest group, or individual volunteer, and (c) as full-time versus part-time (i.e., they facilitate or coordinate fewer than 35 hours per week). For facilitators, we also characterize them as either trained professionals or nontrained, occasional facilitators. Finally, we ask the interviewees to describe in their own words the effectiveness of the facilitator and coordinator. We then coded their responses on a 5-point scale from "harmed the partnership significantly" to "helped the partnership significantly," centered at "neutral." For each partnership, the evaluations were then averaged to generate one overall evaluation for each facilitator and each coordinator.

In addition to the variables characterizing the facilitator and coordinator, we also measured many other factors likely to influence partnership success. These include the age of the partnership (months since inception to the time of our interviews), funding (received by or on behalf of the partnership since inception), interpersonal trust, level of ideological diversity, and degree of local versus nonlocal leadership.

Trust was cited as a key to success in 15 of the reviewed studies. Trust is measured using a scale constructed from survey data, as described in Box 8-2. The scale is calculated for each respondent, and then partnership-level indicators are created by averaging the scores for all respondents from a given partnership.

Conflict among stakeholders was cited as a potential obstacle to success in 13 of the reviewed studies. We focus here on ideological diversity, which we measure by first creating a scale to measure resource management ideology and then by calculating the standard deviation of scores on the scale among respondents from a given partnership. The scale is described in Box 8-3.

Eight of the reviewed studies concluded that it is important for stakeholder partnerships to be led or initiated locally. The opposite extreme would be a partnership led by officials and executives from outside the watershed conspiring to impose policies on an unwilling local community. To measure

Box 8-2. Scale for Interpersonal Trust

How many of the participants

a. are honest, forthright, and true to their word? ($\rho = 0.76$)
b. have reasonable motives and concerns? ($\rho = 0.76$)
c. are willing to listen and sincerely try to understand other points of view? ($\rho = 0.83$)
d. reciprocate acts of good will or generosity? ($\rho = 0.83$)
e. propose solutions that are compatible with the needs of other members of the partnership? ($\rho = 0.82$)

Notes: On a 5-point Likert scale: 1 = none, 2 = few, 3 = half, 4 = most, and 5 = all. The scale is the mean of the five items. Chronbach's $\alpha = 0.88$. ρ = nonparametric correlation between item and scale.

Box 8-3. Scale for Management Ideology

a. People were intended to rule over the rest of nature. ($\rho = 0.73$)
b. Plants and animals exist primarily for use by people. ($\rho = 0.74$)
c. Environmental regulations should not be promulgated unless the proponents can prove that the monetary benefits will exceed the costs. ($\rho = 0.78$)
d. The best government is the one that governs the least. ($\rho = 0.74$)
e. A first consideration of any good political system is the protection of private property rights. ($\rho = 0.78$)
f. Government laws and regulations should primarily ensure the prosperity of business because the health of the nation depends on the well-being of business. ($\rho = 0.76$)
g. Government planning almost inevitably results in the loss of essential liberties and freedoms. ($\rho = 0.74$)
h. Decisions about development are best left to the economic market. ($\rho = 0.62$)

Notes: On a 7-point Likert scale: 1 = strongly disagree and 7 = strongly agree. Scale is the mean of the eight items. Chronbach's $\alpha = 0.88$. ρ = nonparametric correlation between item and scale.

local leadership, we first assume that leaders are those respondents who work on partnership activities at least 10 hours per month, which places them in the top third of all respondents in terms of hours. We then measure the proportion of leaders in each partnership who reside in the watershed.

Results and Discussion

Of the 50 partnerships, 20 have a facilitator but no coordinator, 15 have a coordinator but no facilitator, and 15 have both. In other words, all partnerships have either a facilitator or a coordinator or both.

Of the 35 facilitators, 18 are trained professionals, 8 are disinterested, 16 are hired, and 4 are full-time. All disinterested facilitators are trained and hired—reflecting the necessity of paying for the services of someone who, by definition, would have little other incentive to participate. Among the 30 coordinators, 5 are disinterested, 10 are hired, and 4 are full-time. All disinterested coordinators are hired.

The effectiveness scores for both facilitators and coordinators range from −1 to +2, with a mean and median of 1 on a scale where −2 indicates "harmed the partnership significantly" and +2 indicates "helped the partnership significantly."

Bivariate Models of Success

Table 8-1 displays bivariate correlations between the four measures of success and the traits of facilitators and coordinators. The perceived effectiveness of facilitators is correlated fairly strongly with each of the four measures of success. The perceived effectiveness of coordinators is correlated with only two: perceived impacts on human and social capital and perceived impacts on the watershed. Surprisingly, trained facilitators are negatively correlated with the "level of agreement reached", and hired facilitators are somewhat negatively correlated with "level of agreement reached" and "perceived impacts on the watershed." Being disinterested appears to be important only for coordinators, for which there is a mild association with "restoration projects" and "perceived impacts on human and social capital."

Bivariate correlations must be interpreted with caution for a number of reasons. Most notoriously, the observed correlations may be spurious rather than resulting from a cause-and-effect relationship. Spurious correlations occur when a third, unobserved variable causes the two observed states.[18] Similarly, unobserved variables can mask true cause-and-effect relationships. Such confounding variables are almost certainly present whenever one studies complex social phenomena, such as watershed partnerships.

These problems can be ameliorated through statistical techniques that simultaneously account for the intercorrelations among multiple variables.

TABLE 8-1. Correlation between Measures of Success and the Traits of Facilitators ($n = 35$) and Coordinators ($n = 30$)

	Level of agreement reached	*Restoration projects*	*Perceived impacts on human and social capital*	*Perceived impacts on the watershed*
Facilitator traits				
Perceived effectiveness	0.25	0.39*	0.62**	0.45**
Trained	−0.39*	−0.03	−0.24	−0.19
Disinterested	−0.04	0.09	−0.03	−0.15
Hired	−0.28	0.06	−0.13	−0.27
Coordinator traits				
Perceived effectiveness	0.07	0.04	0.51**	0.40*
Disinterested	0.15	0.28	0.33	0.04
Hired	0.25	0.20	−0.06	−0.22

*Correlation is significant at the 0.05 level (two-tailed).
**Correlation is significant at the 0.01 level (two-tailed).

Multivariate regression is one technique that produces estimates of the true relationship by statistically "controlling" for the influence of other variables included in the analysis. It is a second-best approach, used in lieu of an actual experiment in which confounding variables are controlled physically.[19] Lacking the luxury of experimental control over the partnerships in our study, we rely upon the statistical approach. Multivariate regression models of partnership success are presented below.

Multivariate Models of Success

Regression models[20] for each of the four success measures are shown in Tables 8-2 and 8-3. Out of necessity, we have separated the data into two overlapping groups, and we present separate analyses for each. Table 8-2 includes only those partnerships that include a facilitator, and Table 8-3 includes only those partnerships that include a coordinator. For each measure of success we present one model consisting of control variables only, another model consisting of facilitator traits or coordinator traits only, and a third model combining traits and controls. Together, the juxtaposed models indicate how much explanatory power (i.e., adjusted R^2) the facilitator and coordinator traits contribute above and beyond that which can be explained with control variables alone.

In general, we find that facilitator traits are important for explaining "level of agreement reached," whereas coordinator traits are important for explaining "perceived impacts on human and social capital." Facilitators and coordi-

TABLE 8-2. Impacts of Facilitator Traits and Control Variables on Measures of Partnership Success ($n = 35$)

	Level of agreement reached			Restoration projects			Perceived impacts on human and social capital			Perceived impacts on the watershed		
Adjusted R^2	0.39	0.08	0.64	0.60	0.02	0.57	0.47	0.29	0.46	0.74	0.10	0.70
(Constant)	NA	NA**	NA#	NA#	NA	NA	NA**	NA	NA**	NA**	NA	NA
Age of partnership	0.45**	—	0.60**	0.65**	—	0.69**	0.42**	—	0.42*	0.17#	—	0.25*
Grant funding	0.16	—	0.12	0.33**	—	0.32**	-0.09	—	-0.11	0.19*	—	0.18#
Interpersonal trust	0.27#	—	0.27#	0.21#	—	0.15	0.54**	—	0.38*	0.69**	—	0.72**
Local leadership	-0.29*	—	-0.35**	-0.23*	—	-0.19	0.02	—	0.07	0.07	—	0.06
Ideological diversity	0.02	—	-0.04	0.17	—	0.17	0.07	—	0.04	-0.23*	—	-0.23*
Facilitator												
Effectiveness	—	0.13	-0.33*	—	0.39*	0.08	—	0.58**	0.22	—	0.44*	-0.13
Trained	—	-0.34#	-0.34**	—	0.00	0.03	—	-0.07	-0.12	—	0.01	0.00
Disinterested	—	0.26	0.44**	—	0.04	0.18	—	0.02	0.21	—	-0.06	0.11
Hired	—	-0.27	-0.47**	—	0.07	-0.16	—	-0.05	-0.08	—	-0.19	-0.17
Full-time	—	0.09	0.04	—	0.18	0.10	—	0.02	-0.05	—	-0.06	-0.01
Coordinator present	—	-0.05	-0.05	—	0.08	0.12	—	-0.15	-0.06	—	0.07	0.05

Note: Ordinary least squares regression. Standardized coefficients. #$p < 0.10$; *$p < 0.05$; **$p < 0.01$. NA = not applicable; — indicates that the variable was omitted from the model.

TABLE 8-3. Impacts of Coordinator Traits and Control Variables on Measures of Partnership Success ($n = 30$)

	Level of agreement reached			Restoration projects			Perceived impacts on human and social capital			Perceived impacts on the watershed		
Adjusted R^2	0.07	-0.13	-0.05	0.44	-0.08	0.38	0.24	0.32	0.43	0.74	0.17	0.72
(Constant)	NA	NA**	NA#	NA#	NA*	NA#	NA**	NA**	NA**	NA**	NA**	NA**
Age of partnership	0.45*	—	0.47*	0.57**	—	0.57**	0.27	—	0.21	0.14	—	0.09
Grant funding	0.23	—	0.27	0.25#	—	0.31#	-0.09	—	-0.14	0.18#	—	0.19#
Interpersonal trust	0.10	—	0.17	0.28#	—	0.35#	0.53**	—	0.40*	0.87**	—	0.82**
Local leadership	-0.11	—	-0.09	-0.21	—	-0.21	0.04	—	0.09	0.07	—	0.06
Ideological diversity	0.13	—	0.06	0.43**	—	0.36*	0.11	—	0.12	0.00	—	0.00
Coordinator												
Effectiveness	—	0.10	-0.04	—	0.04	-0.14	—	0.51**	0.38*	—	0.46*	0.09
Disinterested	—	-0.05	-0.08	—	0.22	0.13	—	0.41#	0.34#	—	0.12	0.11
Hired	—	0.25	0.36	—	0.04	0.17	—	-0.23	-0.09	—	-0.33	-0.05
Full-time	—	0.03	-0.03	—	0.07	-0.06	—	-0.11	-0.22	—	0.11	-0.06
Facilitator present	—	-0.10	0.00	—	-0.15	-0.04	—	-0.23	-0.06	—	-0.30	-0.13

Note: Ordinary least squares regression. Standardized coefficients. #$p < 0.10$; *$p < 0.05$; **$p < 0.01$. NA = not applicable; — indicates that the variable was omitted from the model.

nators do not appear to be important for explaining "restoration projects" or "perceived impacts on the watershed" in the multivariate analyses.[21] Additionally, each of the control variables contributes to the fit of one or more models. A detailed summary of the results follows.

In every model where they appear, the variables "age of partnership" and "interpersonal trust" are the most important variables in terms of having the largest standardized coefficients, β. For the subset of partnerships with facilitators, age is significant for all four measures of success. For partnerships with coordinators, age is significant for the two objective success measures. Interpersonal trust is significant in almost every model, although it drops out of some models for the two objective measures of success. Across all 50 observations, the trust variable alone accounts for 64% of the variance of "perceived impacts on the watershed."

Funding from grants is stongly associated with "restoration projects," as would be expected, considering that such projects usually require intensive inputs of labor and materials. Funding is also associated with "perceived impacts on the watershed"—a result which may be largely attributable to the role that funding plays in enabling tangible, visible restoration projects.

Local leadership is also important, but in a way that contradicts much of the literature on watershed partnerships. Local leadership is *negatively* associated with the "level of agreement reached" and "restoration projects," at least for the subset of partnerships with facilitators. We can only speculate about why. Watershed boundaries are rarely coterminous with political boundaries,[22] and a partnership that draws its leaders strictly from watershed residents may well lack sufficient participation from government officials (local, state, or federal), who hold the permitting authority, technical expertise, and purse strings that lubricate conflict resolution and sustain project implementation.

Ideological diversity regarding conservative environmental management is another variable that behaves in an unexpected way. One would expect that, in partnerships with highly diverse ideological perspectives, conflict would be greater and conflict resolution more difficult. However, we find no association between ideological diversity and "level of agreement reached." We do find a *positive* association with restoration projects—significantly so for the subset of partnerships with coordinators. One plausible explanation is that medium or high levels of conflict are necessary to motivate the stakeholders to persevere through the arduous process of crafting and implementing projects and watershed management plans.

Ideological diversity is negatively associated with "perceived impacts on the watershed." In ideologically diverse partnerships that have attempted to strengthen environmental regulations or enforcement, resource users may view these efforts as threats to their property rights and may give the partnership poor marks, lowering the "perceived impacts on the watershed" score.

Ideologically consistent groups that are relatively liberal would have fewer discontents, while ideologically consistent conservative groups are unlikely to pursue such strategies in the first place.

Facilitator Traits

Controlling for the variables discussed above, substantive agreements are more likely in partnerships with facilitators who are disinterested, untrained, in-kind or volunteer, and who are viewed as being relatively *in*effective according to the three to five stakeholders whom we interviewed from each partnership (Table 8-2). Two traits that are not significant in any model are whether facilitating the partnership is a full-time job versus part-time and whether a separate coordinator is present. Thus, the data as modeled substantiate Hypothesis 3 and contradict Hypotheses 1, 2, 4, and 5 as they pertain to facilitators.

Hypothesis 3 predicts that being disinterested, rather than a party to the dispute, is advantageous because it imparts the facilitator with a greater degree of perceived legitimacy and objectivity. The hypothesis is supported, and yet, all disinterested facilitators in our study were also trained and hired—detrimental traits according to the model. Thus, the ideal combination of traits does not occur in the sample. To compare the net effects of various combinations of traits that do occur, we looked at the interaction effects for several of the trait dummy variables. These auxiliary regression models (available from the authors) suggest that stakeholder facilitators perform at least as well as disinterested facilitators, as long as they are neither trained nor hired. (i.e., as long as they are untrained and in-kind or volunteer). The most inopportune combination of traits is a stakeholder facilitator who is both trained and hired.

Again, we can only speculate about the mechanism underlying the observed patterns. From our interviews, we know that farmers, ranchers, and other business people frequently resent having to sacrifice personal time and money whenever they attend a midday partnership meeting, whereas agency officials can participate "on company time." The feelings of resentment may be even stronger when those who are "paid to be there" also wield a large degree of control over the process, as facilitators. Occasionally, stakeholders we interviewed overtly expressed concerns about the facilitator being "in it for the money" or "getting rich off the process." If being hired engenders resentment toward the facilitator, the participants would be distracted from the actual work of negotiation, and agreements could be impeded. Because every disinterested facilitator is also hired, the positive trait and the negative one offset one another, and therefore we find that disinterested facilitators are no more effective than stakeholder facilitators who are volunteer or in-kind, rather than hired. Hired stakeholder facilitators would be perceived as being motivated by both an agenda and personal financial gain. The data suggest that this is the most disadvantageous combination of traits.

The negative association between "level of agreement reached" and the facilitator being trained is also curious. Hypothesis 4 predicts that training or ample professional experience would be a clear asset, imparting a larger tool-box of skills, techniques, and experiences from which to work. It is conceivable that the skills and techniques of the facilitation profession actually impede progress on tangible agreements. Facilitators are trained to focus on the partnership structure and process. As a result, facilitated partnerships often spend many months discussing the process before they begin to address actual issues. Besides delaying the actual negotiations, these procedural deliberations run the risk of dissipating valuable energy that typically accompanies a new collaborative endeavor. A drawn-out drafting of the bylaws and ground rules could permanently alienate stakeholders such as farmers, ranchers, and small business owners, who participate on their own time and who tend to be product-oriented, impatient, and have little tolerance for new layers of bureaucracy. Even after a partnership is up and running, an overly structured and managed process can overwhelm participants, some of whom complain about being "processed to death." A highly structured process can also create unintended barriers to consensus building. As one county livestock advisor (and untrained facilitator) told us, "Every ground rule represents another opportunity for someone to try to manipulate the process on technicalities if they think they aren't going to get their way."

The final facilitator trait, perceived effectiveness, is also *inversely* associated with "level of agreement reached." Why would highly rated facilitators be associated with partnerships that are less successful at reaching consensus? One clue and potential explanation comes from the literature on psychotherapy evaluation. Researchers observe that the patients with the most serious clinical problems tend to form the highest affinities for their therapists, viewing them as highly competent and essential to their recovery, even though these patients may in fact experience smaller improvements in psychological well-being relative to other patients (who hold their therapists in lower regard) according to objective measures of progress.[23] Similarly, stakeholders in dysfunctional partnerships may place greater faith in their facilitators, out of some psychological necessity.

Are Facilitators Perceived as More Effective in Difficult or Easy Situations?

An alternative explanation is that partnerships that face the most intractable conflicts tend to seek out more skillful facilitators, thereby generating the inverse relationship between perceived facilitator effectiveness and the objective measure of partnership success, "level of agreement reached." It is

TABLE 8-4. Impacts of Facilitator Traits and Measures of Issue Intractability upon Perceived Facilitator Effectiveness ($n = 35$) and Perceived Coordinator Effectiveness ($n = 30$)

	Perceived effectiveness of the facilitator			Perceived effectiveness of the coordinator		
Adjusted R²	0.44	–0.04	0.52	0.00	–0.07	–0.08
(Constant)	NA	NA**	NA	NA	NA**	NA
Ideological diversity	–0.06	—	–0.07	–0.02	—	–0.06
Interpersonal trust	0.48**	—	0.42*	0.49*	—	0.51#
Norms of compromise	0.33*	—	0.36*	–0.26	—	–0.27
Severity of the initial crisis	–0.20	—	–0.36*	–0.10	—	–0.07
Size of the watershed	–0.22	—	–0.29#	–0.03	—	–0.05
Number of participants	0.26#	—	0.22	0.09	—	0.16
Trained	—	–0.26	–0.25#	—	—	—
Disinterested	—	0.21	0.35*	—	0.26	0.28
Hired	—	–0.13	0.09	—	–0.17	–0.09
Full-time	—	0.05	0.13	—	0.04	–0.07

Note: Ordinary least squares regression. Standardized coefficients. #$p < 0.10$; *$p < 0.05$; **$p < 0.01$. NA = not applicable; — indicates that the variable was omitted from the model.

important to remember that the facilitators in our study were not randomly assigned to partnerships. Naturally, each partnership selects its own facilitator according to the partnership's needs and means. To examine this choice of facilitator and coordinator effectiveness, we can model effectiveness as a function of several other facilitator and coordinator traits as well as variables indicating the difficulty of the consensus-building challenge confronting a particular partnership (Table 8-4).

Two of these difficulty variables were discussed above: ideological diversity and interpersonal trust. We also assess whether the stakeholders are culturally inclined toward cooperation and compromise. Some people relish confrontation and are motivated primarily by a desire to win. Such personalities may be poorly suited to consensus-based conflict resolution. To measure "norms of compromise" we ask survey respondents to agree or disagree that "it is essential to find solutions that are satisfactory to all members of the partnership" (on a 7-point Likert scale). Another indicator of difficulty is constructed by asking participants whether they agree or disagree that "when I first joined the partnership, the problems in the watershed had reached a state of crisis" (on a 7-point Likert scale). This measure has the welcome property that it reflects the level of crisis at or near the inception of the partnership, before the partnership or its facilitator could have influenced the crisis severity. Finally, the size of the watershed is an indicator of the number and complexity of the issues the partnership faces, whereas the number of participants is an indicator of the complexity of the interpersonal relationships, bargaining positions, and coalitions that shape behavior within the partnership.

If partnerships select facilitators whose caliber is proportional to the difficulty of the task, we would expect to see negative coefficients for interpersonal trust and norms of compromise and positive coefficients on ideological diversity, severity of initial crisis, size of watershed, and number of participants. As it turns out, none of these expectations can be conclusively corroborated by the multivariate analysis (see Table 8-4).[24] As predicted, the number of participants is positively related to perceived facilitator effectiveness, but the effect is not statistically significant. In the model of perceived coordinator effectiveness, the fit to the data is too poor (adjusted $R^2 = -0.08$) to draw any affirmative conclusions. On the other hand, several of the expected relationships can be conclusively ruled out in the model of perceived facilitator effectiveness for which the fit to the data is good (adjusted $R^2 = 0.52$). Perceived effectiveness is higher in partnerships with greater trust and stronger norms of compromise that initially confront a less severe problem or crisis. Facilitators also receive higher marks in smaller watersheds. Each of these relationships is statistically significant. In other words, perceived facilitator effectiveness is greater—not lower—in partnerships facing relatively easy tasks.

Coordinator Traits

In multivariate models, coordinator traits are important only for explaining "perceived impacts on human and social capital" (see Table 8-3). Positive impacts are associated with disinterested coordinators and with positive perceptions of coordinator effectiveness. Hypotheses regarding other traits of coordinators are either unsupported or inconclusive.

Coordinators influence human and social capital by organizing activities that build interpersonal networks and that educate participants about the watershed and each other. For example, coordinators often bring in guest speakers to partnership meetings or lead the partnership out into the community through field trips to restoration sites. Coordinators are also often charged with reaching out to potential participants to ensure that the membership is as inclusive as is feasible. If the fence-sitting stakeholders are hesitant to participate due to suspicions about the partnership and its participants' motives, then disinterested coordinators ought to be more persuasive than stakeholder coordinators, who have an agenda of their own, by definition.

Conclusions

The importance of effective facilitation is widely accepted in the literature on consensus-based watershed planning. Anyone with substantial experience as a participant in facilitated disputes has probably been impressed by some facilitators and disappointed by others. Most would probably agree that an incompetent facilitator or coordinator—or one whose personality, ethics, or style is poorly matched to those of the participants—can derail an otherwise promising partnership. But what does "effective" mean? A major goal of this chapter has been to develop a list of attributes of successful facilitators and coordinators and to assess the importance of these two roles relative to other factors.

Overall, facilitator traits are important for explaining the "level of agreement reached," and coordinator traits are important for explaining the partnership's "perceived impacts on human and social capital." Facilitators and coordinators do not appear to be important for explaining "restoration projects" or "perceived impacts on the watershed." Control variables such as interpersonal trust and the age of the partnership are important for modeling all four measures of success, and, as a whole, they are more important than the traits of the facilitators and coordinators.

For both facilitators and coordinators, it is better to be disinterested, all other factors being equal. However, non-disinterested stakeholders can be effective facilitators—especially if they facilitate on an in-kind basis or as volunteers rather than being paid to facilitate through a grant administered by (or on behalf of) the partnership. Despite their best intentions and abilities, paid facilitators, like attorneys, may evoke feelings of resentment among those whose affairs are being facilitated for financial gain. Surprisingly, professional training in the arts of consensus building also appears to be a detriment, all other factors being equal. Training may lead facilitators to devote excessive amounts of time and attention to "getting the process right," thereby delaying substantive negotiations and dissipating stakeholder enthusiasm for collaboration. Watershed stakeholders may also be skeptical of overly polished or excessively managed processes.

These findings have several implications for financial decisionmaking within watershed partnerships and for the current debate over whether public funding for facilitators and coordinators should be expanded. The overarching conclusion is that, in California and Washington state, hiring professional facilitators and coordinators is not always the most judicious use of the finite amounts of public funding earmarked for watershed restoration. Agencies and foundations should not automatically presume that funding such services is the most effective way to support collaborative resource management. Similarly, partnerships with limited financial resources should not automatically invest in hiring a trained, professional, third-party facilitator. A viable option for many cash-poor partnerships is to recruit a layperson facilitator from within one of the participating agencies or interest groups, as long as that person's services are provided on an in-kind or volunteer basis.

Conceptually, funding for a professional facilitator or coordinator is justified only if the marginal value of the resulting consensus agreements and improved social capital exceeds the opportunity cost of diverting the funds away from some other type of intervention, such as a tangible restoration project. When estimating this marginal value, one should inventory existing levels of organizational capacity within the partnership. If no agency or interest group is willing or able to bear the costs of initiating and sustaining collaboration, external assistance may be justified. Once this determination is made, however, the funding organization should stipulate that the person hired must be a neutral third party. Our findings suggest that funding is detrimental if it goes toward hiring a party to the dispute.

We close the chapter by reflecting upon some of the strengths and limitations of this study and its implications for future research. First and most obviously, our study considered only a handful of facilitator and coordinator characteristics, and future research could add both breadth and detail in this area. Particularly absent were measures of intangible traits, such as personality, style, and ethics. Second, the data represent only a snapshot of each partnership at one moment in time. By revisiting each partnership at five-year intervals, one could measure actual changes in social capital over time, rather than relying on "perceived impacts." A repeated-measured design would also allow more sophisticated modeling and stronger causal inferences. Another useful extension of the research would be to replicate the methods in other states where the ambient levels of organizational capacity are lower than in California and Washington state, which could make professional facilitation and coordination more important.

Finally, we encourage future researchers to recognize the complexity of facilitated environmental disputes and to reflect this complexity in their approaches to data collection and analysis. For example, the multivariate regression models revealed patterns in the data that were not apparent in the simple bivariate correlations. In some cases, relationships that were present in the bivariate analyses disappeared when control variables were introduced. Other relationships appeared where previously there were none. And one relationship (between "perceived facilitator effectiveness" and "level of agreement reached") switched from being moderately positive to decisively negative after controlling for interpersonal trust. The mirror that we have turned upon the profession is beveled and bowed. As in all research, it is useful to view the subject from multiple perspectives and in varying lights.

Notes

Special thanks to Chris Weible for assistance with analyzing and interpreting the data. The field research was carried out by the authors and Kate Reza, Beth Cook, Jared Ficker, Maryann Hulsman, Tamara LaFramboise, Erin Klaesius, Steve Kropp, Neil Pelkey, Martha Turner, and Chris Weible. Funding was provided by the National Science Foundation's "Decision Making and Valuation for Environmental Policy" program (grant 9815471), the U.S. Environmental Protection Agency's "Science To Achieve Results" (STAR) program (grant R82–7145), and the David and Lucile Packard Foundation's "Conserving California Landscapes Initiative."

1. William D. Leach and Neil W. Pelkey, "Making Watershed Partnerships Work: A Review of the Empirical Literature," *Journal of Water Resources Planning and Management,* vol. 127, no. 6 (2001), pp. 378–385.

2. William D. Leach, "Surveying Diverse Stakeholder Groups," *Society and Natural Resources,* vol. 15, no. 7 (2002), pp. 641–649.

3. See also Gail Bingham, *Resolving Environmental Disputes: A Decade of Experience* (Washington, DC: Conservation Foundation, 1986); Susan L. Carpenter and W.J.D. Kennedy, *Managing Public Disputes: A Practical Guide to Handling Conflict and Reaching Agreements* (San Francisco: Jossey-Bass, 1988); "Top 10 Watershed Lessons Learned" (Washington, DC: Environmental Protection Agency, Office of Water, 1996); Judith E. Innes, "Evaluating Consensus Building," in Lawrence Susskind, Sarah McKearnan, and Jennifer Thomas-Larmer, eds., *The Consensus Building Handbook: A Comprehensive Guide to Reaching Agreement* (Thousand Oaks, CA: Sage Publications, 1999); Lawrence Susskind, Ole Amundsen, and Masahiro Matsuura, *Using Assisted Negotiation To Settle Land Use Disputes: A Guidebook for Public Officials* (Cambridge, MA: Lincoln Institute of Land Policy, 1999); Lawrence Susskind, Sarah McKearnan, and Jennifer Thomas-Larmer, eds., *The Consensus Building Handbook: A Comprehensive Guide to Reaching Agreement* (Thousand Oaks, CA: Sage Publications, 1999).

4. Jo Clark, *Watershed Partnerships: A Strategic Guide for Local Conservation Efforts in the West* (Denver: Western Governors' Association, 1997).

5. Michael L. Poirier Elliott, "The Role of Facilitators, Mediators, and Other Consensus Building Practitioners," in Lawrence Susskind, Sarah McKearnan, and Jennifer Thomas-Larmer, eds., *The Consensus Building Handbook: A Comprehensive Guide to Reaching Agreement* (Thousand Oaks, CA: Sage Publications, 1999); Ann Moote, *Partnership Handbook: A Resource and Guidebook for Local Community-Based Groups Addressing Natural Resource, Land Use, or Environmental Issues* (Water Resources Research Center, College of Agriculture, University of Arizona, 1995).

6. Jesse A. Gordon and Timothy T. Jones, *Monitoring and Evaluation of Selected Rural Watershed Councils in the Continental United States* (Fayetteville, AR: Buffalo River Stewardship Foundation, 1998); Marjorie M. Holland, "Ensuring Sustainability of Natural Resources: Focus on Institutional Arrangements," *Canadian Journal of Fisheries and Aquatic Sciences,* vol. 53 (1996); Betsy K. Landre and Barbara A. Knuth, "Success of Citizen Advisory Committees in Consensus-Based Water Resources Planning in the Great Lakes Basin," *Society and Natural Resources,* vol. 6 (1993); M.A. Moote, M.P. McClaran, and D.K. Chickering, "Theory in Practice: Applying Participatory Democracy Theory to Public Land Planning," *Environmental Management,* vol. 21, no. 6 (1997); Betsy Rieke and Douglas Kenney, "Resource Management at the Watershed Level: An Assessment of the Changing Federal Role in the Emerging Era of Community-Based Watershed Management. Report to the Western Water Policy Review Advisory Commission, Denver, Co." (Boulder: Natural Resources Law Center, University of Colorado School of Law, 1997).

7. Julia M. Wondolleck and Clare M. Ryan, "What Hat Do I Wear Now? An Examination of Agency Roles in Collaborative Processes," *Negotiation Journal,* vol. 15, no. 2 (1999), pp. 117–133.

8. Rosemary O'Leary and Susan Raines, "Lessons Learned from Two Decades of Alternative Dispute Resolution Programs and Processes at the U.S. Environmental Protection Agency," *Public Administration Review,* vol. 61, no. 6 (2001), p. 665.

9. Landre and Knuth, "Success of Citizen Advisory Committees in Consensus-Based Water Resources Planning in the Great Lakes Basin."

10. President's Council on Sustainable Development, "Lessons Learned from Collaborative Approaches" (Washington, DC: New National Opportunities Task Force, 1997).

11. California partnerships were identified through several means. First, a brief questionnaire was mailed to a random sample of district conservationists with the Natural Resources Conservation Service, directors from local resource conservation districts, field personnel of the California Department of Forestry and Fire Protection, and University of California Cooperative Extension specialists. Second, we searched the Natural Resource Projects Inventory, a database housed in the University of California—Davis Information Center for the Environment. Third, we used Internet search engines to find relevant Web pages. Fourth, we asked interviewees from each partnership to name other partnerships in their regions. Finally, in cases where there was any doubt about whether a suspected partnership satisfied our operational definition, we called a partnership representative to obtain further information. A similar process was used for Washington state.

12. We partitioned California using Hydrologic Unit Code (HUC) watersheds defined by the U.S. Geological Survey. There are 160 HUCs in the state, ranging from 35 to 9,000 square miles.

13. We partitioned Washington state using the 62 Water Resource Inventory Areas, which range from 140 to 3,000 square miles.

14. Innes, "Evaluating Consensus Building."

15. Robert D. Putnam, *Bowling Alone: The Collapse and Revival of American Community* (Simon & Schuster, 2000); Robert D. Putnam, Robert Leonardi, and Raffaella Y. Nanetti, *Making Democracy Work: Civic Traditions in Modern Italy* (Princeton University Press, 1993).

16. James S. Coleman, "Social Capital in the Creation of Human Capital," *American Journal of Sociology,* vol. 94 (1988), pp. 95–120.

17. The 12 problems were impaired water quality, threatened species or habitat, lack of open space, population growth, inadequate water supply, risk of damaging floods, threat of catastrophic fire, lack of economic prosperity, severe regulation or taxes, threats to property rights, threats to tribal or treaty rights, and conflict among stakeholders.

18. For example, the retail prices of peanut butter and jelly are highly correlated over time—not because demand for one creates demand for the other—but because of general inflation in the economy.

19. In such an idealized experiment, the researcher manipulates a single variable for one of two identical groups of subjects. Any observed differences between the treatment group and the control group must be caused by the one manipulated variable.

20. Regression models for each measure of success consist of an explanatory variable multiplied by a coefficient and summed together. The regression procedure produces estimates of these coefficients. For each explanatory variable in a model, the tables display standardized versions of the coefficients, such that variables with larger

coefficients can be interpreted as having larger effects on the success indicator. For each coefficient, we also report p values, which correspond to the probability that "true" value of the estimated coefficient is significantly different from zero. In other words, the smaller the p value, the greater the confidence that the variable has an actual effect that cannot be attributed to random or chance patterns in the data. Finally, we report the R^2 statistic, which measures how well the model fits the data. The R^2 can be interpreted as the proportion of the variation in the success measure that is accounted for by the model. Thus, $1 - R^2$ is the "unexplained variation." Larger R^2 statistics indicate more accurate models.

21. In the case of "perceived effectiveness" variables, the null findings could be partly attributable to the truncated range and somewhat skewed distribution of the data. The scores range from −1 to +2, with a mean and median of 1, on a scale where −2 indicates "harmed the partnership significantly" and +2 indicates "helped the partnership significantly."

22. Doug S. Kenney, "Historical and Sociopolitical Context of the Western Watersheds Movement," *Journal of the American Water Resources Association*, vol. 35, no. 3 (1999), pp. 493–503.

23. Kirsten von Sydow and Christian Reimer, "Attitudes toward Psychotherapists, Psychologists, Psychiatrists, and Psychoanalysts: A Meta-Content Analysis of 60 Studies Published between 1948 and 1995," *American Journal of Psychotherapy*, vol. 52, no. 4 (1998).

24. We also attempted to specify logistic regression models for each of the other facilitator traits (trained, disinterested, hired, full-time) as indicators of caliber, using the various measures of "difficulty" as explanatory variables. These efforts uniformly failed, yielding models with poor fit to the data and no significant coefficients. These results further contradict the hypothesis that partnerships select the caliber of their facilitator in proportion to the difficulty of the situation.

PART IV

Downstream Environmental Conflict Resolution at the State and Federal Levels

This section moves us from theory to application, from recognition to lessons learned, from transforming people to transforming programs. Setting the stage is Andy Rowe, a professional evaluator, who explains evaluation methods, opportunities, and challenges in environmental conflict resolution (ECR) in Chapter 9. Rowe maintains that the evaluation of ECR programs is constrained by two challenges: first, establishing the incremental contribution of dispute resolution compared to alternative processes, and second, measuring the effectiveness of dispute resolution. Rowe presents his own model for evaluating environmental and public policy dispute resolution programs.

Specific program evaluations at the state and federal level are the foci of the remaining four chapters in this section. In Chapter 10, Kirk Emerson and Christine Carlson examine an ongoing effort to design and implement a self-administered program evaluation system for state and federal ECR programs. Specifically, current evaluation efforts at the U.S. Institute for Environmental Conflict Resolution, the Massachusetts Office of Dispute Resolution, and the Oregon Dispute Resolution Commission are explained and analyzed. Lessons learned that will be useful to other federal and state ECR agencies are presented.

A research team from Florida State University presents its analysis of Florida's current dispute resolution efforts in Chapter 11. Frances Berry, Bruce Stiftel, and Aysin Dedekorkut focus on two areas: first, managerial attitudes toward the use and usefulness of dispute resolution techniques and processes, and second, the results of their survey of participants in Florida dispute resolution efforts. Citing many environmental examples, the authors conclude that mediation can be used effectively to improve process and outcomes in a significant portion of administrative agency disputes.

A recent ECR pilot project in the Oregon federal trial court is the focus of Lisa A. Kloppenberg in Chapter 12. Kloppenberg overviews the goals and parameters of the Oregon project and explains the role of judicial attitudes toward alternative dispute resolution (ADR) in the outcome of the program. Kloppenberg presents invaluable lessons learned, including the need for an early assessment of the case docket to ascertain whether ADR is appropriate, and the need for an ADR administrator.

Rosemary O'Leary and Susan Summers Raines analyze two decades of federal ECR efforts in Chapter 13. After highlighting the 20-year history of the programs at the U.S. Environmental Protection Agency (EPA), the authors present their results of a comprehensive survey as well as 10 lessons learned that are applicable to other governmental organizations contemplating using ADR. Among these lessons learned is the recommendation that EPA evaluate its ADR efforts continually.

When examined as a whole, the five chapters in this section offer considerable insights for policymakers, public managers, and concerned citizens interested in improving the evaluation of environmental and public policy conflict resolution programs and policies.

9

Evaluation of Environmental Dispute Resolution Programs

Andy Rowe

Evaluation is a vehicle for acquiring quality information about the performance of programs and activities. Systematic evaluation is an integral element of reflective practice and an invaluable contribution to improved practice and theory of dispute resolution. As a unified field of practice, program evaluation, like dispute resolution, is relatively young, and we are still developing the knowledge and practice necessary for high levels of performance. Evaluation offers a mechanism for systematically gaining valid and reliable information about dispute resolution—information that can successfully undergo critical scrutiny from internal and external stakeholders.

Because some of the claims and statements I will make in this chapter could be surprising, I believe that it is important for readers to know my background in both evaluation and working with dispute resolution programs. I have a long and successful track record in evaluation and limited recent experience developing evaluation systems for two different programmatic settings—the U.S. Institute for Environmental Conflict Resolution (USIECR) and the Oregon Public Policy Program, of which the Oregon Dispute Resolution Commission is one of three state agency partners. My evaluation experience spans almost 20 years of practice in diverse public program areas, including environmental and resource programs, all areas of health and human services, arts and culture programs, and international development. Over this period I have employed most of the many approaches evaluators have in their toolkits. I was president of the Canadian Evaluation Society (CES) for two years and have been privileged to serve at national levels in both the CES and the American Evaluation Association (AEA). One thing that has struck me over the relatively brief period in which I have been involved with dispute resolution programs is the gap between professional program evaluation and evaluation of dispute resolution. From anecdotal and personal observation, it seems that many who evaluate dispute resolution programs come from some association with the practice area itself and have limited evaluation experience outside these programs. As indicators of this, when I asked attendees at the invigorating and successful conference associated with this volume how many were members of the American Evaluation Association or had attended AEA meetings, only a few indicated that they fitted the bill. Likewise, an important state dispute resolution program held a set of invited meetings on evaluating dispute resolution programs the same week as the AEA annual meetings. Both examples demonstrate the gap between professional evaluation and the evaluation of dispute resolution programs. Both practices could benefit from closer collaboration among practitioners.

Because of this observed gap, my initial focus in this chapter is to communicate what evaluation can offer to dispute resolution, and then I assess the options for evaluating environmental dispute resolution programs. The chapter concludes with a summary of the key opportunities and challenges. Before embarking on that agenda, it is important to set evaluation in its contemporary context.

Contemporary Context for Program Evaluation

The expectations and requirements of public and nonprofit funders changed fundamentally in the 1980s and 1990s. This period saw the enactment of federal and state legislation requiring that programs focus on and achieve success in attaining their mandated results. The movement began in the United States, Australia, and New Zealand, but by the end of the period could be seen in all industrialized and many developing countries. Foundations and other nonprofit funders also enacted similar requirements. In the United States, the Government Performance and Results Act (GPRA) has changed forever the expectations and requirements of accountability for federal agencies and programs. Most states have also enacted accountability and performance requirements for state agencies, many before GPRA—for example, the Oregon legislature requirement of performance measures from all state agencies.

This move to a results focus and accountability has raised the profile and utility of evaluation for state, federal, and community agencies. It has also provided a useful structure around which evaluators can organize their work. In 1997, John A. Koskinen of the U.S. Office of Management and Budget described GPRA this way:

At its simplest, the Government Performance and Results Act (GPRA) can be reduced to a single question: what are we getting for the money we are spending? To make GPRA more directly relevant for the thousands of federal officials who manage programs and activities across the government, GPRA expands this one question into three: what is your program or organization trying to achieve? How will its effectiveness be determined? How is it actually doing? One measure of GPRA's success will be when any federal manager anywhere can respond knowledgeably to all three questions.

The evaluation systems I developed for the Oregon Public Policy Program and the U.S. Institute for Environmental Conflict Resolution directly address these three questions.

Evaluation and Dispute Resolution Programs

Many readers of this volume will not be familiar with the categories and terms used in program evaluation. I will outline a few terms that are basic to evaluation of dispute resolution programs.

Evaluators have traditionally grouped evaluations as *summative* or *formative*.[1] Summative evaluations are about judging the merit or worth of a program or service; the purpose of formative evauation is to support program improvement. Evaluation is reaching a level of maturity where the distinction between formative and summative evaluation is of less value. At one time, summative evaluation was taken to refer to assessment of outcomes, whereas formative evaluation assessed processes. This was never really the case and today is clearly not correct; the difference has more to do with the types of decisions made using the evaluation information than with evaluation methods or focus.

The agenda for evaluation of dispute resolution programs is both summative and formative. It includes summative considerations of whether to fund these programs or use more traditional mechanisms of addressing disputes (including doing nothing), and it is formative because it provides information identifying means of improving the effectiveness of particular dispute resolution programs, or indeed, the practice of dispute resolution. Most existing evaluations of dispute resolution programs appear to have a formative intent; however, legislatures and foundations are now asking questions and considering decisions that are best addressed with summative evaluations. Ideally, programs should first have the benefit of formative evaluation to aid them in achieving higher levels of performance and as a resource for any summative evaluations that might be requested.

This points to a second consideration regarding the purpose of evaluation. Evaluation seeks to acquire information that is good enough to support current and likely future decisions about the program. This is the guiding principle for the quality of evaluation information and can be contrasted to the standards for applied research that would engage peer review, publication, and hopefully, use. Referring to the program decisions that are likely to be made using evaluation information provides a useful general guide to the appropriate design of an evaluation.

Before initiating an evaluation, an *evaluability assessment*[2] should always be undertaken. Not all programs should be evaluated; for example, programs in their early stages are unlikely to be performing near their potential, and it would be appropriate to use evaluation approaches for immediate program improvement purposes, but inappropriate to use evaluation to judge the merit or worth of the program at this stage. In assessing a program's evaluability, evaluators should take into account

— whether the program has established clear and observable outcomes,
— if adequate information either exists or can be obtained with a reasonable level of effort,
— if there is good evidence that key stakeholders will use the evaluation information, and
— if sufficient key stakeholders can be engaged in the evaluation process.

Dispute resolution programs appear no more or less likely to be evaluable than programs in general. The three programs included in the collaborative evaluation initiative reported on elsewhere in this volume[3] easily passed an evaluability assessment.

As evaluation practice and theory developed, the concept of a *program theory* has gained importance and today is central to many approaches to evaluation. Program theory is the theoretical and practice knowledge underpinning a program. When designing evaluations, evaluators increasingly want to establish the conceptual strength of the program's claims at an early stage. Surprisingly, many programs are launched and continue without ever articulating the underlying program theory, with the near-certain effect of reduced effectiveness and efficiency. Unfortunately many evaluators still undertake an evaluation without first clarifying the program theory, thereby enhancing the likelihood that the evaluation will expend resources assessing program activities that could have been made more efficient if the underlying program theory had been clarified. Program theory is also indispensable in allowing evaluators to offer evaluation judgements about the wider contributions of the programs. For example, many international development programs are funded from sources that have the goal of poverty reduction, but the funded programs themselves do not contribute directly to poverty reduction. Improved environmental infrastructure (such as water, sewage, and solid waste disposal) in slums does little to reduce poverty directly, so we use program theory to provide the link. So long as the environmental infrastructure program successfully achieves the program theory, evaluators can make the statement that the program "is likely to" contribute to poverty reduction, based on good contemporary and relevant knowledge from practice and theory.

The strength of the program's claims can first be assessed via the program theory, and this also provides a valuable guide to the quality of the information the evaluation requires: what is good enough? Lee Sechrest, a professor of psychology at the University of Arizona, uses an effective and simple illustration. For some diseases, inoculation has been established as the optimal treatment approach. This is based on extensive epidemiological research, itself testing hypotheses resting on well-established physiological, pharmacological, and other disciplinary knowledge. From this we know that as long as we monitor inoculations and the conditions under which they are provided, we can be reasonably sure that the desired outcome will be achieved. Consider an alternative treatment—the "laying on of hands." The theoretical and empirical foundations of this treatment are not as well established and do not engender the same level of confidence about program success that we associate with inoculation. Thus, the required quality of the evidence to evaluate an inoculation program is modest. For this we track outputs: who was inoculated where and with what? The claims made by "laying on of hands" can be regarded as quite exceptional, requiring exceptional evidence to evaluate program effects. This would likely require complex experimental designs sustained over a long period and conducted in a variety of settings for a range of ailments.

When I have stated that dispute resolution has a reasonably robust program theory, the response has ranged from outright disagreement to grudging, "Okay, but it is not really a program theory." Perhaps with the advantages of a fresh perspective I have been able to see the practice of dispute resolution more clearly—or perhaps my experience has been too limited, and I have not yet encountered the programs and dispute resolution practices that would put the lie to my claim. Illustrated below are the primary outcomes that dispute resolution seeks to achieve.[4] These were developed with Susan Brody and Mike Niemeyer of the Oregon Public Policy Program (Oregon PPP) and are highly congruent with the outcome structure developed with the U.S. Institute for Environmental Conflict Resolution (USIECR). The differences between the Oregon PPP and the USIECR structures lie in program emphasis, delivery, and accountability, not in the underlying program theory. There is also a tight match between these two outcome structures and the program conception developed by David Fairman of the Consensus Building Institute with the Massachusetts Office of Dispute Resolution. All three were developed independently, as part of a collaborative effort providing credence to the claim that the dispute resolution practice pursues a fairly homogeneous set of outcomes.

When one considers the strength of the underlying knowledge from literature on organizations, psychology, and negotiation, it seems reasonable to state that there is a fairly homogeneous program theory. This does not by any means imply that practice is homogeneous; that is clearly not the case. The program theory states the outcomes that diverse practice approaches seek to achieve.

Figure 9-1 groups process and agreement outcomes separately and distinguishes two further subsets of process outcomes: alternative dispute resolution is used appropriately, and party interactions are constructive. Both process and agreement outcomes are recognized and valued in this program theory, and there is an assumption that successful process outcomes are necessary to have successful agreement outcomes. The top gain from the use of dispute resolution processes is that parties and agencies can redirect resources that were allocated to the conflict issues to more productive uses.

The logic model is one means of describing a program and articulating a programmatic theory. There are many variants of a logic model. The one pictured here is partial but typical, but there are many different ways that evaluators describe outcomes, impacts, and results. Contemporary evaluation places considerable emphasis on the right-hand side of this figure. The key terms in a logic model are activity, output, outcome, and some term that indicates higher-level outcomes that are beyond the ability of the program to affect alone. Figure 9-2, which also includes impacts, uses the following terms: *activities* are what the program does (facilitate resolution of disputes), *outputs* are the products of the activities (number of disputes facilitated), *outcomes* are what has changed as a result (parties collaborate, no unresolved issues), *results* are the biggest outcome(s) for which the agency is accountable, and *impacts* are how society benefits from the program (fewer resources expended on disputes).

Logic models have become much in vogue—public and foundation funders now frequently require fundees to develop a logic model. This is a welcome step, but like most developments that come into vogue, form often becomes more important than content.

Evaluation Methods and Environmental Dispute Resolution Programs

To achieve the goal of judging a program's merit or worth, evaluation must be able to make valid and reliable statements about the program's effects compared to what would have occurred without benefit of the program. I refer to this as the *incrementality* of the program. Evaluation uses the entire

ADR is successful (that is, time and resources spent in disagreement and conflict are now redirected to more constructive purposes)

↑

GOOD AGREEMENTS

Durable and implementable agreement is reached using ADR. (e.g., changes in weather conditions and unanticipated events, terms, and conditions can be accommodated within the schedule)	Agreements reached with ADR are complete; no hard issues are left or deferred. (e.g., nothing is left unaddressed that is likely to derail agreement because it is critical to parties and/or it involves a high level of conflict or tension)	Agency, program mediator, and parties' capacity to use ADR is improved through experience with this case.
		Government decisionmaking is improved through use of the process.

↑

GOOD OUTCOMES FROM PROCESS

All parties involved in an ADR process are satisfied that the process was fair and open.	ADR is more effective (better benefits for the resources expended) than the other options for this dispute

Party interactions are constructive.

The right parties (those affected or who can affect decisions) continue to be engaged, and new parties are added as required.	The use of ADR narrows disagreements (e.g., reduces the number of issues, focuses on priority issues, identifies underlying problems, identifies issues best dealt with another way).	The use of ADR helps adversarial parties collaborate.

ADR is used appropriately.

ADR is used where it is the best approach for this case (based on screening and assessment and on other sources).	Non-ADR processes are used where they are the best approach (based on screening and assessment and on other sources).	An appropriate mediator leads ADR.
		The design of the ADR process is appropriate for the dispute and needs of the parties.

Monitoring and evaluation are used to generate feedback and lessons learned; these are used to improve practices.

FIGURE 9-1. Oregon Public Policy Program Case Management Options

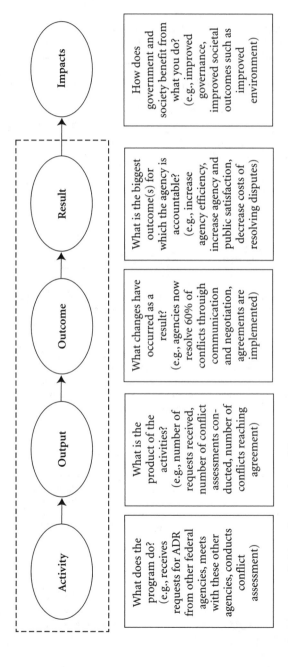

FIGURE 9-2. Core Definitions in a Logic Model

Note: The program is accountable for what lies inside the dotted-line box. In achieving these, the program contributes to impacts, but it is not accountable alone for achievement of impacts.

range of social science research methods, and incrementality can be assessed using qualitative or quantitative approaches,[5] or a combination of both. Dispute resolution is a diverse practice with differences in approaches and in programmatic settings, requiring different evaluation approaches. The Program Evaluation Standards,[6] developed by the Joint Committee on Standards for Educational Evaluation, direct evaluators to ensure that their work is appropriate, useful, of good quality, and affordable.

I have heard a number of stories about external evaluators requiring programs to randomly assign disputes to a dispute resolution treatment group or to a comparison group where dispute resolution services are not provided and the parties use other available options. These experimental and quasiexperimental designs can be appropriate in some settings, but should not be employed where they are not the optimum approach. Experimental and quasiexperimental designs rely on statistical techniques that require members of the program population to have an equal or known likelihood of being in the treatment group. The treatment and nontreatment groups are then both monitored for program effects, and the presumably positive differences are attributed to the program. The general requirement for this approach is a relatively large number of fairly homogeneous disputes and either homogeneous program treatments, or, at least, known differences in program treatment along with knowledge of the disputes' distinguishing characteristics. This approach is the most direct means of identifying incrementality, it has a high level of validity, and it is a well-established approach in social and natural research. I have used these designs frequently and to good advantage. If such methods were appropriate, I would usually employ a quasiexperimental design with a group of comparison projects, rather than an experimental design with random assignment of cases to the treatment or nontreatment groups. For programs such as dispute resolution, random assignment has no identifiable benefit over comparison groups and runs the risk of reducing program effectiveness because cases are often assigned to dispute resolution on the basis of some explicit or implicit screening to ensure a match between the case and the treatment applied.

Environmental disputes are typically quite heterogeneous because of important characteristics of the disputes and the parties. Moreover, the context within which dispute resolution is considered can significantly alter disputes; for example, a court's mandate to use dispute resolution likely alters how parties approach the resolution process. To date, the only settings I have encountered where the requirements of a quasiexperimental design might be met are in family courts, human resource complaints, or similar large-scale programs with fairly homogeneous cases, where sufficient and appropriate program data exists for all disputes in the study population. Given the complexity and heterogeneity of environmental disputes, it is extremely unlikely that they would ever meet the requirements for this approach. This means that another evaluation design is required to evaluate environmental dispute resolution cases. Where quasiexperimental design is not suitable, many evaluators look for comparators. For example, we might seek to match participants in an employment program to nonparticipants with similar characteristics. For environmental disputes, we would seek to find similar disputes that were addressed by a reasonably good alternative, such as a judicial settlement conference, rulemaking, arbitration, litigation, or doing nothing. Some studies ask parties and neutrals to compare their dispute to other alternatives and use this information to make evaluative judgements about the incrementality of dispute resolution. This information is not sufficiently valid or reliable to support these claims—parties rarely have sufficient experience and knowledge of alternatives, and even when they do, the comparisons they make are unlikely to be to similar disputes. Without comparators, it is difficult to credibly state the incremental contribution of dispute resolution. This is a significant challenge for the evaluation of environmental dispute resolution processes, which I will return to in the next section.

Regardless of the means by which evaluators establish the incrementality of the program, many research methods can be used. Good case studies are appropriate for some settings, while others are better served through systematic surveys, anthropological methods, or collaborative and empowerment approaches. No approach has inherent superiority—good evaluators will use the approach that provides credible information that is good enough for the decisions that are likely to be made.

Evaluators must be careful to ensure that their evaluation efforts do not get in the way of the program. This is particularly important in dispute resolution programs, where success is highly dependent on the relationship between the neutral[7] and the parties and the relationships among the parties themselves. The presence of evaluation could potentially disturb this relationship. It is my current view that evaluation should "stay out of the room" unless it is certain that it will not affect the dispute resolution process. Collaborative and empowerment evaluation approaches offer the best options for getting evaluation "inside the room" (see opportunities below).

Many existing evaluations of dispute resolution programs are best described as case studies from a legal or process perspective, using interviews, document reviews, and court information, or are limited reviews of party satisfaction and program implementation. There have also been efforts to use retrospective interviews with parties and others to obtain information about the dispute and the resolution process. My limited review of existing evaluations was neither systematic nor thorough, so it is with caution that I offer the observations that although these studies have been useful and important, the incremental contribution of dispute resolution has not been established, and the quality is often not good enough for the usual types of decisions that would use evaluation information.

Challenges and Opportunities

Systematic evaluation of dispute resolution programs offers valuable inputs for improving the performance of these programs and a means of responding to the increasing accountability requirements of public and foundation funders. Many of the benefits from evaluation will be realized early in the evaluation process. Both the Oregon PPP and the USIECR report benefit from their early efforts to clarify program outcomes and accountabilities. In addition, analysis of the early data from the evaluation has already identified some potential areas for improving overall performance of the programs and neutrals. Prospects are strong that the evaluation systems will also identify aspects of dispute resolution practice that can be improved. However, there are some important challenges in evaluating dispute resolution programs and environmental dispute resolution programs.

Challenges

The main challenge in evaluating environmental dispute resolution programs—because of the difficulty in identifying and obtaining information from comparators—is demonstrating the incremental benefit of the program. The second challenge applies to dispute resolution programs in general: obtaining valid and reliable information on the effectiveness of the program.

Incrementality. Assessing a program's incremental contribution is at the core of evaluation and presents a serious challenge for programs like environmental dispute resolution that feature extreme heterogeneity and complexity. The key is to assess the incremental benefit of the dispute resolution approach compared to other good alternatives. For this we require good information about what would have happened if one of those alternative approaches had been used instead of dispute resolution.

I have briefly described some of the approaches evaluation uses to do this. However, individual environmental disputes appear to be sufficiently unique that comparators are difficult to find. Whereas there are many water disputes, and indeed many water disputes involving similar classes of parties, they can still differ critically in terms of the parties' characteristics, the approach, the way that the neutrals approach the parties, the process actually used, the contextual setting for the dispute, and many other ways. Still, if we had good information on a large number of water disputes with similar parties, the importance of these differences would diminish, but we do not.

However, all is not lost. Early information from the Oregon PPP suggests that parties consider nonlitigative options such as legislation and administrative rulemaking as preferred alternatives to dispute resolution. Furthermore, in some jurisdictions (such as Oregon), attorneys from the Department of

Justice estimate the likely settlement and the internal costs of the state litigating larger cases and compare those estimates to calculated alternative dispute resolution costs. Thus, there are places to start looking for comparator alternatives to environmental dispute resolution. The requirement is for typical types of disputes and disputes addressed through the most likely means, and it is likely that for environmental disputes, the number of cells in this matrix of exemplars would not be that great. However, without applied research to obtain this information, evaluation is constrained in estimating the incrementality of dispute resolution approaches, something with which most external stakeholders such as legislators and funders are very concerned.

Measuring the Effectiveness of Dispute Resolution Programs. Dispute resolution has sometimes been promoted as "faster, better, and cheaper." Although I am told that these claims are no longer made as often or as enthusiastically as they once were, they still have considerable intuitive appeal to legislators and other funders. Indeed it is reasonable to expect any program, when employed appropriately, to generate a combination of greater benefits (better) and greater efficiency (faster and cheaper) compared to the alternatives. This is what we mean by program effectiveness—more benefits per unit of input. Programs can increase their effectiveness by improving the amount or quality of the benefits they achieve for participants, reducing the amount of input required by improving efficiency in the provision of the programs, or a combination of the two. Thus the statement "faster, better, and cheaper" could be restated as, "Dispute resolution, when used appropriately, is more effective than the alternatives." This is a reasonable claim that bears assessment.

It is difficult to measure the effectiveness of dispute resolution. On the cost side, there are serious measurement challenges, some general to estimating the types of costs involved in disputes and some that arise from the duration of disputes. It is also difficult to measure the benefits of dispute resolution.

Conceptualizing and Measuring Costs. When we measure costs of a given dispute addressed with dispute resolution processes, we want to do it in a manner that we can replicate: similar disputes should enable comparison of the different program alternatives' effectiveness. This is essential for assessing the effectiveness of dispute resolution initiatives. On a larger scale, this will allow us to assemble a body of knowledge about ways to reduce the costs to parties participating in dispute resolution processes, as well as increasing accountability. There are two deep challenges in this. First, disputes, and often dispute resolution processes, occur over a long period of time. Disputes can exist for years, even decades or longer, before engaging dispute resolution or another alternative. Moreover, it can take many years for all the benefits and costs of a dispute to occur. To properly measure the costs of a dispute, we want to compare the cost before the use of dispute resolution processes to the cost after all of the benefits occurred (regardless of whether an agreement was reached and implemented). This difference, plus the direct costs of resolving the dispute, constitutes the net change in costs attributable to the program intervention. We assume that the net change should be negative— the costs associated with the dispute should go down, and the fall in costs should be greater than the expense of the dispute resolution process itself.

The longevity of disputes is only one of the challenges. Another challenge arises when we try to conceptualize the costs of disputes. Figure 9-3 provides a partial enumeration of the types of costs that could be associated with the dispute. It is challenging to conceptualize many of these sufficiently to design a standardized information-gathering approach.

There is also a serious measurement problem with implicit costs, such as staff and volunteer time, and with estimating the efficiency and effectiveness of an organization. My own previous timekeeping experiments with consulting staff, for example, indicated that they systematically underestimate billable time by 15% when recording time at the end of the day in 15-minute intervals. I found that their records were even more inaccurate on a weekly basis.

These are important challenges for evaluation, but they are not insurmountable. Remember that evaluation information only needs to be good enough for the decisions that are likely to be made. Evaluation needs are often satisfied with information from applied research. For example, when I was evaluating a geological mapping program as part of a mineral development initiative, with the benefit of known ratios applicable to the area that had been mapped, I was able to roughly estimate the likely future value of economic benefits. The key was the empirically established ratio of 1 mine per 25 exploration sites. Geologists were able to forecast the high and low numbers for exploration sites from the actual results of the mapping, and I used economic input–output tables to estimate the economic benefits of the mines and hence of the mapping. The quality of the forecast was good enough for us to estimate the economic benefits of the geological mapping, but clearly would not be sufficient for publication in an economics journal. Similarly, dispute resolution needs a few known and credible coefficients from which implicit costs can be reliably estimated.

Like the geological mapping example, we can use the extensive applied research on environmental impacts and other applications to generate good enough estimates of the costs (and benefits) to the environment that would arise from most agreements. Likewise, we can estimate other economic effects, such as employment and earnings, that are likely to arise from resolution of a dispute.

The problem is that we do not have the benefit of applied research that would provide us with the necessary coefficients to estimate with sufficient

Upstream Costs	Transaction Costs	Downstream Costs
Examples of Explicit Costs	**Examples of Explicit Costs**	**Examples of Explicit Costs**
Fees of attorneys and other professionals	Fees of attorneys and other professionals	Fees of attorneys and other professionals
Charges attributable to the dispute	Charges attributable to thedispute resolution process	Charges attributable to implementation of an agreement, continuation of a more limited dispute, or the entire original dispute.
Examples of Implicit Costs	**Examples of Implicit Costs**	**Examples of Implicit Costs**
Time of staff and volunteers	Time of staff and volunteers	Time of staff and volunteers
Reduced organizational efficiency and effectiveness because of engagement in the dispute	Outcomes that do not occur because the organization is involved in resolving the dispute	Changes in organizational efficiency and effectiveness
Gains such as earnings that do not occur because of the dispute		Any changes in gains (positive or negative)
Costs of environmental damage, including costs to parties other than the disputants		Costs of environmental damage, including costs to parties other than the disputants

FIGURE 9-3. Classification of Costs of a Dispute

standardization the costs of a dispute to individual parties. This leaves evaluators of environmental dispute resolution programs in a quandary. We can expect that respondents will systematically underestimate current indirect costs if asked and that this will be more serious when we seek information over longer time periods. This creates a bias against dispute resolution being measured as efficient.

Estimating Benefits. The benefits of dispute resolution are described in Figure 9-1 for the Oregon PPP and can be grouped into dispute-specific benefits and party benefits. Dispute-specific benefits are improved collaboration among disputing parties; narrowing of the number of disputed issues; parties' satisfaction with the process, which is more effective than the alternatives; improved capacity of parties to engage in collaborative processes; complete, durable, and implementable agreements reached; and improvement of government decisionmaking. According to the Oregon PPP program theory, all of these outcomes are desirable, and both process outcomes and agreement outcomes are valued. The Oregon PPP has three additional outcomes that the use of alternative dispute resolution must provide in addition to those identified in the program theory described in Figure 9-1: decreased costs of resolving disputes, increased efficiency of agencies, and increased agency and party satisfaction. These three outcomes result from the enabling legislation and are much on legislators' minds. Together these outcomes articulate the intended benefits from the Oregon PPP.

Both the Oregon PPP and the USIECR distinguish between process and agreement outcomes, and the program theories state that the process outcomes are necessary for achievement of agreement outcomes. The early evidence from the Oregon PPP and the USIECR supports this claim.

We can group the benefits from the dispute resolution process as improved capacity of parties to work together with other parties now and in the future and as complete, durable, and implementable agreements that will result in improved efficiency and effectiveness of the parties. Capacity and agreement benefits can be assessed with information from the parties and neutral(s); however, assessing the effectiveness benefits has the same challenges described above. It is likely that the most important benefit will prove to be improved organizational effectiveness, enabled by a reduction in the number and intensity of disputes requiring senior and middle management attention.

Summary of Challenges to Evaluating Dispute Resolution Programs.
Every program should be able to substantiate that it is the most effective option when used appropriately, and dispute resolution is not different. However, there are some serious challenges to obtaining the necessary cost and benefit information for disputes. These challenges arise both for general reasons and for reasons that are specific to dispute resolution. The general problems arise from the difficulties in getting valid and reliable information on the types of implicit costs and benefits that are involved in dispute resolution. The problems specific to dispute resolution arise from the extremely long time over which costs and benefit information must be obtained. Resolution of this challenge will come from applied research that generates knowledge of the costs and benefits for different classes of parties for the main types of environmental disputes.

The main methodological challenge is assessing the incremental contribution of dispute resolution compared to other good alternatives for addressing a dispute. This challenge is a serious constraint on our ability to respond adequately to summative evaluation questions about the value of dispute resolution. When combined with the additional challenges of measuring effectiveness, evaluators of environmental dispute resolution programs face a potentially daunting task. However, it is a challenge that must be addressed because legislators and other funders have an obligation to fund programs that are the most effective response to a problem. To make these funding decisions in the context of accountability for results, funders require information about the effectiveness of dispute resolution programs compared to reasonable alternatives.

Fortunately, the gap in our current knowledge can be addressed with modest investments in applied research to generate dispute resolution exemplars and comparators for a limited number of types of disputes and classes of parties. In the meantime, evaluation is already showing itself to be a means of generating valuable information that can be used to improve environmental dispute resolution programs.

Opportunities: Collaborative, Participatory, and Empowerment Evaluation

Evaluation information is useful to the parties and neutrals facilitating the dispute resolution process, as well as to program officials, and collaborative and empowerment evaluation approaches are worthwhile experiments to improve the utility of evaluation to key stakeholders, neutrals, and parties. These approaches engage the program participants (parties and neutrals) in all aspects of the evaluation as primary stakeholders. Use of these approaches would require capacity building with the neutrals and parties and offers the promise of collaboratively set and implemented monitoring of the process and outcomes of dispute resolution. The information that would be obtained would give parties resources to improve collaborative dispute resolution processes in which they participated and would provide ongoing monitoring of any agreements, as well as information they could use when considering whether to engage in dispute resolution or other options for future disputes. These evaluation approaches need to be embedded in the program processes themselves. They offer the potential to enhance the benefits of evaluation to parties while obtaining better information about the dispute resolution process and its effectiveness, because collaborative and empowerment approaches would bring evaluation "inside the room" and inside the dispute resolution process itself.

Conclusions

This chapter has approached evaluation of environmental dispute resolution programs from the vantage of the professional evaluator. Good evaluation of environmental dispute resolution processes would clarify dispute resolution programs and practices, as well as the mechanisms to gain information about programs and practices. The mechanisms can be used to identify and plan improvements in the theory and practice of environmental dispute resolution and to address the accountability requirements of legislators and funders. However, evaluation of environmental dispute resolution programs is constrained by the twin challenges of establishing the incremental contribution of dispute resolution compared to alternative processes and measuring the effectiveness of dispute resolution. Both of these challenges can be addressed with modest investments in applied research. Evaluation, enhanced with this knowledge, can fulfill the potential of not only aiding improvements in environmental dispute resolution practice and theory, but also aiding program managers, dispute resolution professionals, legislators, and other funders to use dispute resolution where it is the best option and ensure that key stakeholders benefit from the process.

Notes

I am grateful to everyone at the U.S. Institute for Environmental Conflict Resolution, the Oregon Public Policy Program, collaborators in the Policy Consensus Initiative, and in particular to Dale Keyes, Kirk Emerson, Mike Niemeyer, and Susan Brody for their patience and sharing their tremendous insights and knowledge. But of course, despite their contributions, I still make mistakes, and they are entirely of my own doing.

1. See Michael Scriven, *Evaluation Thesaurus, 4th ed.* (Newbury Park, CA: Sage Publications, 1991).

2. See Joseph S. Wholey, "Assessing the Feasibility and Likely Usefulness of Evaluation," in Joseph S. Wholey, Harry P. Hatry, and Kathryn E. Newcomer, eds., *Handbook of Practical Program Evaluation* (San Francisco: Jossey-Bass, 1990), pp. 15–39.

3. See Chapter 10 in this volume.

4. Note that further outcomes are nested under the principal outcomes and that there has also been an effort to define and clarify terms, for example, see the websites of USIECR, www.ecr.gov (accessed January 17, 2003) or the Indiana Conflict Resolution Institute, www.spea.indiana.edu/icri (accessed January 17, 2003).

5. The American Evaluation Association has established topical interest groups (TIGs) around evaluation methods and topics. The list of TIGs testifies to the diversity of evaluation practice and concerns and can be found at the AEA website, www.eval.org (accessed January 17, 2003).

6. Joint Committee on Standards for Educational Evaluation, *The Program Evaluation Standards, 2nd ed.* (Thousand Oaks, CA: Sage Publications, 1994).

7. I use neutral as a generic term to refer to the individual(s) who facilitate the dispute-resolution process. Through experience, I am mindful of the minefield on which I tread by doing so. See Scriven, *Evaluation Thesaurus.*

10

An Evaluation System for State and Federal Conflict Resolution Programs

The Policy Consensus Initiative

Kirk Emerson and Christine Carlson

In 1999, the U.S. Institute for Environmental Conflict Resolution (USIECR) and the Policy Consensus Initiative (PCI) began an inquiry into the feasibility of developing program evaluation guidance for state and federal agencies and programs that administer public policy and environmental conflict resolution programs. PCI is a national, nonpartisan organization that works with state leaders—governors, legislators, attorneys general, and state courts—to promote the use of consensus-building and conflict resolution practices to address difficult policy issues and achieve more effective governance. The inquiry started by USIECR and PCI has grown into a collaborative program evaluation initiative involving USIECR, the Massachusetts Office of Dispute Resolution (MODR), and the Oregon Dispute Resolution Commission (ODRC) with support from PCI. The project has also engaged the support of the Indiana Conflict Resolution Institute, Indiana University's School for Public and Environmental Affairs, Syracuse University's Maxwell School of Citizenship and Public Affairs, and the University of Arizona's Udall Center for Studies in Public Policy.[1]

This chapter focuses on the design and implementation of this initiative to create a self-administered program evaluation system for state and federal programs. It provides an overview of this ongoing, multiyear initiative and the lessons we have learned to date. The potential benefits from this collaborative investment for the field of public policy consensus building and environmental conflict resolution are substantial.

State dispute resolution programs are still small and fragile entities. Some of the most well established and productive programs still face challenges in demonstrating their efficacy to state administrative and legislative bodies. Most state programs do not have the necessary resources to contract with experts to develop evaluation systems. Through this evaluation project, PCI has enabled these states to hire the experts they need to establish reliable, valid systems for demonstrating to policymakers the value of their programs. PCI plans to expand the program evaluation project to more states and to develop materials and conduct workshops for other interested states.

This collaborative project represents a considerable opportunity to address both research and program evaluation challenges in this field. By laying the groundwork for broad agreement on success and its measurement, this collaborative endeavor will be of direct assistance to existing public programs and provide invaluable "tracks" for future ECR research to follow.

Project Background

The need for more explicitly reflective practice and for more systematic assessment of public policy and ECR programs has never been greater. After some 30 years of development and refinement, the field is maturing into a profession. Institutions of government and private-sector organizations and interests are increasingly turning to forms of collaborative problem solving and alternative dispute resolution (ADR). New institutions, new resources, and a growing number of enabling statutes provide more incentives to participate in consensus-building and conflict resolution processes.

PCI recently conducted a survey and found more than 175 public ADR programs in states. In addition to ADR programs located in courts, there are programs located in the administrative branch, in offices of attorneys general and administrative law judges, individual state agencies, and state universities serving state government. These programs are currently listed in the Directory of State Dispute Resolution Programs, which is available online at www.policyconsensus.org (accessed January 17, 2003). There has been similar growth in federal ADR programs across executive-branch agencies and federal courts.

With such growth in the field there has come more attention, more exposure, and more scrutiny. Federal and state legislatures and executive agencies, as well as private parties, want to know if their public investment is paying off. And all want to know about the experiences and results from using these processes. Have settlements been reached, agreements breached, or are resolutions generally favorable to their side? Is it cheaper, quicker, and better?

The conceptual and methodological challenges to rigorous evaluation of public policy and environmental conflict resolution processes have been with us for some time (see Chapter 1). There has been considerable discussion within the practice and within the academy about just what are or should be the expected outcomes of such processes, let alone how one might measure them. However, considerable progress is being made through innovative research approaches such as those presented in this book.

A related set of challenges arises when trying to evaluate the public-sector programs that administer public policy and ECR services. Increasingly, such legislation as the federal Government Performance and Results Act (GPRA) requires federal and state agencies to account more accurately for their performance. Program evaluation is fundamentally an attempt to define, measure, and improve performance. To paraphrase GPRA, programs need to answer three questions: What is your program trying to achieve? How will its effectiveness be measured? How is it actually doing?

Program evaluation requires a systematic articulation of expected outcomes and accomplishments as well as a methodology for collecting the information that will reliably track program performance. Analyzing performance requires a cataloguing of program inputs and outputs and rests on certain basic operating assumptions or program theory that connect the two. This is where the rubber of research hits the road of service delivery.

Collaborative Project Design

This collaborative program evaluation initiative has chosen to grapple with these complex research and evaluation challenges in the context of three similar, yet distinct programs that have committed to designing and implementing program evaluation systems. Together, the individuals in these programs have agreed to work toward evaluation systems that will be self-administered to the greatest extent possible and will provide data to inform program improvement and encourage reflective practice. Practically, these systems must also be cost-effective and must generate the reports on program performance expected by oversight bodies and legislative committees. Finally, to the extent that these systems can produce valuable information for other programs and the field at large, these programs intend to make that available. PCI and USIECR also hope that the collective accomplishments of these programs will lead to useful guidance and potential efficiencies for other federal and state programs.

To pursue these multiple intentions, PCI and USIECR have defined five phases of activity to guide this collaborative program evaluation initiative through the next four to five years of development work. They are information gathering and consultation; design and piloting for three programs (USIECR, MODR, and ODRC); implementation; extending the initiative to more states and providing guidance for others; and creating a repository for future research.

Although the initial phases of the project have taken longer than anticipated, it is expected that the pace will quicken as we begin to work through refinements to implementation and expansion. What follows is a summary of the activities related to these five phases.

Phase I: Information Gathering and Consultation

In the spring of 1999, USIECR and PCI initiated an assessment and consultation phase that culminated in a two-and-a-half day gathering in mid-September in Tucson, Arizona (the "Tumamoc Hill meeting"). With the assistance of Juliana Birkhoff (of RESOLVE, Inc.) and Mette Brogden (at the Udall Center for Studies in Public Policy), a literature review was conducted and background materials were assembled on evaluation of public policy and environmental conflict resolution programs. Not surprisingly, the list of available resources was relatively short.

A workshop was convened with leaders from the fields of dispute resolution professionals; administrators of federal and state public policy and environmental conflict resolution programs; and academic researchers in psychology, communications, law and legal studies, urban planning, and public administration. PCI invited two state programs from Massachusetts (MODR) and Oregon (ODRC) with the expectation that they would be involved subsequently in the program evaluation initiative. After the meeting, the Indiana Conflict Resolution Institute agreed to support a future meeting and call for papers that resulted in this edited book.

Phase II: Design and Piloting for Three Programs

Subsequently, USIECR and the two state programs, MODR and ODRC, supported by PCI, began the task of selecting program evaluation consultants to help them with the design of their respective systems. MODR selected David Fairman of Consensus Building Institute in Cambridge, MA. USIECR selected Andy Rowe of GHK International, Toronto, Canada, one of a short list of respondents to a national Request for Qualifications. ODRC chose to work with Dr. Rowe as well.

The activities being evaluated by one or more of these programs include case assessment, case management (both mediation and facilitation), training and roster management, and integration of dispute resolution that includes systems design in government.

There is another interesting element of the design of this collaborative effort: determining how to share information across programs. Three basic common goals have been identified regarding the need to gather more data to answer questions about the efficacy of these processes, increase the pool of data to validate program theory, and help programs learn from each other.

The three programs discussed mechanisms for sharing data across programs, such as aggregating the data in one database or establishing common formats for evaluation across programs. These options were ultimately rejected as too costly and too difficult to implement, given the different kinds of programs. Instead, the three programs are now working to identify core concepts and questions that they will agree to use. PCI and USIECR will then urge other agencies to adopt these same concepts and questions so that data can be compared and possibly aggregated at some future time.

Over the next year and a half, the three programs, their consultants, and PCI met three times and participated periodically in conference calls reporting their progress, discussing issues, exchanging instruments, and critiquing evaluation frameworks. Two panel presentations were made at annual professional conferences reporting on the project's progress to date.

Phase III: Implementation

These programs have now completed pilot testing of their data collection instruments and are in the very early stages of implementation. The consultants are completing analyses on preliminary data. It is anticipated that the three programs will continue to coordinate and share what they learn as implementation of their evaluation systems unfolds. Programs will also work together to deepen and articulate program theory as new information emerges from the project. Plans for additional conference presentations in 2002 are intended to inform others of preliminary findings from their data collection and analysis.

The programs will continue to discuss and work on ways they will use and report the data. Some of the ways they intend to use the data and information are to demonstrate accountability, help in budget development and advocacy, improve program management, and educate parties and policymakers.

Phase IV: Extending the Initiative and Providing Guidance for Others

Concurrent with the implementation phase, efforts are now underway to bring additional states into the project and to develop guidance for federal and state agencies' program evaluation efforts. PCI is planning to assist groups in three additional states: the Florida Conflict Resolution Consortium located at Florida State University; the Mediation and Conflict Resolution Office, located in Maryland's state court system; and the Ohio Commission on Dispute Resolution and Conflict Management. PCI has begun developing guidance materials to share with other states. To assist the three new states that are joining the initiative, a handbook has been developed to assist states in finding, choosing, and working with an evaluator.

USIECR expects to share aggregate evaluation results with other federal agencies to demonstrate collective ECR outcomes. It will also report to individual agencies on ECR outcomes for those projects involving USIECR or its contractors. It may also be possible for USIECR to manage the data collection, analysis, and reporting for agency ECR cases and projects with which USIECR is not necessarily directly involved, using its standard instruments. Alternatively, agencies may use the evaluation instruments independently once the U.S. Office of Management and Budget has cleared them for use. Finally, USIECR may assist other federal agencies in developing their ECR programs for internal performance review requirements.

Phase V: Creating a Repository for Future Research

The public programs involved in this collaborative initiative have an interest in the findings from each other's evaluation programs and from the collective findings from other programs. However, none of these programs has the mandate or the resources to extend their program evaluation efforts in service of other, larger research interests. Nonetheless, they do recognize the broader value of the data being collected to those interests. PCI and USIECR are exploring various options for making these data more accessible to other researchers.

From the beginning of this initiative, there has been discussion about the potential benefits of creating a national data repository for researchers. Given the ever-growing powers of Web-based data management tools and systems, it will soon be possible to link numerous databases from a variety of sites. The need for one repository is hence less essential. Although different research methodologies require varying degrees of consistency in research design, data collection, and analysis, nonetheless, the importance of coordination, quality control, and clarity regarding data management and protocols remains. USIECR is considering how it will make its collected data available more broadly and how it might assist other federal agencies in administering the necessary protocols to ensure appropriate confidentiality of data, maintain essential records, and manage appropriate and timely access to such data in the future.

Case Studies in Massachusetts and Oregon and at USIECR

During this collaborative effort, we have found that it is important for each program to develop its own evaluation framework in the way that makes the most sense to the program. Programs have different histories and cultures, and it is important to match the evaluation with those characteristics. Although these approaches start out in different places, they end up gathering the same data in similar ways.

Massachusetts Office of Dispute Resolution

Massachusetts has had a comprehensive state dispute resolution program—the Massachusetts Office of Dispute Resolution (MODR)—in the executive branch for 15 years. MODR is a state agency providing mediation, facilitation, arbitration, case evaluation, ADR training, and other ADR services. MODR provides these services to the public agencies, municipalities, courts, and citizens of the commonwealth using MODR staff and private-sector neutrals. MODR is organized into two main programs: the Court Program and the Government Program.

Since 1985, MODR has provided mediation services for environmental disputes. MODR has handled more than 200 environmental disputes in Massachusetts involving municipalities and state and federal agencies. In the early 1990s, MODR established an Environmental Mediator Panel of 14 skilled private-sector mediators with backgrounds and expertise in environmental dispute resolution.

Recently, Governor Cellucci issued an executive order entitled "Integrating Dispute Resolution into State Government." It identifies the benefits of ADR and orders state agencies to "work diligently to fully utilize, wherever appropriate, alternative dispute resolution to resolve disputes ..." It requires the head of each agency to designate an ADR coordinator who reports directly to the agency head. MODR is designated as responsible for training and providing consultation and assistance to these coordinators.

Massachusetts decided to approach its evaluation design by identifying the best practices for each of its activities: mediation, facilitation, training, and dispute systems design. From these detailed best practice statements, the state designated all the objectives that would need to occur to accomplish them. From those objectives, the state developed indicators and then survey instruments asking questions that reveal whether the state is meeting those indicators.

Oregon Dispute Resolution Commission

Oregon is one of the most advanced states in the use of consensus-based approaches to governance. Oregon has also developed an extensive infrastructure to assist state agencies with the use of collaborative processes. The Oregon Dispute Resolution Commission (ODRC) was established by law in 1989. The commission's mission is to support the beneficial and effective use of mediation, negotiation, conciliation, and other collaborative problem-solving processes. The commission has three major program areas: the Community Dispute Resolution Program, the Court Referred Mediation Program, and the Public Policy Dispute Resolution Program. Since 1990, ODRC has helped resolve hundreds of complex and difficult public policy controversies.

There is also a Dispute Resolution Steering Committee, which is chaired by the governor's advisor and made up of an ODRC member and representatives from the attorney general's office and the Department of Administrative Services. Oregon was the first state to designate coordinators for agencies. Since 1998, four public policy dispute resolution coordinators have worked full-time with clusters of agencies to help them incorporate dispute resolution practices.

Governor Kitzhaber recently issued an executive order similar to the Massachusetts order. It requires heads of all agencies with more than 50 employees to appoint an ADR coordinator to encourage and facilitate appropriate use of ADR within their agencies. ODRC is providing training and consultation for agency coordinators and assisting them with their needs assessments and the development and implementation of their plans for incorporating ADR practices.

Oregon decided to participate in the evaluation project because its legislature required the state to provide measures of how well it was carrying out its mandates. ODRC and USIECR worked with the same evaluation consultant, and their approaches to program evaluation are essentially the same. As part of the evaluation project, ODRC appointed an advisory committee to provide guidance from an independent perspective.

To develop its program evaluation and narrow its scope and focus, ODRC articulated an outcome structure derived from its mission statements, strategic plans, and enabling legislation. These outcomes are divided into two main parts: case management and ADR systems. Case management outcomes relate to the use of ADR in specific disputes or controversies. ADR systems outcomes focus on systems and procedures ensuring that ADR is available and that it advances the mission of state government. Evaluation instruments were then developed to measure progress in achieving these outcomes.

U.S. Institute for Environmental Conflict Resolution

USIECR is a federal program established by the U.S. Congress in 1998 to assist parties in resolving environmental, natural resource, and public lands conflicts. USIECR is part of the Morris K. Udall Foundation, an independent agency of the executive branch overseen by a board of trustees. USIECR offers impartial, nonpartisan assistance through a Tucson-based professional staff, extended by a national referral system of qualified environmental facilitators and mediators. USIECR promotes nonadversarial, agreement-seeking processes that range from large, multiparty consensus-building efforts to assisted negotiations and court-referred mediation. Its services include case consultation, convening, conflict assessment, process design, facilitation, mediation, training, and dispute systems design.

During its first three years of operation, USIECR has been involved in more than 100 environmental cases and projects across 30 states, as well as in the

District of Columbia. Several projects are national, intergovernmental, or mul-tistate in scope. The issues range broadly, including wildlife and wilderness management, recreational use of and access to public lands, endangered species protection, marine protected areas, watershed management, ecosystem restora-tion, wetlands protection, and urban infrastructure planning. [See www.ecr. gov (accessed January 17, 2003) for more information about USIECR.]

In addition to its service provision, USIECR has sponsored several pro-grams, including the Federal ECR Partnership Program, a biannual national conference; an investigation into a National Environmental Policy Act Pilot Projects Initiative, a demonstration program with the Oregon Federal District Court; and the Federal ECR Roundtable for ADR specialists at federal agencies.

USIECR began developing its program evaluation system at the outset, along with its strategic planning process. In addition to meeting the federal government's requirement to develop performance improvement plans, USIECR wanted to participate in the collaborative program evaluation initia-tive to establish consistency with the prevailing standards and practices in the field of ECR and to ensure the credibility of its own evaluation system.

With the guidance of its selected consultant, and similar to Oregon's approach, USIECR articulated its expected service and program outcomes in several areas: conflict assessment, ECR projects, system design, roster man-agement, training, and meeting facilitation. Working with its full program staff and in consultation with its other partners in the collaborative effort, USIECR determined its operating assumptions or program theory about the specific actions and conditions deemed necessary to achieve its desired out-comes. Specific measures and performance standards were then developed to track these achievements through survey questions designed for the involved staff, neutrals, and parties or participants. These survey instruments have been approved by the U.S. Office of Management and Budget to collect the necessary information.

Lessons Learned to Date

MODR, ODRC, and USIECR have encountered numerous challenges along the way in developing their respective program evaluation systems. For the most part, these have been shared challenges, and the lessons learned by these three programs should be instructive for other programs embarking on their own program evaluation efforts.

Starting at the Beginning

It became quickly apparent to the three programs that program evaluation was not going to be a self-contained module simply slipped in at the end of a given case. The unit of analysis is the program and its constituent parts. To understand what to evaluate and why, one has to start at the beginning with the program definition and its strategic plan: the program goals, service areas, resource priorities, expected outcomes, and performance standards.

To effectively implement an evaluation system, it has to be integrated into both the professional culture and behavior of the staff and into the operating mechanics of project management systems. This means that new programs should be thinking about evaluation during their start-up, and more mature programs must identify the appropriate time or opportunity to integrate into or retrofit existing personnel and operating systems. At MODR and ODRC, both relatively long-lived programs, new directors, executive orders creating mandates for state agencies to incorporate ADR, and new legislative mandates for program accountability provided the impetus for investing in the development of program evaluation systems. At USIECR, the design of the evaluation system paralleled the program's start-up and early strategic planning processes.

Finding Program Evaluation Experts

All three programs took much longer than anticipated to identify competent program evaluation consultants. There is a growing number of academic researchers studying the field of public policy and ECR from a great variety of disciplines—communications, social psychology, political science, public administration, legal studies, economics, and natural resource management, to name a few. There are, however, relatively few program evaluation experts familiar with ADR, ECR, and public-sector dispute resolution programs. A working familiarity with ADR or ECR may be less important, however, than experience in evaluating public sector programs generally.

To make the search for program evaluation consultants easier in the future, PCI has developed some guidance for programs in seeking such services. In addition, the Request for Qualifications developed by USIECR to solicit candidate program evaluation experts may be helpful as a model for other federal agencies.

This collaborative initiative had the benefit of working with two quite different consultants: one was an economist with limited substantive knowledge of the field, and the other had an extensive background in ECR. Their approaches were different in several respects, but they were equally valuable to their client programs. They worked with a synergy that validated and reinforced the overall direction of the program evaluation initiative.

Taking Our Own Process Advice

When working internally with program staff, the process of developing an evaluation system requires a collaborative approach. Not surprisingly, the basic principles for sound process design apply in this context as well. The need and importance of gaining upfront buy-in and support cannot be over-estimated. The commitment and participation of the agency leadership are essential ingredients for ensuring that the evaluation's direction is on target with agency priorities and plans, for communicating the importance of the process to the staff, and for enabling ready access to decisionmaking.

Active involvement of the entire program staff as key stakeholders in the evaluation system is also central to improving the quality and appropriate-ness of the evaluation design, as well as for ensuring the staff's meaningful participation in implementation of the evaluation system later. Engaging advisory boards or councils can also be of considerable value, as Oregon has found, particularly in articulating the questions that outside funders, legisla-tive committees, and executive oversight agencies need answered.

In addition to an in-house evaluation project coordinator, it is important to think about who should facilitate internal discussions and decisionmak-ing. This might be the same person serving as the evaluation project man-ager, or this role might be shared with the outside evaluation consultant, or it might become an additional responsibility of the evaluation consultant or an organizational development expert. Depending on the needs of a given agency, its organizational culture, and its staff makeup, the choice will vary. However, acknowledging the need for such facilitation of in-house discus-sions will be useful because the identity of the program and management's expectations for staff performance and its measurement are at stake.

Controlling Collective Curiosity

Designing a program evaluation system provides the opportunity to ask important questions. It also provides the enormously tempting chance to ask everything you ever wanted to know about what makes these processes work. Work hard to control that impulse. Focus on your own program design or theory first. What is the program mandated to accomplish or deliver; what are the underlying operating premises that support the services being pro-vided; and what factors or inputs can your program control to deliver the desired outcomes? Interestingly, the three programs discovered that whereas their mandates and missions varied somewhat, their working assumptions about best practices and process outcomes were quite consistent. What they needed to know to improve their performance was quite similar, particularly once they whittled down their expectations to the sphere of accomplishments for which they were ultimately responsible.

There are at least three good reasons for being circumspect about the extent of the information sought through a program evaluation system. First, less may be more, because the number of questions asked no doubt influ-ences the response rate to questionnaires. In addition, the more responses across diverse participants, the more valid the analysis will be. Second, whereas it may appear that the marginal cost of adding another question to a given form is small, the subsequent processing and analysis time may not be. And finally, be careful what you ask for, as one consultant advised—you might just get the answer. This is not intended as a cynical obstruction of important inquiry, but rather to guard against the inadvertent collection of information that, if made public, could be harmful to an individual or a process. In the end, balancing what you would like to know with what you need to know will make the program evaluation system more economical and more useful.

Database Design and Management

The issues surrounding development and implementation of the database turned out to be more difficult, costly, and time-consuming than anyone anticipated. The scope of the project is large, and it must rest on a good foundation in terms of the technology employed as well as the communications among all necessary participants. Issues such as whether to use a staff person or an outside consultant as a database designer and decisions about the scope of an automated system, what information needs to be tracked, and the volume of information all require careful consideration.

Because of the scope of this project, programs needed assistance from someone knowledgeable about systems design, rather than simply computer coding. The programs have found that day-in and day-out contact between the agency and the designer is critical to understanding the scope of the project. The programs used a combination of in-house staff and consultants to create their databases.

The programs again learned how important it is to involve others in the organization to gain their buy-in. Based on experience, they suggest designating an in-house database manager at the beginning of the project. It is beneficial to have the designer work with the database manager from the beginning of the project rather than to "back fill" knowledge about the creation of the database. To ensure the reliability of the information, attention must also be given to who will do the actual data entry. In Oregon, for that reason, the commission decided to have one staff member do all the data entry. The support provided while that person was learning the ropes has paid dividends.

The programs have found that there is no "off-the-shelf" system or product that shortcuts the necessary work. Though all three programs are using Microsoft Access software, there is some sentiment that Access is good for generating reports but not as flexible for purposes of analysis. Some of the project participants think that programs should start with paper and simple Microsoft Excel spreadsheets before finalizing the database design. This is a topic that will be given further attention as we look toward extending the project to other federal and state programs.

Implementation Challenges

As the three programs embark on their implementation phase, they have discussed several common issues they will be facing.

Confidentiality Issues. One of the most concerning issues ahead for these programs is the extent to which the information they collect should or can be protected as confidential information. There is an understandable tension between the value of the information collected for independent analyses and further research and the need to appropriately protect the confidential nature of some processes and the privacy of individuals responding to questionnaires. USIECR has published the following statement, which accompanies its preliminary request for comment on its proposed information collection request:

> To encourage candor and responsiveness on the part of those completing the questionnaires, the U.S. Institute [USIECR] intends to report information obtained from questionnaires only in the aggregate. The U.S. Institute intends to withhold the names of respondents and individuals named in responses. Such information regarding individuals is exempt from disclosure under the Freedom of Information Act (FOIA), pursuant to exemption (b)(6) (5 U.S.C. Section 552(b)(6), as the public interest in disclosure of that information would not outweigh the privacy interests of the individuals. Therefore, respondents will be afforded anonymity. Furthermore, no substantive case-specific information that might be confidential under statute, court order or rules, or agreement of the parties will be sought.
>
> The U.S. Institute is committed to providing agencies, researchers and the public with information on the effectiveness of environmental conflict resolution (ECR) and the performance of the U.S. Institute's programs and services. Access to such useful information will be facilitated to the extent possible. The U.S. Institute is also committed, however, to managing the collection and reporting of data so as not to interfere with any ongoing ECR processes or the subsequent implementation of agreements. Case-specific data will not be released until an appropriate time period has passed following conclusion of the case; such time period to be determined. FOIA requests will be evaluated on a case-by-case basis.[2]

Each state has its unique approach to the issue of confidentiality. In the case of Oregon, state agencies are authorized to make mediation communications confidential. Agencies can limit the discovery and admissibility of mediation communications in subsequent proceedings if they adopt, with the approval of the governor, mediation confidentiality rules developed by the attorney general.

Incorporating Findings To Improve Program Management. Another area for further discussion and development is the manner in which programs will use the information to improve program delivery. Ultimately, if the information collected is not useful or is not used, the evaluation system will probably not be maintained for long. One way to ensure its utility is to formally draw on information from the evaluation system for staff performance reviews. Most important, however, will be the analyses that can identify conditions or practices that either contribute to or impede successful outcomes, leading to strategic program improvements. In addition, it will be important to manage the data analysis so that case- or project-specific information will be available to project staff in a timely fashion. Furthermore, aggregate analysis should be made on a periodic basis for the full staff and oversight entities to track the overall performance of the program.

The three programs anticipate using, in some manner, the information collected concerning the performance of contracted neutrals, but no specific plans have been proposed at this point. Over time, experience with the management of roster systems for professional neutrals has grown. Increasingly, roster managers voice concern over issues of qualifications, quality control, and mechanisms for providing feedback and mentoring. Careful attention to these issues is warranted, and further reflection is needed.

Dealing with Staff Turnover. Finally, among many other implementation issues still to be addressed is a caution about staff turnover. As those involved in the early design and implementation of these systems move on and new staff are brought on board, programs will have to orient them to the importance of the evaluation system and their role in making it work. To the extent that evaluation is successfully integrated into the way programs do business, this will be less of a challenge and more of a platform for staff development and training.

Conclusions

Public-sector dispute resolution programs generally have small staffs and little administrative overhead, yet their activity level can be quite high, particularly for those administering professional rosters and small grants programs. As discussed at the outset of this chapter, the need for effective program evaluation has never been greater, but it must be attainable at a realistic price.

There is no question that programs must make an upfront investment to realize the ongoing benefits of a program evaluation system. PCI and USIECR hope that the accomplishments of this collaborative initiative will benefit other programs and the field by fostering effective guidance, models, tools, and systems that also reduce the costs of program evaluation.

Notes

We extend special thanks to the members of the collaborative program evaluation initiative cosponsored by PCI and the USIECR, including Susan Jeghelian, Harry Manasewich, Israela Brill Cass, and Freddie Kay (all from the Massachusetts Office of Dispute Resolution); Susan Brody (Oregon Dispute Resolution Commission) and Mike Niemeyer (Oregon Department of Justice); Andy Rowe (GHK Associates); David Fairman (Consensus Building Institute); Juliana Birkhoff (PCI consultant); and Dale Keyes (USIECR).

1. Early theory discussions at the Tumamoc Hill meetings in the fall of 1999 were instrumental to the collaborative program evaluation initiative, as was additional guidance provided by University of Arizona researchers Mette Brogden (Udall Center for Studies in Public Policy), Bonnie Colby (Agriculture and Resource Economics), and Lee Sechrest (Department of Psychology). Lisa Bingham (Indiana University) and Rosemary O'Leary (Syracuse University), codirectors of the Indiana Conflict Resolution Institute, were early supporters of the initiative and brought their institutional resources to bear in sponsoring the conference in Washington, D.C., in spring 2001, the results of which have been translated into this book.

2. Further information regarding USIECR's evaluation program and instruments can be obtained at www.ecr.gov/techdoc.html (accessed January 17, 2003). USIECR would welcome feedback on this proposed approach.

11

State Agency
Administrative Mediation
A Florida Trial

FRANCES BERRY, BRUCE STIFTEL,
AND AYSIN DEDEKORKUT

Throughout the country, and particularly in Florida, the use of mediation
has grown dramatically in recent years. Mediation is now being employed
to resolve disputes of all types involving family law, environmental issues,[1]
criminal law, neighborhood disputes, and personal injury. In some states,
including Florida, courts are now mandated to require mediation of certain
types of disputes and are permitted to mandate mediation in other dispute categories as well.[2] Court-connected mediation, both mandatory and voluntary,
has typically been well received. Mediation users are generally satisfied with the
dispute resolution technique and perceive that it saves time and money while
offering a fair way to resolve disputes in an amicable, lasting fashion.[3]

Advocates have urged that mediation can also be a valuable alternative to
administrative processes and can serve as an alternative to court.[4] Examples
of administrative applications of mediation include resolving complex environmental disputes,[5] drafting administrative rules,[6] and resolving more simple licensing or land use matters.[7] At the federal level, the Administrative
Dispute Resolution Act[8] was passed in part to encourage administrative agencies' use of mediation. Though less frequently used in the administrative
process context, mediation nonetheless has often received positive reviews.[9]
Public dispute mediation has received credit for increasing both the
frequency[10] and quality of settlements,[11] as well as for reducing the time and
costs necessary to achieve such settlements.[12]

After the success of mediation in Florida's court system, the Florida legislature amended the Florida Administrative Procedures Act (Chapter 120
Florida Statutes) in 1996 to encourage alternative dispute resolution (ADR)

as an element of agency case management. The 1996 act, and the uniform rules that implemented the act (28 *Florida Administrative Code*), said that agencies should inform parties in administrative disputes when mediation is available to them and should explain established procedures for using facilitated and negotiated rulemaking. Two years later, the legislature authorized the State Administrative Dispute Resolution Pilot Project, administered by the Florida Conflict Resolution Consortium. The project's purpose was to demonstrate and evaluate the implementation and potential effects of the ADR amendments.

The project included an assessment of barriers and incentives to state agency institutionalization of ADR and a program evaluation, both conducted by a study team from Florida State University. This chapter, drawn from the work of that study team, focuses on two areas: (a) managerial attitudes toward the use and usefulness of ADR in administrative procedures, including barriers to and use of mediation, and (b) a survey of pilot program mediation participants to ascertain their assessment of the ADR process and outcomes.

National Context

Mediation has grown as an alternative to formal adjudication before administrative law judges and hearing officers. It is widely perceived that litigation has become too slow, too costly, and too formal to be the sole mechanism for effectively and efficiently resolving disputes. In reaction to this perception, federal and state laws have been amended to encourage mediation.

In 1990, Congress amended the federal Administrative Procedures Act, requiring federal agencies to explore and use ADR techniques in the adjudicative context. Although the 1990 act contained a "sunset provision," in 1996 Congress permanently reauthorized administrative ADR (P.L. 104–320). The Clinton administration subsequently issued directives encouraging agencies to use ADR. In a 1996 executive order aimed at "improv[ing] access to justice for all persons who wish to avail themselves of court and administrative adjudicatory tribunals," President Clinton instructed federal agencies and their litigators to resolve claims informally when feasible and to use ADR when appropriate.[13] More recently, in 1998, President Clinton issued a memorandum promoting greater use by agencies of ADR and negotiated rulemaking.[14] The memorandum authorizes creation of a working group, composed of representatives from cabinet-level departments and the attorney general's office, to facilitate agency use of these procedures through means such as employee training, development of ADR programs, and development of procedures that allow agencies to obtain the services of neutral persons on an expedited basis.[15]

Many states authorize the use of mediation in administrative adjudication, as well as in the preadjudicative stages of agency decisionmaking.[16] As with the federal system, most states require state agencies or administrative law judges (ALJs) to evaluate and consider ADR but do not impose ADR on regulatory agencies or regulated parties. Specific legal authorization for agencies to use mediation and other ADR techniques is important because without it agencies or parties may doubt the legality or propriety of ADR.[17] For this reason, many states address mediation in specific regulatory programs, such as environmental or public service commission programs.[18]

Some states, including Florida, have authorized agency mediation more generally as part of their administrative procedures acts (APAs) or other general mediation statutes. In many, if not most, states, ALJs operate under the auspices of a central panel. Because all ALJs in central-panel states operate under a single umbrella, the central panel works as an identifiable and discrete dispute resolution system; by contrast, ALJs in the federal system are spread among dozens of key agencies, each of which has its own dispute resolution issues. Thus, mediation may be less costly to implement and yield more immediate benefits in central-panel states than in the federal system. Among the states authorizing mediation more generally—by statute, rule, or executive order—are California, Minnesota, New Jersey, and Texas, all of which are, to one degree or another, central-panel states. Although none of these states impose mediation in regulatory matters, some of these states require a central panel to offer mediation where the parties both agree to it.

Florida Agency Use of Mediation and Managerial Attitudes

To gain information on existing use of mediation and informal negotiation in administrative case management, project staff conducted interviews with the three primary project pilot agencies—the Department of Insurance (DOI), the Department of Transportation (DOT), and the Public Service Commission (PSC)—and with the Division of Administrative Hearings (DOAH) in the Department of Management Services. These interviews found that there is limited direct recording or analysis of case settlement practices.[19]

The project's independent study team, in consultation with project staff, conducted a survey to get a better understanding of how agencies responded to controversial or challenged agency decisions that are subject to Chapter 120 of the Florida Administrative Procedures Act. Based on the Florida DOAH 1997 statistics, the study team selected 10 agencies that made up more than 80% of the 3,733 non–Baker Act cases for its survey of existing agency practices.[20] The agencies the study team selected for the survey included the Agency for Health Care Administration, Department of Busi-

TABLE 11-1. Categories of Contested Cases under Florida's Administrative Procedures Act

Categories	Number of agencies
Licensure	9
Permitting	5
Enforcement actions	9
Bid protests	7
Contract disputes	8
Rulemaking	10
Other	6

ness and Professional Regulation, Department of Community Affairs, Department of Education, Department of Environmental Protection, Department of Health, Department of Insurance, Department of Revenue, Department of Transportation, and the Public Service Commission. The study team interviewed 19 respondents from these 10 agencies (using the survey for baseline questions) to get a "snapshot view" of the agencies and program practices as they existed in 1999.[21]

Types of Action under the Florida Administrative Procedures Act

Survey respondents were asked which categories of cases are the most contested under the Florida APA in their agencies. The responses demonstrate the wide variety of actions that Florida agencies handle under the APA; see Table 11-1.

Agency representatives indicated a strong reliance on an established practice of prehearing conferences and settlement meetings to resolve cases before going to the DOAH or to court. The agencies use the processes in Table 11-2 for resolving disputes.

Use of and Need for Mediation

There is limited agency use of administrative mediation and alternative rulemaking techniques under Chapter 120 mediation provisions. Although eight of ten agencies surveyed indicated that they notify parties of mediation opportunities, only four of the agencies directly indicate that mediation may be available as required by Chapter 120.573 of the Florida Statutes. Survey respondents were asked to describe the overall level of mediation use in their agencies or programs. Seven indicated no use or very little use, and five indicated some use. No agencies indicated extensive use of mediation. All of the respondents indicated a perceived need in their agencies for mediation as a

TABLE 11-2. Processes Used by Florida Agencies To Resolve Disputes under the Administrative Procedures Act

Processes	Number of agencies
Mediation	3
Arbitration	2
Special master	0
Hearing officer	5
Other	1

means of resolving Chapter 120 contested cases.[22] Five respondents (or 30%) perceived a fairly high level of need for mediation (ranking the perceived need as a 4 or 5 on a 5-point scale, with 5 being high perceived need), whereas 11 respondents (65%) rated the need as a 2 or 3 on a 5-point scale. One respondent saw no perceived need for the agency to use mediation in resolving Chapter 120 cases. The average rating was 2.94.

Respondents who indicated only a limited need most often said that agency staff understood their current system of informal negotiation and the system worked well. Other respondents cited limits to the use of mediation based on statutes, need for enforcement, and lack of understanding of how to use ADR in the context of their agencies' needs. Despite those concerns and differing levels of perceived need, 14 respondents (82%) indicated that they would support increased use of mediation; three respondents (18%) supported the same level of use, and none supported less use of mediation.

When specifically asked why they would not use mediation, respondents cited concerns regarding the possible loss of control by the agency, concerns that enforcement actions were not suitable for mediation, a general lack of interest and understanding of ADR by parties outside the agencies, and time and budget constraints. Respondents were then asked to suggest the most important actions that could be taken to encourage broader use of mediation in their agencies or programs. An executive mandate or statutory requirement was the most commonly cited action needed to encourage use of mediation, and the next most common action cited was additional education, particularly education tailored to how ADR could be effectively implemented within the respective agency for its type of cases. Several respondents felt that the Florida Division of Administrative Hearings needs to have an accessible mediation system for citizens and state agencies to use.

Whereas agencies have made limited use of the formal mediation processes authorized in the 1996 ADR amendments, agency attorneys and program managers make extensive use of *informal* negotiation in both Chapter 120.57, contested case hearing process, and Chapter 120.54, rule development. In addition to APA mediation provisions, within Florida state government there are other well-established examples of administrative ADR in program-specific mediation authorized by statutory provisions. These provisions include the following:

— The Department of Business and Professional Regulation uses a mediation program for lot rental and service disputes under the mobile home statute.
— The Department of Insurance has developed nationally recognized mediation programs, including an emergency program for postdisaster homeowner claims and statutory mediation programs for automobile and property insurance claims up to $10,000.
— The Florida Commission on Human Relations has a mediation program for cases filed pursuant to the state's antidiscrimination laws.
— The Department of Community Affairs and the Department of Environmental Protection have conducted successful pilot programs for mediation of selected enforcement and compliance cases.

Individual agencies have developed innovative approaches to resolve or focus matters without proceeding to litigation. Some examples include the following:

— The Department of Environmental Protection makes use of interdisciplinary negotiation teams.
— The Public Service Commission staff convenes "pre-pre" hearing meetings with parties to discuss evidentiary issues, anticipated testimony, and possible limitation of issues to be heard.
— The Department of Transportation has established an informal hearing process before designated individuals in the interest of resolving matters without formal hearing procedures.
— The Public Employees Relations Commission has a settlement officer who assists parties in evaluating claims and issues.
— The Department of Business and Professional Regulation has established a program to train its field inspectors in mediation techniques.

Ranking of Constraints and Obstacles to Wider Use of Administrative Mediation

There are many reasons why managers and lawyers believe barriers exist to using mediation. Our study sought to understand the views of Florida agency managers about the constraints and obstacles that they perceive limiting the use of mediation in their agencies. Respondents were asked to rate 26 issues that act as constraints or obstacles to the use of mediation to resolve administrative cases in their agencies. Each issue was rated on a 5-point scale ranging from 1 (no obstacle) to 5 (a significant obstacle); the 19 responses were averaged to get a mean response for each issue. (See Table 11-3 for a full list of the obstacles and benefits listed on the survey and the respondents' average rankings for each.)

TABLE 11-3. Issues Rated by Managers as Constraints or Obstacles in State Agencies

Constraint or obstacle[a]	Average rating by respondents[b]
Current settlement practices are satisfactory	2.9
Time constraints (i.e., statutory deadlines)	2.6
Concern that negotiating representatives lack sufficient authority (as agents) and that principals will not agree to the settlement	2.6
Budget constraints (no funds to pay for mediation)	2.5
Belief that mediation may mean compromising nonnegotiable standards of compliance (including federal program requirements)	2.3
Belief that mediation is inefficient and takes too long in solving disputes	2.3
Potential high costs of mediation	2.2
Belief that the probability of success in mediated agreements is remote (the other party will never agree)	2.2
Lack of adequately trained staff	2.2
Belief that mediation is ineffective (does not result in better resolutions than litigation or administrative hearings) in resolving disputes	2.1
Lack of awareness of mediation and how it works	2.1
Concern about precedent and effect on cases previously settled (e.g., a good deal to one party may result in other parties asking to renegotiate a past settlement)	2.0
Concern that mediation is a delay tactic	2.0
Concern that if mediation becomes popular, the agency may not be able to handle the mediation caseload (e.g., personnel demands, administrative costs)	2.0
Lack of confidence in the enforceability of negotiated agreements	1.9
Lack of legislative support for mediation	1.8
Concern that "sunshine" requirements will render mediation ineffective	1.6
Existence of an antimediation culture in our agency	1.4
Belief that entering mediation signifies admission of personal or agency failure	1.4
Concern that matters discussed during mediation will not be excluded as evidence in subsequent administrative hearings	1.4
Belief that collaborative processes are illegitimate methods of making public decisions	1.3
Perception by staff that attorneys should not be involved in mediation	1.3
Existence of a reward system in the agency that favors successful litigation rather than mediation	1.2
Negative attitude of senior management toward mediation	1.2
Prior negative experience with mediation in [our] agency	1.1

Note: 1 = no obstacle, 3 = somewhat of an obstacle, and 5 = a significant obstacle.

[a]Listed from highest-rated obstacle to lowest-rated obstacle.

[b]Each respondent was asked to rate each issue on a scale of 1 to 5.

Overall, managers did not perceive any of the issues as significant obstacles to increased use of mediation. The four constraints or obstacles that received the highest scores were the following:

— the feeling that current settlement practices are satisfactory;
— time constraints (i.e., statutory deadlines);
— concern that negotiating representatives lack sufficient authority (as agents) and that principals will not agree to the settlement; and
— budget constraints (no funds to pay for mediation).

The first issue indicates that managers do not perceive their standard operating procedures as inadequate. This may reflect lack of knowledge about what mediation is and what it offers to an agency. (Managers ranked "lack of awareness and how it operates" as a slight obstacle—it received 2.1 out of 5 points—but did not see lack of knowledge as the major reason they are not using more mediation.) However, if there is not a perception that procedures need to change, given the general strength of the saying, "If it ain't broke, don't fix it," status quo practices will remain, and managers will need outside assistance (or intervention) to use mediation more widely.

The other constraints point to some concrete issues for managers. They believe that there is not enough time to pursue mediated settlements, given the legislatively mandated time spans for resolving issues, and that no budget allocation can pay for mediators. Both of these issues are major barriers that require legislative attention to fix. Managers' final concern about a possible lack of authority to reach settlement agreements points to the need for each agency to decide how this issue can be addressed.

Whereas those four issues were rated "somewhat of an obstacle" to agency use of mediation, a variety of issues were rated as slightly significant obstacles, including lack of adequately trained staff, high costs of mediation, inefficiency and ineffectiveness of mediation, and lack of awareness of mediation and how it works. These issues show that there is not a clear perception that mediation works well and will lead to positive outcomes, although few things are viewed as significant obstacles to the use of mediation for administrative cases in state agencies. Finally, a series of possible constraints were seen as nonfactors in Florida agencies; there is not a negative attitude toward or past negative experience with mediation, and staff members do not believe the agency favors litigation over mediation.

Ranking of Benefits and Incentives

After they rated constraints, respondents rated a list of possible benefits of, and incentives for using, mediation based on the importance each would have to their agencies. (See Table 11-4 for the full list.) The scale ranged from 1 (of no importance) to 5 (of great importance). The incentive cited as most

TABLE 11-4. Issues Rated by Managers as Benefits or Incentives of Using Mediation

Benefit or incentive[a]	Average rating by respondents[b]
Policy directive from the agency head favoring mediation as a technique to resolve disputes	4.1
Mediation increases the likelihood of settlements	3.8
Mediation may cost less than litigation or other settlement practices	3.8
Mediation increases the quality of settlements	3.6
Mediation decreases the time required to reach settlement	3.6
Voluntary compliance is more likely due to the parties' acceptance of the settlement	3.6
Track record of success with prior mediations, positive experiences, or case studies	3.6
Improved relationship with affected parties	3.5
Employee rewards or recognition for resolving administrative cases through negotiated settlement rather than through litigation	2.5

Note: 1 = no obstacle, 3 = somewhat of an obstacle, and 5 = a significant obstacle.

[a]Listed from highest-rated to lowest-rated benefit or incentive.

[b]Each respondent was asked to rate each issue on a scale of 1 to 5.

important was "a policy directive from the agency head (secretary, board of commission) favoring mediation as a technique to resolve disputes. Clearly managers are not going to start using mediation without encouragement and "permission" from their agency directors. Key benefits cited as being most important for encouraging use included mediation costing less than litigation, increasing the likelihood of settlements, and reducing the time needed to reach settlement. These factors show that managers are looking for new practices to handle contested administrative cases that work effectively in leading to a timely, high-quality, documentable settlement.

Survey of Pilot Program Mediation Participants

The Florida State Administrative Dispute Resolution Pilot Project included test mediations of 31 disputes involving 10 state agencies over an 18-month period from July 1998 to December 1999. Members of the study group evaluated 11 of these cases. These 11 cases were a universal sample of all pilot program cases that had reached conclusion at the time data collection began in mid-1999. All of the cases fall under the jurisdiction of Florida Administrative Code, Chapter 120.57, the Administrative Procedures Act provisions for contesting state agency decisions. The cases came to the pilot program through six state agencies: the departments of Business and Professional Regulation,

Community Affairs, Education, and Transportation; the Loxahatchee River Environmental Control Board; and the Public Service Commission.

The Survey

Through a telephone survey, we sought to interview all pilot program case mediators and at least one representative of each party participating in each of the cases, which had been completed at the time the survey was conducted. When a negotiator for a party was not the responsible decisionmaker for that party, we sought to interview both the negotiator (or agent) and the decisionmaker (or principal). The 81 individuals involved in 11 cases were interviewed following a one-hour protocol, which included questions on case background, outcomes, incentives and obstacles to mediation, mediation practices, settlement evaluation, cost, and respondent demographics. Our objectives were to understand the nature of the cases mediated and respondents' views of the success of mediation, as well as to uncover, to the extent practicable, the reasons underlying successes and difficulties. In particular, we were interested in identifying the contextual variables that make a case appropriate for mediation, as well as the mediation practices that are best linked to success in given circumstances.

Respondents

The 81 respondents included 21 principals (25.9%), 48 agents (59.3%), and 12 mediators (14.8%). Among the agents, 36 (75.0%) were attorneys. All 11 pilot program cases were represented in the sample, with as few as 3 to as many as 15 respondents. All mediators involved were interviewed. For all cases, at least one representative of each party participating in the mediation was interviewed.

About 29% of respondents work for state government, 21.7% work for local or regional government, 14.5% work for private corporations or represented themselves in the negotiations, and 10.1% work for nonprofit organizations. The remaining 14.5% were professionals in private practice, generally either attorneys or engineers. Respondents identified their occupations as follows: 44.4% in law, 24.7% in administration and management, 19.8% in another technical profession, 4.9% in business, 4.9% in elected public posts, and one respondent was self-identified as a homemaker.

The majority of respondents had used mediation in the past (56.7%); 39.7% said that they knew about mediation but had never used it. Only 4.4% did not know about mediation before the case in question. This response contrasts with 72.5% of respondents who had previously participated in administrative hearings, 21.7% who knew about hearings but had never used them, and 5.8% who did not know about hearings.

The median age of respondents was 51; the oldest was 73, and the youngest 27. Men made up 82.7% of the respondents. Some 74.1% of respondents had completed graduate degrees; 21.0% had bachelor's degrees, and the remaining 4.9% had high school diplomas.

Case Background Information

Taken as a set, the disputes involved in the pilot program were multiparty disputes concerning actions to be taken. Lead authority at the negotiating table varied widely, but the majority of the negotiation teams did not have the authority to make decisions at the negotiating table. The majority also had at least a general sense of their bottom lines and thought that the case would settle. In almost half of the cases, an affected party may have been absent from the negotiations.

About 18% of cases involved two parties, 36.4% involved three parties, 18.2% involved four parties, 18.2% involved six parties, and 9.1% involved 12 parties. Some cases in question had been in dispute for a long time; 9.9% of respondents said that the dispute was more than four years old, and 39.0% said that the dispute was between one and four years old.

Respondents were asked to identify the nature of the conflicts involved in the mediation. About 47% named dispute over actions to be taken or prohibited; 27.4% named disagreements over the projected impacts of actions; 24.7% named interpretation of laws, rules, or policies; 20.6% named disagreements over the facts of the case; 11.0% named disagreements over who is responsible for proposed actions; 11.0% named how resources will be distributed; and 6.9% named basic values, principles, or perspectives. Only one respondent (1.4%) identified interpersonal relationships as an issue in the dispute.

Respondents said that the lead negotiator for their party was most often a senior staff person from their organization (39.7%), followed by a staff attorney (29.4%), and hired negotiator (22.1%). About 6% of respondents said that their party had no lead negotiator. One respondent represented him or herself.

As shown in Table 11-5, 35.4% of respondents said that final decision-making authority for their party rested with a collegial body (such as an elected commission). Higher officials of the agency in question had final authority in 44.6% of the cases. The respondent had final authority 20.0% of the time.

A substantial fraction of respondents (42.5%) said that they thought there might have been affected parties who did not participate in the mediation. In the vast majority of cases (91.1%), no statutory time limit constrained the mediation.

About 29% of respondents said that they had a clear idea of the bottom line in their organizations as they negotiated; an additional 40.0% had at

TABLE 11-5. Negotiator Authority: "Within Your Organization, Who Had the Final Decisionmaking Authority?"

Authority	Number	Percent
Collegial body	23	35.4
Higher official of the organization	29	44.6
Respondent (negotiator)	13	20.0
Total	65	100.0

least a general sense of their bottom lines, and 30.8% had, at best, a vague notion of a bottom line. More than two-thirds of respondents (71.9%) thought, as they were going into mediation, that the case would settle.

Outcomes

Arguments in favor of public policy mediation generally assert that, when compared with more traditional forms of dispute resolution, mediation results in better agreements, is faster and cheaper, and improves relationships among parties. A high fraction of pilot program cases settled. In general, respondents were quite pleased with the settlements and think that they were better than results that administrative hearings would have produced. They also think that the mediations were quicker and cheaper than hearings would have been.

Using settlement as the indicator of success, the mediation pilots were highly successful, as shown in Table 11-6. About 89% of respondents reported that their cases settled.[23] This compares with a settlement rate of 90.1% for all of the sample cases and a settlement rate of 87.1% for all cases that went to mediation under the pilot program, as shown in the project team's records.

We asked a variety of questions aimed at gauging respondent comparisons between mediation and the more traditional administrative hearings that would have often been used to settle these disputes. About 75% said that the mediated outcome would not have been likely if the case had gone directly to an administrative hearing. About 82% of respondents said that the mediation took less time than an administrative hearing probably would have.

Respondents consistently said that the settlement reached would not have been likely in an administrative hearing. They often focused on mediation's ability to find a creative settlement that might not have been possible in a more contentious setting. Administrative hearings were, by comparison, said to produce largely "up or down" decisions without flexibility or exploration of underlying facts and perspectives. The confrontational nature of administrative hearings was often said to reduce the likelihood of joint problem solving among the parties and to build defensiveness.

TABLE 11-6. Settlement: "What Was the Outcome of the Mediation?"

Outcome	Number	Percent
Settled at mediation	57	77.0
Settled after mediation	9	12.2
Continued conflict with administrative law judge or court	8	10.8
Total	74	100.0

About 87% said that staff and consultant costs were less in mediation than in administrative hearings. Almost two-thirds of the respondents (63.6%) reported that the case was in formal mediation sessions for no more than one day; an additional 18.2% reported two days of mediation (the highest report was six days). About 53% of respondents said that the case in question never appeared before an administrative law judge; reports of appearances before ALJs ranged from 1 to 11 days. Private parties in the cases estimated that their savings ranged from $2,250 per case to $85,000 per case. State agencies had a harder time estimating costs relative to other settlement means, but in one case, an agency official suggested that the case in question cost the agency more than $700,000 less than a similar case that had been recently litigated. Respondents attributed cost savings to eliminating the need for formal discovery, court reporters and officers, and expert testimony, as well as to shorter proceedings. In one case, reduced cost was said to have been important to the involvement of a large number of homeowners.

Respondents reported that mediator fees ranged from $600 to $5,600 for the cases in question. The median response was $1,600, and the mean was $2,335. About 78% of respondents reported that the parties paid the cost of the mediator. In other cases, the pilot program paid the mediator's fees.

About 55% of respondents said that they were very satisfied with the mediation process, and an additional 25.0% said that they were moderately satisfied. Only one individual said that she or he was very dissatisfied.

Respondents were asked to what degree the mediation resulted in the achievement of their party's goals. Just under half (49.1%) said that their party achieved almost all of its goals, and another 38.6% said that their parties achieved at least some of their goals.

Effects on the relationships among the parties were positive but less dramatic. About 42% of respondents reported moderate improvement in relationships, but 35.4% reported no effect on relationships.

Incentives and Obstacles to Mediation

In an effort to understand the extent to which mediators and negotiators in the pilot program cases faced the barriers to using mediation, we asked a number of questions about incentives and obstacles the parties faced. In general, although the levels of potential barriers were considerable, these barriers were not seen as having derailed the mediation. There were important exceptions, however.

The first barrier examined concerns the parties' authority to settle the matter and, in particular, the authority of the negotiators involved in the mediation to commit the parties to settlement. Almost two-thirds of the respondents (66.2%) reported that the regulatory agency involved in their case had sole jurisdiction over the matter and that it did not share authority with any other government agency. However, only 36.5% of respondents thought that the negotiators representing their parties had enough authority to settle the key issues.

In several cases, respondents cited the absence of decisionmakers from the negotiating table as a serious impediment to progress. This appears to have been more of a problem when the amount of authority varied among the parties. Therefore, when one party's decisionmakers were at the mediation, they were frustrated, even angered, that although they could commit to settlement, another party could not. In one particular case, this problem led to perceptions of bad faith negotiation on the part of a state agency. In another case, private parties referred to the presence of a high-ranking state agency official at the mediation as key to the case's resolution.

The second barrier concerns the balance of expertise, resources, and political influence among the parties. We expect that when expertise, resources, and influence are in rough parity among the parties, mediation will work better than when these factors vary widely among the parties. Respondents widely described expertise and financial resources as unimportant to the outcome. As was previously stated, in one case the less expensive nature of mediation relative to administrative hearings facilitated the participation of individual homeowners. Lack of parity in political influence was a concern to certain parties in certain cases. They described the ultimate power of certain parties to veto the outcome as critical to the settlement reached.

Rigidity in regulation is often seen as having the potential to undermine the flexibility necessary in mediation. Only 18.2% of respondents thought that such rigidity existed in their cases, but this minority was quite forceful in their beliefs. These respondents said that agencies were unwilling to differentiate among petitioners who come before them, applied narrow regulatory categories to cases, had a fixed mentality, operated under statutory mandates that allowed little wiggle room, took issues off the table, and refused to recognize the ambiguity of certain regulations.

The potential for precedents that an agreement establishes to constrain future actions by the parties is also often seen as a threat. This was a concern of 24.7% of respondents, who thought that the potential for precedent was an obstacle to settlement. This minority was quite sure that the potential for affecting future cases was a major consideration to at least some parties in their cases. Both agency personnel and private-party representatives spoke of the need for consistency and the potential for cumulative effects of many similar decisions. Others claimed that fear of precedent experienced in a case was more perceived than real. Other respondents, who evaluated precedent as not having been an obstacle to settlement, still saw the potential of precedent as a difficulty they had needed to overcome.

Political interference is often cited as undermining policy mediation processes. This was an obstacle to settlement according to 45.6% of respondents. The specifics these respondents discussed were wide ranging, however, including such diverse concerns as the following:

— the necessity of later approval by an elected body,
— public pronouncements by elected officials that constrained the possible positions taken by the negotiators for their side,
— elected officials who took positions different from their negotiators,
— statutory constraints of federal law,
— the effects of the weight of public opinion, and
— efforts to tie the case in question to larger political agendas or longer running interjurisdictional conflicts.

Florida's so-called "sunshine laws," which require extensive public information about and access to the deliberations of government officials, are sometimes thought to impede the frank exchange of information that is necessary for successful mediation. However, only 10.3% of respondents thought that open government laws were obstacles to mediation efforts. One mediator was quite sure that the inability to ensure confidentiality impeded progress in the case. Another respondent said that sunshine laws impeded candid discussions. A third indicated that the list of mediation participants was specifically chosen to avoid open meeting requirements. Others saw openness as a threat that never really materialized, including parties who were concerned about maintaining strategic options for later litigation. In one case, however, thorough and perceptive press coverage was cited as important to building support for the process that led to settlement.

Clarity in understanding the issues, both generally and in terms of the parties' objectives, can be a barrier to effective mediation. About three-quarters (75.3%) of respondents, however, said that they thought there was total or substantial clarity about the issues in dispute. A full 93.5% of respondents said that they thought their organizations knew what outcome they wanted from the mediation.

Parties accustomed to traditional litigious methods of dispute resolution may seek to hide or distort information, as they believe this to give them a tactical advantage. However, only 15.8% of respondents thought that this was an obstacle to settlement in their cases. One party was charged with distributing false information outside the mediation sessions. One was said to have withheld information. Multiple representatives of one party were said to have made conflicting statements about factual matters. Several parties were said to have misrepresented their bottom lines. One respondent shrugged off this sort of behavior as ordinary negotiating behavior that is unavoidable.

Finally, parties that work together regularly are generally imagined to have better chances for finding mutually advantageous ways to meet each other's interests. In part, this is because they can trade concessions across cases as well as within the case in question. About one-half of the parties involved with the pilot cases had never negotiated, mediated, or litigated with each other, according to respondents. Remaining pairs of parties ran the gamut in terms of prior joint disputes, with about one in 10 pairs having negotiated, mediated, or litigated with each other on a regular basis. In one case, respondents indicated that familiarity was a liability, as animosity from a difficult prior dispute carried over into negotiations for the case in question.

Mediation Practices

Strategies and styles vary widely among mediators. Nationally, this is mirrored by debate over which mediation approaches are most effective in which situations. Respondents were asked a series of questions designed to gauge mediators' relative expertise, experience, and performance in a variety of areas.

Mediators' substantive knowledge of the issues in dispute varied widely, with about one-quarter of responses falling in each of the categories of "very good" (24.6%), "good" (27.7%), and "fair" (27.7%). About one of every 10 (9.2%) responses graded the mediator's substantive knowledge as "excellent," and about one in 10 (10.8%) graded it as "poor." The mediator was often seen as lacking specific knowledge about the policy issues under discussion, but respondents who expected mediators to have process knowledge, not substantive knowledge, seldom saw this as a problem. Respondents indicated that they often had to explain things to the mediator the way that they would to an "ordinary citizen on the street."

Mediators' group process knowledge was reviewed more positively: 33.8% of respondents said that the mediator had "excellent" group process knowledge, 55.4% "very good," 7.7% "good," 1.5% "fair," and 1.5% "poor."

Mediators' involvement before and after mediation sessions can be critical to success. The majority of respondents (51.3%) said that they had no contact with their mediators before formal mediation sessions. An additional 8.8% said that before mediation, they discussed only the time, place, and context of their sessions with the mediators. The remaining 40% discussed some issues (22.5%), all issues in general terms (15.0%), or all issues in detail (2.5%) with the mediators before mediation. Participants in several cases

TABLE 11-7. Mediator Effectiveness

	Very effective		Moderately effective		Moderately ineffective		Very ineffective	
	No.	Percent	No.	Percent	No.	Percent	No.	Percent
Verbal skills	41	63.1	20	30.8	3	4.6	1	1.5
Avoiding bias	59	90.8	4	6.2	2	3.1	0	0.0
Developing new solutions	29	45.3	28	43.8	7	10.9	0	0.0
Reaching agreements	35	55.6	22	34.9	5	7.9	1	1.6
Reducing tension	37	59.7	18	29.0	4	6.5	3	4.8

described mediators' failure to follow through. This included the lack of expected postmediation communications and a mismatch between expectations and postmediation written drafts.

Several respondents referred to the voluntary or nonbinding nature of the mediation process as important to its success. Motivation to participate was described as critical to engaging in the work of the process.

Respondents rated their mediators in five areas:

1. verbal expression, persuasiveness, and communication with the parties;
2. avoiding appearance of bias;
3. helping parties to develop new solutions;
4. helping parties to reach agreements; and
5. reducing tension among the parties.

Responses are given in Table 11-7. In all five areas, the "very effective" category received the largest number of responses, ranging from a high of 90.8% of respondents on avoiding bias to a low of 45.3% on helping develop new solutions. At most, 4.8% of respondents graded their mediators as "very ineffective" in any of the five areas, ranging from a high of 4.8% for reducing tension to a low of 0% for avoiding bias and for developing new solutions. Three to 10 percent of respondents saw their mediators as "moderately ineffective" across all five categories.

The mediators' relatively lower ratings on helping to develop new solutions appears tied to the perception that mediators lacked detailed substantive knowledge in the policy areas under discussion. Mediators were seen as guiding the process and not as having enough contextual knowledge to suggest innovative or additional alternatives or options.

More than a few respondents voiced belief that the use of independent mediators not affiliated with the regulatory agency involved in the dispute was important to the success of the process. Several referred to prior experience with agency-based mediators who had been less satisfactory.

Summary of Findings

Our sample of the 10 state agencies that provide more than 80% of the cases that go to ALJs in Florida (in 1997) found some striking results. Agencies do not widely use mediation under the APA procedures, and managers do not perceive a clear need for its use. The majority of managers believe that current settlement practices are satisfactory and that agencies face real obstacles to using mediation more widely.

For mediation to become more widely used, timelines for resolving disputed administrative cases may need to be modified, or at least clarified. Many programs must meet a built-in clock of 30 to 60 days for resolution to be in legal compliance. If mediation is used as a first process, there may be no time left after unsuccessful mediation to go to ALJs.

Agency budgets do not have built-in staff or consultant costs to allow regular use of mediation, and few agencies have much in-house expertise for the APA mediation process to work through.

Finally, managers do not have a highly positive view of mediation; they did not generally agree with statements that mediation offers a faster, more efficient, or higher quality method of resolving administrative disputes. Accordingly, greater use of mediation would seem to require staff education on the basics of mediation, when it can be used most productively, and skills for conducting mediations.

Despite these obstacles, more than 80% of the respondents said that they would support increased use of mediation in their agencies if more resources and other problems were dealt with. They felt that the most successful way to increase agency use of mediation was through overt support from agency leadership, an executive mandate, or statutory requirement.

Evaluation of the project cases generally supported the positive assessments of mediation found in literature. A high fraction of pilot program cases settled. Respondents were quite pleased with the settlements, with more than 80% either very or moderately satisfied with the mediation process. About three-quarters of the participants believed that the mediated outcome would not have been likely had the case gone to administrative hearing. They often focused on the ability of mediation to find specially tailored, creative solutions to problems that less flexible administrative hearings might not craft. A high proportion (more than 80%) also think the mediations were quicker and cheaper than hearings would have been, estimating savings to their organizations in the wide range of $2,000 to $700,000. Nearly all participants (more than 90%) felt that the mediations led to achievement of at least some of their goals.

The most common barriers to mediation were generally not seen as obstacles to settlement in these cases. The regulatory agency usually had sole jurisdiction. Organizations knew what they wanted. Negotiators were not often seen as distorting or hiding information. Issues were perceived as clear.

Even though barriers were not widely seen as settlement obstacles, some issues complicated mediations. Participants often had to overcome regulatory rigidity, the potential for setting precedents, and open government laws. Almost a quarter of respondents felt that the potential for precedent coming from the mediation was an obstacle to settlement. Imbalance in expertise and financial resources, imagined by the researchers as potential threats to the process, were widely described by respondents as unimportant to the outcome.

Notable exceptions where barriers *were* obstacles were the absence of authority to settle on the part of negotiators at the mediation and the fact that many of the parties had never negotiated with each other before. Political interference was widely seen as an obstacle to settlement, but the specific concerns incorporated in this label were quite diverse.

The practices mediators used generally received quite high ratings. In particular, mediators got high marks for group process knowledge, avoiding bias, and verbal skills. Areas where mediators received somewhat lower evaluations were substantive knowledge in the areas of dispute, involvement before and after formal mediation sessions, and help with developing new solutions.

Conclusions

The prospects for wider use of mediation to resolve state agency administrative disputes are strong. Project cases showed high rates of settlement, lower dispute resolution costs, short timelines to conclusion, and the fact that relationships among disputants were not adversely affected. It would seem, in fact, that all the major advantages cited for mediation in the ADR literature[24] were evidenced in this pilot program. This is in contrast to challenges raised by Amy,[25] and by Kakalik and others.[26]

Lack of familiarity with mediation among agency personnel, along with statutory procedural requirements that may demand fast timelines for processing of disputes, seem to be the strongest obstacles to increased use of mediation. Wider discussion of experiences to date, among both administrative and legislative leadership, may therefore be important to broader adoption. Florida's national position as a leader in the use of both environmental mediation and court-annexed mediation[27] suggests that if familiarity is low in Florida, it must be very low in many other states. There is likely to be broad value from the assessment and discussion of mediation practices nationwide.

Some other obstacles troubled mediations but did not prevent or undermine them. Among these were the absence of authority to settle on the part of negotiators at the mediation and a grab bag of government leadership influences, which we have called political interference. Both of these obstacles derive from fundamental characteristics of government, including built-in checks and balances, the multiobjective nature of agencies, cyclic electoral selection of top decisionmakers, and the attempts in place to distinguish policy and administrative functions.[28] They are problems for government negotiation, litigation, and mediation. The fact that the sample of mediations examined here suffered little from these obstacles suggests hope that mediation may be a procedure that can reduce the importance of these government structure-derived obstacles.

Indeed, the generally low level of impact of expected barriers to mediation is a surprising finding of this research. Efforts of the Policy Consensus Initiative,[29] the University of Texas Law School's Center for Public Policy Dispute Resolution,[30] and others[31] have led us to expect that mediation would confront a regular and steady diet of obstacles in the administrative context. The contrary, promediation findings must be tempered by the realization that the pilot project cases make up a small subset of all cases processed by the agencies in question. Whereas project staff sought uniform procedures to offer mediation to all parties in wide classes of cases, there may well have been strategic efforts within agencies to select cases for mediation based on nonrandom criteria. Even if this is the case, however, the strong successes in project cases suggest that mediation can be used effectively to improve process and outcomes in at least a significant portion of administrative agency disputes.

Notes

The authors acknowledge two other study team members: James Rossi, College of Law, Florida State University, for his research on federal and state mediation procedures, and Jean Sternlight, Law School, University of Missouri, for her work on background, obstacles, and incentives to the use of mediation. We also acknowledge the professional expertise and contribution of the other pilot project staff to this study's write-up: Jonathan Davidson, Anne Clausen, Hal Beardall, Robert Jones, and Tom Taylor, all of the Florida Conflict Resolution Consortium.

1. See Rosemary O'Leary and Tracy Yandle, "Environmental Management at the Millennium: The Use of Environmental Dispute Resolution by State Governments," *Journal of Public Administration Research and Theory,* vol. 10 (2000), pp. 137–155 for information on state programs for environmental dispute resolution.

2. See, e.g., Florida Statute §44.102(2)(b) (1997) ("upon a finding that there is a dispute circuit courts *shall* refer to mediation … parental responsibility issues"); see Florida Statute §44.102(2)(c) (1997) ("circuit courts *may* refer to mediation matters relating to dependency or a child or family in need of services"). Most states require mediation for at least some types of cases. See the discussion in Nancy H. Rogers and

Craig A. McEwen, "Employing the Law To Increase the Use of Mediation and To Encourage Direct and Early Negotiations," *Ohio State Journal on Dispute Resolution,* vol. 13 (1998), pp. 831–864. In addition, most states use some form of discretionary, court-connected mediation; see Peter S. Chantilis, "Mediation U.S.A.," *University of Memphis Law Review,* vol. 26 (1996), pp. 1031–1079.

3. See, for example, Craig A. McEwen and Richard J. Maiman, "Small Claims Mediation in Maine: An Empirical Assessment," *Maine Law Review,* vol. 33 (1981), p. 237. For additional sources, see A. Gutherie and G. Lewin, "A 'Party Satisfaction' Perspective on a Comprehensive Mediation Statute," *Ohio State Journal of Dispute Resolution,* vol. 13 (1998), pp. 885–906.

4. See Rogers and McEwen, "Employing the Law To Increase the Use of Mediation and To Encourage Direct and Early Negotiations," p. 833.

5. See Neil Sipe and Bruce Stiftel, "Mediating Environmental Enforcement Disputes: How Well Does It Work?" *Environmental Impact Assessment Review,* vol. 15 (1995), pp. 139–156; Bruce Stiftel and Neil Sipe, "Mediation of Environmental Enforcement: Overcoming Inertia," *Journal of Dispute Resolution,* vol. 1992, no. 2 (1992), pp. 303–324 for in-depth studies of Florida environmental dispute assessments; see Peter Steenland and Peter A. Appel, "The Ongoing Role of Alternative Dispute Resolution in Federal Government Litigation," *University of Toledo Law Review,* vol. 27 (1996), pp. 805, 815, discussing the advantages of mediation with respect to petitions filed under the National Environmental Policy Act; also see Ann L. MacNaughton, "Collaborative Problem-Solving in Environmental Dispute Resolution," *Natural Resources and Environment,* vol. 11 (1996), p. 3; Allen Talbot, *Settling Things* (Washington, DC: Conservation Foundation, 1983) on evaluating the use of mediation in a wide variety of environmental disputes and concluding that it is effective.

6. See Philip J. Harter, "Fear of Commitment: An Affliction of Adolescents," *Duke Law Journal,* vol. 46 (1997), p. 1389; Cary Coglianese, "Assessing Consensus: The Promise and Performance of Negotiated Rulemaking," *Duke Law Journal,* vol. 46 (1997), p. 1255; William Funk, "Bargaining toward the New Millennium: Regulatory Negotiation and the Subversion of the Public Interest," *Duke Law Journal,* vol. 46 (1997), p. 1351.

7. See, e.g., Florida Statutes §163.3181 (1997), requiring that when a local government denies a landowner's request to amend the local comprehensive plan, the local government must provide the landowner an opportunity to use alternative dispute resolution to resolve the dispute.

8. U.S. Code Section 571.5 *et seq.*

9. See Bruce Stiftel, Robert M. Jones, and Thomas A. Taylor, *Statewide Offices of Dispute Resolution: The Florida Experience* (Tallahassee: Florida Conflict Resolution Consortium, 1998); Lawrence E. Susskind and Jeffrey Cruikshank, *Breaking the Impasse: Consensual Approaches to Resolving Public Disputes* (New York: Basic Books, 1987).

10. See, for example, Marcia Caton Campbell and Donald W. Floyd, "Thinking Critically about Environmental Mediation," *Journal of Planning Literature,* vol. 10 (1996), pp. 235–236; Nancy Kubasek and Gary Silverman, "Environmental Mediation," *American Business Journal,* vol. 26 (1988), pp. 533, 541; Gail Bingham and Leah V. Haygood, "Environmental Dispute Resolution: The First Ten Years," *Arbitration Journal,* vol. 41 (1986), pp. 3–4.

11. See, for example, Gail Bingham, *Resolving Environmental Disputes: A Decade of Experience* (Washington, DC: Conservation Foundation, 1986), pp. 130–132; Jay Folberg and Allison Taylor, *Mediation: A Comprehensive Guide to Resolving Conflicts without Litigation* (San Francisco: Jossey-Bass, 1984); Lawrence E. Susskind and Denise Madigan, "New Approaches to Resolving Disputes in the Public Sector," *The Justice System Journal*, vol. 7 (1984), p. 179; Scott Mernitz, *Mediation of Environmental Disputes* (New York: Praeger, 1980), p. 47.

12. See, for example, Lawrence E. Susskind and Alan Weinstein, "Toward a Theory of Environmental Dispute Resolution," *Boston College Environmental Law Review*, vol. 9 (1980), pp. 311–312; Mernitz, *Mediation of Environmental Disputes*.

13. Executive Order 12,988, 61 *Federal Register* 4729 (1996).

14. See "President's Memorandum on Agency Use of Alternative Means of Dispute Resolution and Negotiated Rulemaking," 34 Weekly Compilation of Presidential Documents 749 (May 1, 1998).

15. Ibid.

16. Chantilis, "Mediation U.S.A.," compiled a survey of mediation statutes from numerous states that demonstrates their breadth. A few examples of court-based mediation are provided here. In Alaska, court-ordered mediation can be mandated on a case-by-case basis. In Georgia, the courts are allowed to establish their own ADR programs and to dictate the types of cases referred to ADR and the type of ADR used (p. 1048). In California, contested child custody cases must be sent to mediation (p. 1041). Texas allows courts to order mediation for medical malpractice claims (Texas Civil Practice and Remedies Code §88.003(d)(1999)) and for divorces (Texas Family Code §6.602(a)(1999)).

17. See Michael Asimow, "The Influence of Federal Administrative Procedure on California's New Administrative Procedure Act," *Tulsa Law Journal*, vol. 32 (1996), pp. 319–320.

18. For a summary of program-specific mediation in Arizona, California, Iowa, Minnesota, New Mexico, and Oklahoma, see Brian D. Shannon, "The Administrative Procedure and Texas Register Act and ADR: A New Twist for Administrative Procedure in Texas?" *Baylor Law Review*, vol. 42 (1990).

19. The DOI keeps aggregate statistics on its statutory mediation programs for automobile and homeowner disputes between policyholders and their insurers. At the DOT, administrative counselors track the disposition of cases that are referred to them and are integrating negotiated settlements and mediation into their management system. The PSC has a sophisticated case-tracking system that integrates case settlement and mediation results. The DOAH maintains an annual list of cases filed, which is broken down by agencies and program areas. The DOAH, however, does not provide mediation as an option for its cases.

20. Florida's Baker Act provides for hearings on involuntary civil commitments. For reasons that include confidentiality requirements, the project did not address these cases.

21. The agencies chose the appropriate respondents to represent them. In most cases, the general counsel's office assisted in identifying the most appropriate person to give an agencywide perspective. The agencywide respondent then assisted in identifying appropriate division or bureau chiefs to also respond to the survey. The survey

and interviews were conducted from April to August 1999 and represent agency personnel assessments of their agencies at that time.

22. Most agencies do not collect information on the contested cases they face under Chapter 120.573 of Florida Statutes.

23. More than half of respondents (53.8%) said that the settlement included a written agreement. About 26% reported that all provisions of the agreement were incorporated into a final order issued by the regulatory agency. About 43% of respondents reported provisions in the agreement that concern monitoring, enforcement, or both. One respondent reported an appeal to an administrative law judge of an issue resulting from a mediated agreement. The one case that did not settle was described as having been convened hastily, with neither the mediator nor the parties coming to the negotiations properly prepared.

24. See Neil G. Sipe, "An Empirical Analysis of Environmental Mediation," *Journal of the American Planning Association*, vol. 64 (1990), p. 275; Caton Campbell and Floyd, "Thinking Critically about Environmental Mediation"; Bingham, *Resolving Environmental Disputes.*

25. Douglas J. Amy, *The Politics of Environmental Mediation* (Columbia University Press, 1987).

26. James S. Kakalik, Terence Dunworth, Laural A. Hill, Daniel McCaffrey, Marian Oshiro, Nicholas M. Pace, and Mary E. Vaiana, "Just, Speedy, and Inexpensive: An Evaluation of Judicial Case Management under the Civil Justice Reform" (Santa Monica, CA: RAND Corporation, Institute for Civil Justice, 1996).

27. O'Leary and Yandle, "Environmental Management at the Millennium"; Robert M. Jones, "Faster, Smarter, Cheaper? Assessing the Barriers to Delivering on the Promise of Administrative Dispute Resolution in Florida," *American Bar Association Dispute Resolution Magazine* (Summer 2001).

28. Bruce Stiftel, "Can Governments Bargain Effectively? Lessons from a Waste Transfer Station Location," paper presented at the World Planning Schools Congress, Shanghai, China, July 11–15, 2001.

29. Andrew Bowman, "Literature Review: Barriers to the Use of ADR Processes at the State Level" (Santa Fe, NM: Policy Consensus Initiative, 1999).

30. Center for Public Policy Dispute Resolution, *Report of Survey Results of Alternative Dispute Resolution Use in Texas State Agencies* (University of Texas at Austin, School of Law, 1999).

31. Kenneth J. Arrow, Robert H. Mnookin, Lee Ross, Amos Tversky, and Robert Wilson, eds., *Barriers to Conflict Resolution* (New York: W.W. Norton, 1995).

12

Court-Annexed Environmental Mediation
The District of Oregon Pilot Project

Lisa A. Kloppenberg

This chapter details a unique pilot project designed to promote the use of mediation to resolve selected environmental cases in the federal trial courts in Oregon. I survey some important factors to consider in environmental conflict resolution, discuss the goals and parameters of the Oregon pilot, and consider some tentative lessons to be drawn.

Environmental Conflict Resolution and Oregon Federal Filings

The pilot involved cases filed in the Oregon federal trial court, not conflicts at a prelitigation stage or in an agency setting. Peter Adler's study of the prospective use of alternative dispute resolution (ADR) for environmental cases contains relevant findings for trial courts in the Ninth Circuit. Discussions with judges, other court employees, and mediators familiar with prior cases in the Oregon federal court confirmed the presence of many similar factors in Oregon environmental cases.

Environmental cases make up only about 2% of the civil docket in Oregon. From 1990 to 2000, between 23 and 35 environmental lawsuits were filed annually. These cases involved claims regarding water use, fishing rights, endangered species, Native American sacred remains, water and land pollution, hazardous waste cleanup, impacts of proposed development, timber theft, and timber sales. The handful of cases that proceeded to mediation through this pilot primarily centered on water and ground pollution, with federal agencies involved.

Although few in number, the cases are important in several respects. Few proceed to trial, but motion activity is significant. Judges, law clerks, and administrators report that the cases are time-intensive for them. Some are resource-intensive for the court because they involve complex legal and factual issues. Many are paper-intensive (one judge referred to "endless paper piles"), and some involve multiple appeals. Because some environmental cases are resolved by consent decrees, they linger on the docket and can entail more of a managerial, long-term role for the court. They often return to judges more than other civil cases due to the need for monitoring compliance and the potential for changed circumstances. Nevertheless, some judges like these challenging, complex, and fascinating cases.

In the litigation process, environmental cases are resource-intensive and slow-moving, in part because they tend to involve scientific uncertainties. Parties sometimes spend significant amounts of money for data they do not use. Some cases require expertise on economic and sociological issues. Judges cited the presence of scientific experts with adamant viewpoints, which tend to harden adversarial positions. It is difficult to find credible neutral scientists. Some Oregon judges believe that the focus on "dueling experts" is a barrier to greater party involvement and exploration of the parties' broader interests. One judge noted that it is difficult for clients to let lawyers "take off advocacy hats and put on reconciliation hats" in this context.

Additionally, environmental cases often present difficult questions of public interest. Some are politically charged, with strong emotions and opposing views on significant legal and factual questions. People or entities that are not parties in the lawsuit can have a significant interest in the conflict or in barriers to implementing solutions. Some are high-stakes cases in terms of precedent, media attention, and public significance. A wide variety of interests participate, including tribes, federal entities, states, local political subdivisions, public interest groups, private entities, industry, and commercial and sports fishing interests. The "big picture problems" in some of these disputes—involving long-term, multiple interests—require durable solutions, not just swift disposition.

Environmental cases are frequently resolved on procedural issues or narrow grounds, so that courts do not reach the merits of underlying scientific or legal claims. Federal court review of federal agency action, for instance, is limited. Parties sometimes request narrow legal relief, which will not address the root problem or prevent future rounds of conflict among the parties. Sometimes conflicts among federal entities dominate environmental lawsuits.

Rather than revolving around retrospective harms (e.g., an appropriate amount of damages), these cases often center on future problem solving. It is challenging for courts to structure remedies that can account for future environmental and financial uncertainties (e.g., changing habitat, water levels, species levels, and political settings). Moreover, court solutions are not always sufficiently flexible. Judges spoke of cases that need a balancing of interests (e.g., water flow levels) versus others that "cried out for legal resolution."

The U.S. attorney in Oregon noted that the office's efforts in using ADR had been quite successful in all areas *except* in environmental cases. It was much more difficult to get the parties involved in the environmental cases to use mediation. The U.S. attorney's office expressed interest in understanding and changing this phenomenon and worked with sponsors as the pilot proceeded.

Whereas the sponsors realize the countless challenges presented when courts encourage environmental mediation, they believe that the potential gains make the effort worthwhile. A mediated agreement, in contrast with adjudication, could expand the parties' ability to bring important underlying issues to the table, to include necessary nonparties, and to provide more flexible and durable solutions, preventing some future conflicts among the parties and others.

The Goals and Parameters of the Oregon Mediation Pilot Project

In 1998, Congress and the president enacted the Alternative Dispute Resolution Act, requiring federal trial courts to promote the use of ADR by implementing an ADR program. The act required districts with existing ADR programs to evaluate them and designate a judicial officer or court employee to handle certain ADR-related functions. As part of implementing this mandate, Judge Proctor Hug, Jr., then chief judge of the Ninth Circuit, and Judge Dorothy Wright Nelson, chair of the Ninth Circuit ADR Committee, expressed critical support for this pilot and similar projects throughout the Ninth Circuit.

The pilot's sponsors included the U.S. District Court for the District of Oregon, the U.S. Institute for Environmental Conflict Resolution (USIECR), the Western Justice Center Foundation, and the ADR Program of the University of Oregon School of Law. Oregon was the first district to initiate an environmental mediation pilot, in large part because of the interest of the District's chief judge, Michael Hogan, and other Oregon judges in promoting ADR. The other sponsors helped keep the potential pilot on the judges' agenda. Resources devoted by sponsors, particularly the USIECR's funding of a part-time project coordinator position, were also critical to getting the pilot underway.

This voluntary environmental mediation pilot is designated "court-annexed" because the court is giving access to cases (either through the docket or by judicial referral), is supportive of a neutral project coordinator who contacts parties in environmental cases to see if they will consider mediation, and is trying to learn how it can best work to support the use of voluntary external mediation.

Goals for the Pilot Project

The sponsors first developed goals for the pilot and case criteria. They agreed on multiple goals for the Oregon pilot. They sought to mediate five to ten environmental litigation cases. The mediators might reduce the number of contested technical and scientific facts, help streamline or resolve legal issues, and create opportunities for integrative and mutual gains negotiations. The pilot would use highly experienced and well-assigned mediators. Sponsors wanted to discover better ways to match the skills and experience of prospective mediators with the needs of litigants and lawyers, as well as to facilitate connections between the court and potential mediators. Sponsors also sought to help develop capacity locally in environmental mediation. Mediation might expedite cases by saving the court and the parties time and money, by increasing the pace of issue resolution, and by beating the average time to disposition for environmental cases. Mediation might also provide a fair and satisfying dispute resolution process by improving the quality of communication and information exchange and reducing the emotional friction involved in these cases. Sponsors hoped that mediation would produce more stable, enduring, and implementable outcomes, thereby reducing the chance that issues would return to litigation and improving the relationships between parties who would need to work together in the future. Finally, sponsors agreed that the law school would evaluate the outcomes of the pilot project and publish the results. This could lead to a compendium of settlement tips, tools, and techniques for those involved in environmental litigation and encourage the further use of environmental mediation in the District of Oregon, with potential applications for other courts.

Notably, the basic aims of decreasing costs and resolution time for the court and the parties were balanced with potentially opposing goals such as creating stronger, more durable outcomes (which could broaden the issues on the table), bringing better scientific and technical information to solutions (which could require significant expenditures), and fostering stronger working relationships that allow for better postlitigation implementation and compliance as well as collaboration in future conflicts (which could involve more extensive mediation, increasing time and cost). An important assumption of many parties involved—from sponsors to individual judges to members of the mediator panel—was that ADR is not a panacea. It is not a complete replacement for existing court processes and cannot completely address the competing health and resource problems involved in environmental disputes.

Case Selection Criteria

The sponsors agreed that the environmental cases selected for mediation would involve natural resources, pollution and toxics, or public land. The pilot sought cases with parties in continuing relationships, with some over-lapping interests, and a possibility for mutual gain. Over time, the USIECR emphasized the goal of selecting cases with possibilities for integrative gains rather than pure cost allocation problems. In part because the District of Oregon already has experienced adjudicators and settlement judges available at no additional cost to litigants, the pilot's focus was on issues that are not easily resolved by a judicial decision or a traditional evaluative settlement conference. Thus, the pilot encompassed only cases that would require estab-lishing a new legal precedent for resolution. Because of USIECR involvement, cases had to involve a federal agency or a federal statute, rule, or policy deal-ing with the environment. Cases with scientific and technical complexity were preferred.

The consent of the parties and the court to mediate a particular case was an important criterion. Parties had to be willing to mediate in good faith and to bring the right people to the table. Parties were required to pay the media-tion fees. All mediators involved in the pilot agreed to reduce fees if necessary to get cases into mediation. Parties entering mediation also had to agree to cooperate with the evaluation of the pilot. Sponsors designed a sample medi-ation agreement containing these criteria. Thus, the voluntary nature of par-ticipation in the project was one of its critical features.

Unless otherwise agreed by the parties and the court, no substantive com-munications or reports could be made from a mediator to a trial judge about a particular mediation. Mediators could coordinate their efforts with a settle-ment judge assigned to a case, unless the parties precluded such communica-tion. The parties, attorneys, and mediators agreed to participate in a program evaluation.

Operations during the Initial Phase of the Pilot

Significant time and resources were devoted to building relationships between the sponsors and Oregon judges during the initial phase of the pilot. A series of meetings was convened to design the pilot, and approval was sought from the Ninth Circuit and the Oregon District Court. The pilot commenced in January 2000. Sponsors anticipated that it would continue for 18 months or until five to ten mediations were complete.

The Role of the Project Coordinator

With support from the Oregon court, the USIECR designated and funded a local project coordinator, Elaine Hallmark, to start the project. As a lawyer and experienced environmental mediator, Hallmark served as a liaison among the court, parties and counsel, and potential mediators. She helped build and maintain support for the program, reported to the sponsors regularly on her activities and progress, and talked with numerous parties in cases. Her duties included the following:

— Fostering the relationships and continuing communications among the project sponsors, particularly the USIECR, the judges, and the University of Oregon; this included developing or refining the protocols for various aspects of the project, such as the case selection criteria and the development and administration of the panel of mediators.
— Establishing, orienting, and communicating with a panel of experienced environmental mediators; this included coordinating an orientation and advanced environmental mediation workshop for the mediators.
— Finding appropriate environmental cases and developing voluntary commitments from the parties in litigation to participate in mediation, to pay for the mediators, and to participate in the evaluation process for the pilot project.

In all cases, Hallmark explained the pilot project and informed parties about the range of approaches to environmental mediation. In some instances, she negotiated premediation agreements on the substance of the dispute and fee-sharing agreements among the parties. In some cases, she helped to "keep a conversation going" about expeditious and early resolution, which may have spurred some attorneys and litigants to undertake earlier or more serious settlement negotiations on their own.

Because of earlier commitments, the project coordinator was unable to begin assessing cases before April 2000. By early 2001, Hallmark had spent more than 350 hours on the pilot, and by the time her activities concluded in September 2001, she had devoted 535 hours to the pilot project.

Developing Information on the Environmental Docket

It was fairly difficult to get data on the court's environmental docket. The court did not keep detailed records on types of dispositions in environmental cases, time to disposition, ADR efforts, or other data. More specifics on how such cases are processed would be extremely useful. The clerk's office reported that from 1990 to June of 2000, between 23 and 35 environmental (category "893") cases were filed per year. For 1998, 27 of 34 cases were closed by June 2000. Of those 27, 11 settled, 16 were disposed of by motion, and none had gone to trial yet. For 1999, 18 of the 28 files were closed by June 2000. Four of those settled, nine were disposed of by motion, four were disposed of in another fashion, and none had gone to trial yet. It is unclear what role lawyers for the parties, assigned judges, settlement judges, or external conflict resolution experts played in the settlements.

At the time of the pilot, little was recorded about time spent in judicial settlement conferences or results achieved. Whenever the Oregon judges had a case that they believed could benefit from judicial settlement efforts, they usually sent a notice to their colleagues, who volunteered for the duty.

Judicial Attitudes toward ADR and External Mediation in Environmental Cases

At the outset of the pilot, sponsors encountered mixed views among Oregon federal judges about the appropriateness of ADR for environmental cases, as well as about the need for encouraging external mediation. Some judges in the District of Oregon, including the chief judge, believe that environmental cases are well suited to ADR because of the number of parties, the complexity of the disputes, and for other reasons mentioned above. Other judges thought that cases involving legal issues or matters of public concern should not be routed to private ADR settings. A few were troubled by the cost of external mediation. Based on prior experience, a few judges expressed worries about the quality of external mediation and the court's obligation to ensure a high-quality experience when engaging in court referral of cases.

Some judges also had ideas, based on their experience with environmental disputes, about the types of cases appropriate for ADR versus adjudication. For example, some judges suggested that some Endangered Species Act (ESA) cases are clearly appropriate for adjudication because they require a legal ruling on the listing of or failure to list a species. Judges also said that some cases are inappropriate for mediation because they have high precedential value, or there are reasons to wait for agency action, such as ESA listing determination. A party involved in numerous ESA cases, however, opined that although some judges think that ESA cases only involve legal questions, some ESA cases may be amenable to settlement because of constraints on agency resources in handling listing matters.

Each federal judicial district is unique. Oregon is a small court (with fewer than a dozen Article III and magistrate judges). The judges place a premium on not having a long backlog, and some aggressively promote settlement efforts. Some of the Article III and magistrate judges are experienced settlement judges who like this work. Thus, some judges did not see a strong need for further ADR options. Some were concerned about the additional cost and delays for parties. Other judges expressed hopes that external mediators could offer more time and a "different persona" than federal judges could. For example, when judges engage in settlement conferences or encourage ongoing negotiations between parties in environmental cases, they find that multiple conferences and long-term negotiations are frequently needed. As one experienced settlement judge noted, whereas most settlement conferences in civil cases take a day or part of a day, environmental cases take conference after conference. Most judges were not familiar with many environmental conflict resolution practitioners, despite the strong cadre of experienced environmental mediators in Oregon, and did not regularly refer such cases to external mediators. The judges were most familiar with mediators traditionally connected to courts in civil cases (e.g., former U.S. attorneys and retired judges). The pilot served at least to inform judges about a broader pool of experienced environmental mediators in Oregon.

In addition to judicial attitudes, parties' and attorneys' perceptions about their options within a court influence their decisions about conflict resolution options. The Oregon judges' track records are often well known to parties in environmental cases. Some public interest organizations are concerned about being forced into settlement in general, or into settlement discussions with a limited set of issues on the table. Repeat players' skepticism of some judges or past experience with judicial settlement efforts may make them less disposed to resolve cases early rather than await a judicial ruling, appeal to the Ninth Circuit, or settle on their own.

The judges had different views on the appropriate timing for ADR. Some viewed an early referral to mediation or a settlement conference as worthwhile in most environmental cases, whereas others expressed reservations about choosing the right timing for ADR. They differed on whether a consistent court ADR process should be used or case-by-case flexibility should be retained. One judge noted that a host of factors must be examined and that experienced judges are skilled at conducting such assessments. One attorney told the project coordinator that repeated reminders about mediation could be useful if a lawyer or party is awaiting the next stage of litigation or another event. In one potential pilot case, the parties chose to await a summary judgement ruling, but that ruling did not offer complete victory to either side.

One judge expressed concern that referrals to external mediators might lead to delays if judges put cases on hold during an unproductive or stalled mediation process; this judge suggested a reporting requirement. Sponsors devised a proposed order that individual judges could use to protect mediators. The order also established a mechanism for keeping judges informed, in general terms, about mediation progress.

The Oregon judges, like lawyers and mediators, expressed divergent views on whether to seek mediators with subject matter or other expertise, or whether the primary value a mediator brings is process skills. In their own adjudicatory work, some judges think environmental expertise and technical backgrounds are helpful, whereas others see more value in retaining generalists who can be educated by adversarial experts like lawyers and scientists or by credible neutrals. There is little familiarity with the role that premediation assessment, convening, and process design play in the likelihood of a successful mediation in environmental, natural resource, and public policy cases.

Little is known about use of ADR in environmental cases filed in the Oregon federal trial court before the pilot. All information is anecdotal, drawn from interviews with the judges and without the perspective of parties and lawyers involved. Judges have used settlement conferences to resolve some environmental cases. One federal and state judge teamed up to resolve a dispute with complex scientific and technical issues about a landslide and used agreed-upon neutral experts to assess remedial options. In a few cases involving water usage, judicial settlement conferences resulted in agreements, but the issues could return to court if water shortages occur. One case selected binding arbitration but eventually settled as delay ensued and arbitration expenses escalated. Judges have referred a few cases to external mediation. Several judges have been innovative in using court-appointed scientific experts. For example, in one case involving a resource management plan, the parties agreed to use a respected scientist to educate the judge, but the expert was not available for adversarial discovery. The scientist was an advisor, not a special master to whom the adjudicator delegated work.

Referrals

The pilot sponsors initially assumed that most referrals would come from judges who had identified cases on their dockets that were "ripe" for mediation. The project was also open to self-referred cases and cases identified in other ways. Cases would be referred to the project coordinator, who would then gather information on the cases, contact the parties' attorneys and help them assess the potential for mediation, resolve any initial issues including fees, help them select a mediator, and secure the agreement to mediate and the confidentiality agreement.

Chief Judge Hogan requested that judges provide a list of cases to refer to the project. He compiled a list of 10 cases for the initial referrals. The project coordinator tracked down information from the dockets and the law clerks on those cases and obtained contact information for the parties. One of the cases was ready to go to mediation but subsequently chose to use a settlement judge. It was soon discovered that 70% of these initial cases were "distributive," primarily cost allocation among private parties for cleanup costs. Almost all had settled or were near settlement by the time that the coordinator contacted the attorneys. Discussions with the sponsors clarified that the pilot project should not be heavily weighted with that type of case because those cases were settling on their own or with judicial settlement conferences. Thus, the project coordinator had to obtain referrals for more complex environmental cases that were not immediately seen as being amenable to mediation.

The project coordinator and the evaluator held individual meetings with each of the judges to discuss the pilot and the potential of external mediation in environmental cases and to seek further referrals. When an intern became available during the summer, she was able to secure the full docket list of open environmental cases and contacted the judges or their law clerks to learn the status of the cases and their potential for mediation. As the project became more widely known, some attorneys requested information, and some judges referred newly filed cases. In January 2001, the project coordinator increased her focus on identifying cases that had not been referred and that involved the United States as a party. The local assistant U.S. attorney provided a list of cases being handled by the Justice Department in Washington, D.C., and the project coordinator also identified cases directly from the latest docket list.

As of August 15, 2001, the following summary statistics had been compiled: The project coordinator screened 75 cases during the pilot. This represented almost the entire docket of environmental cases in the District of Oregon. The entire docket was reviewed, but only those cases that appeared to have some potential for mediation were screened.

Of the 75 cases, 21 were referred directly by the judges (10 from the chief judge initially and 11 in interviews or direct calls); 38 were identified by the coordinator from the docket; 6 were requests from attorneys; and another 10 were referred by the U.S. attorney who handles environmental cases in the district.

In 13 cases, parties agreed to participate in mediation (including one case that was not filed). Seven of those cases were primarily distributive (i.e., the key issues were distributing costs among parties). Six of the cases might be termed integrative because resolution of the issues was likely to require more joint problem-solving approaches.

Outcomes in Mediation

The seven distributive cases were primarily cleanup cases, involving allocation of costs or recovery of costs. For example, one case was an enforcement case involving the amount of a fine for violations of the Clean Air Act. One was a National Environmental Policy Act (NEPA) case involving the amount of a fine for violations of the Clean Air Act. One was a NEPA case involving a federal agency, and one was an enforcement case under the Clean Water Act involving a federal agency and a pro se party. Two of the seven distributive cases resolved in mediation (including the nonfiled case). One case remained pending as of August 15, but there are indications that it will settle without an actual mediation session. Two cases returned to the litigation process after mediation (in one of these, parties had selected a settlement judge rather than an external mediator from the pilot project). One case went to mediation without success, then used litigation to resolve a major issue, then returned to mediation to structure an arbitration process to resolve the remaining issues. One case proceeded to a mediation session, using a legal expert as an advisor to the mediator, but then parties continued with further discovery. The coordinator reports that the mediation may resume later.

The six integrative cases include a range of environmental issues: a NEPA–federal agency action; a Clean Water Act enforcement case; a timber sale; an

ESA case involving contested time for establishing critical habitat designations; and two agency permit–private party actions. Of these, three cases settled through mediation. In one case, parties withdrew from mediation before even finalizing the choice of the mediator. The parties tried to negotiate on their own, but proceeded to trial. In one case, mediation was still pending as of August 15. One case did not settle in mediation. Although the substantive issues appeared to be resolved, a dispute over attorney fees remained. The coordinator reported that some negotiations may have continued in that case.

Other Dispositions

Of the 75 cases reviewed, 23 settled on their own without using mediation. One of these involved parties from an earlier mediation within the pilot, and they negotiated settlement using a similar approach. One case settled in mediation outside the pilot. Two cases settled using a settlement judge, and parties in one case were before a settlement judge as of August 15. In another case, a major convening–case assessment effort was being considered as of August 15.

Thirty-five cases continued in litigation. In two of these, parties tried to resolve issues using a settlement judge, but settlements were not achieved. One of the original referrals had not yet been filed as a case and may be developed into another pilot project after August 15, 2001. That process might include designing an ADR process for a large number of similar cleanup cases likely to be filed in the district in the future. Another pilot may involve a collaborative approach to looking at the best use of resources for addressing the myriad filings of ESA petitions. A settlement judge was working on several suits arising from drought and curtailment of irrigation issues. Depending on the outcome, a longer range solution may be considered on a broader basis and with some ADR process assistance provided to the court. Another potential special project being considered as a result of discussions connected to the pilot project involves problems related to multiagency permitting. It may be possible to combine state and federal issues within a mediation on site-specific issues.

An Assessment Note

In one of the mediated cases within the pilot, the project coordinator engaged in significant preliminary work to assist parties with negotiations around entering into mediation. The parties eventually negotiated a premediation agreement relating to substantive issues and agreed to modify an existing injunction. They then negotiated their fee and mediation agreements and ultimately did resolve the case through mediation. This suggests that in some environmental cases, the premediation conflict assessment and subsequent convening work are necessary precursors to actually engaging in mediation. This pilot, however, was not designed to include this kind of conflict assessment for every case. If the attorneys reported that a case was not ready for mediation or that the parties did not want to mediate, the project coordinator either tried to explore the issues more thoroughly or suggested that she could talk further with the parties to help them evaluate their interests in resolving the case. However, in most cases the attorneys did not see the need for direct discussions with the parties.

Selecting Mediators

Sponsors defined criteria for pilot mediators relying on standards from the Society of Professionals in Dispute Resolution, the American Bar Association, the American Arbitration Association, and the U.S. District Court. The mediators had to be experienced in environmental conflicts and credible to the parties, the court, and the USIECR. The project coordinator (who agreed not to serve as a mediator in pilot cases) worked with the USIECR and the court to develop a panel of neutrals committed to working with the project. She used mediators already on the USIECR's national roster and added others suggested by the court, plus other well-known environmental mediators in Oregon. The panel of about 15 mediators includes lawyers and nonlawyers, men and women, and people with expertise in many different subject matter areas within environmental law. A preference for Oregonians was given to enhance relationships among the court, attorneys, parties, and local environmental mediators and to strengthen capacity in Oregon for environmental mediation.

For certain highly technical cases, sponsors hoped to encourage mediation teams using a mediator and scientist. No mediation team has been used to date. Although several judges were interested in serving as mediators in the pilot, sponsors decided to focus the pilot on court-annexed mediation by external mediators. In part, some sponsors sought to complement the settlement expertise already offered by judges in the district. They also wanted to concentrate on voluntary mediation rather than settlement conferences, in a district where some attorneys complained that some judges push settlements aggressively.

In June 2000, the sponsors conducted a workshop for the pilot's mediators. In the one-day session, sponsors prepared mediators generally for environmental cases that might be assigned during the pilot and facilitated an exchange of ideas and experience to hone the skills of the panelists. Panelists learned about environmental mediation in the Ninth Circuit, the pilot, and the evaluation component, and were introduced to some judges and court administrators active in ADR. During several portions of the workshop led by Peter Adler, the mediators worked with a hypothetical dispute between a cement company and a neighborhood association. The panelists shared their views and experience on issues such as process design, raw emotions, joint versus separate sessions, and judicial involvement. Mediators brought differing opinions and approaches to many topics, and the hypothetical situation generated useful discussion. Adler also drew on a variety of environmental cases to offer ideas for managing scientific and technical information. Mediators' evaluation of this workshop was positive, and several mediators expressed an interest in continuing such discussions in other workshops or lunch meetings as a form of advanced training.

The project coordinator worked with lawyers to select mediators for particular cases. She asked them what qualities or expertise they wanted in a mediator. She chose three to five mediators and checked on their time and interest. She sent the mediators' resumes to the lawyers and encouraged them to interview some mediators. Within the small group of cases that went to mediation, for distributive cases, lawyers usually sought mediators based on their reputation and legal expertise rather than their process skills. For the integrative cases, lawyers seemed to focus on process abilities and subject matter expertise. In both integrative cases, lawyers and one or more of the parties interviewed potential mediators by phone. No preliminary interviews took place in the distributive cases.

Work with Other Constituencies

As the pilot progressed, the project coordinator found a need to share information about the pilot and environmental conflict resolution with repeat players in the environmental cases filed before the court. For example, the U.S. attorney's office is involved in many pending cases, and the project coordinator also dealt repeatedly with several attorneys who regularly represented public interest groups and private litigants. She met with Oregon's U.S. attorney and subsequently conducted a pilot orientation session for a group of assistant U.S. attorneys who expressed interest in learning more about the pilot and the potential gains of mediation for environmental cases. Dealing with these types of barriers and developing such relationships appears to be important for serious exploration of mediation in environmental cases.

Project Evaluation

The pilot involved a small number of cases, so it has limited utility in terms of identifying patterns and comparisons among environmental mediations. These are unique cases; standardized measurements and random assignments are not possible. Evaluation involved gathering the subjective views of participants and their projections of the cost, time, and utility of alternative methods of resolving a particular dispute. It must be recognized that lawyers and parties often want to justify the alternative they chose, and their ability to predict the outcome and satisfaction with another alternative will be somewhat suspect. Moreover, better data are needed to draw comparisons between litigated and mediated cases and between ADR options such as judicial settlement conferences and external mediation. Potential differences in perceptions among participants may yet prove interesting, and the narratives of how cases were disposed may prove useful for courts, litigants, attorneys, and mediators interested in environmental mediation.

The pilot evaluation includes questionnaires sent to parties, attorneys, and mediators who participated in mediation of pilot cases. Follow-up interviews could develop this information further. The questionnaires and interviews cover both process and outcome evaluation. Participants are asked, for example, about items such as their satisfaction with mediation sessions, the mediator's approach, and their own conduct during mediation. They are surveyed on outcomes such as settlements or partial settlements, time to settlement, streamlining of factual or legal issues, the perceived effect of mediation on long-term relationships of parties, and the cost-effectiveness of mediation. Further evaluation could include longitudinal work on participants' satisfaction with the process and outcome over time, changing perceptions or conditions, and the durability of agreements reached.

Eventually, the assessments may provide diagnostic tools for parties, lawyers, and courts seeking to implement environmental ADR programs.

Lessons Learned during the Initial Phase of the Pilot

Not all the parties, lawyers, and mediators in mediated cases have been surveyed or have responded to their questionnaires. The sponsors plan to follow up in these cases, and surveys may be sent in cases that did not proceed to mediation through the pilot.

At this juncture, only tentative lessons can be drawn, but several themes are worth noting. First, significant effort is required to design and implement such a pilot, even in a supportive court. Docket assessments and administrative help are critical to focus judges and lawyers on early resolution. Second, the work during the initial phase of the pilot presents a catalogue of potential challenges to implementing ADR programs for court-annexed environmental cases. As more information is gleaned from the attorneys, mediators, and project coordinator, we may learn a good deal more. For example, surveying attorneys who did not counsel their clients to use the pilot mediation option may provide useful insights.

Resource-Intensive Work Is Required

Establishing this pilot mediation program for environmental cases required significant investments of time, effort, and resources. The District of Oregon, like most courts, did not have the resources, complete data on environmental cases, or sufficient personnel to initiate the pilot and conduct the case development portion. The funding of a part-time project coordinator was essential to get the pilot underway and to sustain it. In light of the congressional mandate to promote ADR when it saves time and cost, federal courts face difficult choices about how to implement ADR. Should courts invest in the infrastructure to support ADR programs because of their potential to reduce time and cost for particular cases in the future? Evaluating the court's costs is complicated because though the court will devote significant resources to establishing an ADR infrastructure, such support may result in benefits that are difficult to measure. For example, would mediation result in more durable agreements so that parties do not bring their conflicts back to court for several years? Would future conflicts never be filed in court because the parties improved their working relationships? The pilot may support the conclusion that, left alone, a large percentage of environmental cases, like other cases, appear to settle without additional promotion of ADR. Alternatively, it may support a finding that judicial or project coordinator queries about mediation encourage earlier settlements, whether through the pilot's options or outside the pilot.

The pilot's administrative costs may be high because this is a unique project, and one of the first focusing on federal environmental cases. The costs may be elevated due to the pilot's upfront efforts and its particular goals. For example, work was invested to foster the relationships among the project sponsors, develop or refine protocols, establish a panel of experienced environmental mediators, and initiate some case assessments, with priority given to integrative rather than distributive cases. The costs may be attributable in part to how the particular coordinator performed tasks. For example, helping the parties to negotiate the preagreement required some expense but may have proved worthwhile in that it allowed the parties to enter mediation comfortably and may have increased the likelihood that they would reach agreement ultimately.

Early Docket Assessment Is Important

As noted earlier, environmental cases make up a small portion of the Oregon docket. The goal of including fewer cases that were primarily distributive and more cases that contained integrative possibilities emerged as the pilot proceeded. Yet the coordinator learned that the cases filed in Oregon were mostly distributive. Of the first 10 she examined, 70% were distributive. By the time the coordinator learned about those cases and checked with the attorneys, almost all had settled.

The clerk of the court expressed interest in a screening process that assesses environmental cases sooner because the filings entail court costs beyond the time invested by judicial personnel. The developing field of environmental mediation practice reveals that assessment of the conflict situation is typically necessary to determine whether the conflict is likely to be amenable to mediation, to "get the parties to the table," to identify interests underlying legal positions, and to get a start on the commitments that would structure the process for success. Another court-annexed pilot might consider whether this type of early assessment is needed in pending environmental litigation or whether, once a suit is filed, the parties are already at the table with the structure and issues defined or hardened by the litigation itself.

Future projects might use more flexible case criteria or more intensive assessment work for a broad range of cases. Alternatively, a court might require case assessments in some complex environmental cases. To preserve the goal of offering opportunities for mediation that complement (rather than duplicate) judicial settlement conferences, future projects could involve more than one district to yield a much wider pool of integrative environmental cases. Additionally, some of the initial characterization of cases by lawyers involved may be too narrow. Perhaps some cases that appear to be primarily distributive involve parties who will deal with each other in the future, and they could use mediation to improve working relationships.

Comparisons can be attempted between the approaches of Oregon settlement judges and the panel mediators. Are some cases best handled by judges? When are external mediators valuable? Which types of environmental cases should be subject to a full case assessment? Who should perform assessments? Could some cases be best served by a combination of a settlement judge and an external mediator who can manage the ongoing issues in a multisession mediation that may take months or even years to complete?

Judges can play an important convening role in environmental mediation. One of the attorneys who found the pilot mediation process and outcome quite favorable noted that the parties participated in mediation through the pilot because the judge "ordered" them to do so through a letter suggesting the option. Another attorney used the pilot's mediation option in part because he or she did not want to appear reluctant to cooperate in ADR efforts, which many judges in Oregon value.

Rather than investing personnel time (from an external coordinator or a court administrator) in searching dockets, judges could create incentives for parties to move to mediation or at least go through a case assessment soon after a lawsuit is filed. At the Department of the Interior, for example, administrative law judges order parties to an assessment. The judge then receives a report about the appropriate dispute resolution method. The report does not reveal case facts, underlying interests, or which party declined to use ADR. Such mandated assessment might be necessary to trigger fuller case assessments by parties and attorneys early in litigation. In this pilot, the project coordinator dealt primarily with the attorneys and did not aggressively push to speak with the parties themselves because the project was not intended to do full assessments of each case. Although she tried to educate the attorneys about the advantage of mediation and about the possibility of different models that might fit their case, she did not try to coerce reluctant parties to participate in mediation because of the emphasis on voluntary external mediation, because the parties had to pay for the mediation, and because the mediation was court-annexed.

The Value of an ADR Administrator

Courts are likely to need at least a part-time administrator for such projects. As of early 2001, the coordinator had devoted more than 350 hours to pilot start-up operations; case assessment; formation of the mediator panel; and fostering discussion of dispute resolution options among judges, attorneys, parties, and mediators. As described above, the coordinator devoted extensive time to collecting information about particular cases and continuing conversations about appropriate resolution processes. In one case, she helped the parties negotiate a preagreement before they were ready to proceed with mediation. An administrator who understands litigation can work well with judges (e.g., inquiring about particular cases on their calendars). It is also important that the administrator be knowledgeable about various dispute resolution options. Hallmark's background in environmental cases was useful in dealing with the attorneys during this pilot.

Additionally, there is some value in using a person with sufficient autonomy from the judge adjudicating a case. Because the sponsors placed a high value on obtaining truly voluntary consent before cases entered mediation, the coordinator was not forceful about pushing mediation in the face of resistance. One party wanted a judge to tell the opposing party to go to mediation, in part because the judge had suggested mediation in a separate case involving the first party. In another case, one party felt obligated to explore mediation due to a judicial suggestion, but the project coordinator made clear that mediation was voluntary, and the party did not choose mediation. The coordinator thus served as a buffer when parties or their counsel felt pressure from judges. A court employee could serve the same function, but if the court places a high value on the voluntary nature of the ADR option, that should be made clear to all involved, and the standard for reviewing the employee should reflect that priority (i.e., valuing the voluntary nature more than disposition figures).

An ADR administrator can be in a difficult position because he or she may hear things that judges do not say directly to litigants and lawyers. For example, during one case, a judge gave the coordinator permission to tell a recalcitrant party that the judge was ready to rule against the party. The ruling would not have completely resolved the dispute but would only have instigated another round of appeals or further administrative activity. The coordinator chose not to reveal what the judge said to either party.

The coordinator also created a link between the court and conflict resolution practitioners, helping to expand courts' options in the private ADR sector. State and local rules, plus court rules, about confidentiality and mediator immunity should be canvassed as any new ADR program is implemented.

Challenges to Using Mediation in Environmental Cases

Although this pilot was limited in scope, time, and resources, it is a useful vehicle for reflecting on the challenges courts face in trying to promote ADR options for environmental and some other complex public policy cases.

Perceptions of ADR as Evaluative Settlement Conferences. The project coordinator reported some resistance to mediation in the pilot because of some judges' and lawyers' (and perhaps their clients') perceptions of court-connected ADR. Although the pilot sought to offer opportunities for mutual-gains bargaining and facilitative processes, some lawyers assumed that the pilot involved traditional settlement conferences or primarily evaluative mediation. This is understandable, because many lawyers are not well educated about the range of process options within mediation. Additionally, the differences between judicial settlement conferences and external mediation are not well defined or agreed upon. Indeed, some courts call settlement conferences mediation and some mediators engage in evaluative processes similar to settlement conferences. This perception is particularly understandable in a court where some judges promote settlement in an aggressive manner. Past exposure to mediators and settlement judges, combined with the lack of lawyer and client education about the diversity of mediation approaches, compounds this perception.

During the evaluation, lawyers and clients who did not enter the pilot could be queried about their perceptions and their ADR preferences. The data might reveal little demand for court-annexed mediation programs because settlement judges are available at no cost and can handle distributive cases competently. Alternatively, the data might support the need for education about environmental conflict resolution and possibilities beyond settlement conferences.

The Cost of External Mediation. For some parties, the cost of external mediation was cited as an impediment to attempting it. As noted, judicial settlement conferences are readily available in the District of Oregon at no extra fee. Panel mediators were willing to reduce their standard fees, at least for an initial period, but the cost barrier appeared significant in a few cases. In another pilot mediation, lawyers for both parties reported substantial cost savings because the case did not proceed to a complex trial, necessitating expert witnesses. One attorney estimated that mediation saved a client between $200,000 and $400,000. In another pilot case, attorneys for both parties estimated that the mediation was more expensive than litigation would have been. Nevertheless, the result achieved through mediation was more satisfactory to both. They reported that litigation would have produced a "bizarre" and "completely impracticable" procedural solution for this NEPA case, but through mediation the parties designed substantive relief.

Attorney's fees were raised as an issue in several questionnaires. Fees sometimes presented an obstacle to reaching settlement. Judges and attorneys expressed concerns about how lawyers would get paid when mediation was used to solve environmental conflicts. On the plaintiff's side, attorney fees stemming from statutory mandates need to be resolved as part of a settlement. On the defense side, concerns about billable hours were frequently mentioned. One judge reported that judges need to understand when a case is "ripe" for mediation or settlement talks, being cognizant of such practical issues as whether the lawyers have been able to generate enough fees from a given case before commencing settlement discussions. The project coordinator suspected that some lawyers were "hiding" behind cost justifications as a reason to decline the mediation option. Cost as a barrier could be examined for cases not entering mediation.

Lawyer Control of Communications with Parties and the Need for Case Assessment. The project coordinator offered to speak with parties about the possibilities mediation offered in some pilot cases, but lawyers refused, wanting to control the flow of information to their clients about dispute resolution options. If the attorneys said that the case was not ready for mediation or that the parties did not want to mediate, the project coordinator did not push mediation aggressively. The project coordinator suspects that full case assessments might have produced more cases amenable to mediation, but such assessments are not possible without interviewing parties. Other pilots might consider a judicial order to assessment, with parties included.

Lawyer Perceptions of Cases as Inappropriate for Mediation. In some cases, lawyers concluded that matters should not be resolved by mediation. Sometimes this was simply because they were already involved in negotiations, or negotiations ensued after an inquiry from a judge or the coordinator. In other instances, the lawyers reported that the parties sought adjudication to establish a legal precedent. Sponsors agreed that such cases were not appropriate for the pilot.

At other times, lawyers reported that their convictions were firm, their client was right, and the case contained no issue on which potential for compromise existed. Some lawyers wanted to wait for a judge to tell them that there was a reason to consider settlement. The project coordinator surmised that some of these attorneys were focused on legal issues and excluded other interests that their clients might have had. For example, lawyers sometimes focused on narrow legal issues before court, such as, "Is there a duty for the agency to consult on an endangered species matter?" Lawyers did not seem willing to explore other interests, such as, "What would you really want to see happen for your client in terms of species management?" Judges told of environmental cases in which a party was upset that an agency did not take their views into account. Mediation might afford an airing of their views, as well as decisions about future communication processes. Lawyers often failed to see the role that mediation could play in advancing such interests.

Distrust among Parties. Several Oregon judges noted that communication problems and distrust between governmental organizations and other interests sometimes spur environmental disputes. The project coordinator confirmed this observation, perceiving that serious distrust among some parties prevented them from exploring mediation in a few cases. The parties did not believe that the relationship could improve. Because these cases involved long-term relationships and repeated contact between parties, they might be ideal for mediation. However, past experience among the parties made them hesitant to even enter a dialogue with a mediator or judge about a less adversarial process than litigation. One case in which attorneys exhibited significant distrust did proceed to mediation, but no agreement was reached. The mediator characterized both parties as "enamored of their respective positions" and found at least one party unwilling to put sufficient time into the conflict resolution process.

Delay Is Sometimes a Gain. In environmental cases, delay is helpful to some parties. For example, litigation may be filed with the goal of stalling a development project or government action. If so, parties are reluctant to enter a more expedited process like mediation. The slower pace of litigation and the fact that the parties have less control over that process can be beneficial. The project coordinator concluded that appreciation for delay was one factor in several decisions not to enter mediation during the pilot.

Sometimes a party or lawyer wants to wait for discovery or the judge's ruling on a dispositive motion or an agency decision such as an ESA listing determination. When no firm deadline, imminent decision, or forcing mechanism is present, reluctant parties may simply await more pressure before exploring settlement options. In one potential pilot case, the parties chose to await a summary judgement ruling, but that ruling did not offer complete victory to either side. In another potential pilot case, some parties thought that national elections could affect the lawsuit. Nevertheless, prodded by earlier judicial settlement conferences, the parties reached a tentative agreement conditioned on congressional funding. In another potential pilot case, parties decided to await a judicial determination of whether they needed an environmental impact statement (EIS). Although issues for negotiation often remain when an EIS is ordered, in this case it was thought that ordering an EIS would be so cost-prohibitive that the project would be stopped, satisfying one party's goal.

Future Development of the Pilot and Spin-offs

In addition to the avenues for exploration noted above, other possible directions for court-related environmental conflict resolution projects could include the following:

— Offer training or education about ADR options for judges, mediators, lawyers, and repeat players in environmental litigation.
— Provide fuller docket assessment, including an evaluation of emerging patterns of environmental disputes; in Oregon, for example, assessment of one potential pilot case led to interest in exploring a process for resolution of a group of related future filings.
— Study why cases did not choose to pursue mediation through the pilot. Gather more perceptions on which types of environmental cases are well suited to mediation and what barriers to mediation exist.
— Study whether court-related inquiry (from the judges or the coordinator) spurred private settlement efforts.
— Explore the differences between ADR offered by judges and external mediators on such grounds as persona (including the voluntary nature), approach, and cost. Can judges not devote as much time? Would judicial encouragement prompt the same outcome?
— Study how court and external mediation processes could be effectively combined in environmental cases. Oregon judges expressed a willingness to back up external mediations by coming in to "close the deal" if needed. They offered to bring persons with settlement authority to the table, to hold mock trials if helpful, to hold a settlement conference on a narrow legal issue as a supplement to mediation, or to explore appointing a neutral scientific expert. Such teamwork may offer wonderful means for individual tailoring of dispute resolution options.
— Compile some suggested settlement tools and strategies from particular case studies after further evaluation of external mediations under the pilot and judicial settlements of environmental cases in the Oregon federal court. Thus far, judges and mediators have provided information on using neutral scientists agreed on by the parties, purchasing an insurance policy to cover future contingencies, and setting up a trust for plaintiffs to cover potential long-term damages or contingencies.

— Explore the value of partial settlements of complex cases.
— Provide training on environmental conflict resolution options to court personnel (including judges, administrators, and law clerks), industry groups, public interest groups, other repeat players, and lawyers in U.S. attorney's offices and the U.S. Department of Justice.

Conclusions

The Oregon pilot gives us a glimpse of how challenging it is, even for a receptive court system, to provide public policy mediation for environmental cases. This type of mediation is not amenable to one-day settlement conferences or a limited amount of mediator time at a free or discounted rate such as the common, four-hour pro bono panels used by many courts. To provide an option for public policy mediation within courts is expensive because it requires administrative oversight as well as education and training of court personnel and mediators. It may be worthwhile to do within a large court system (e.g., the federal courts as a whole or within a large circuit) or agency because of the important public issues raised by environmental disputes and the potential of mediation to resolve those disputes with creative, durable solutions. It appears difficult, however, for a smaller court system to create and sustain such a program on its own.

Because many of the filed litigation disputes centered on allocation of costs for pollution cleanup (i.e., primarily distributive issues), the litigation queue might be less costly and more efficient than an early mediation program for promoting settlement without trial. These distributive cases are much more likely to move into settlement or mediation on their own and do not usually require extensive case assessment, convening, and process design, as do many multiparty public policy cases with integrative issues. The Oregon pilot may reduce some of the start-up costs for other courts and agencies wishing to experiment with public policy mediation programs by reviewing some of the issues that need to be considered, educating decisionmakers about the resources required, or simply by providing a sample format and documents to get another project underway.

Perhaps the market for public policy mediation services will grow, lessening the need for court-annexed programs in environmental and other public policy cases. Currently, however, many barriers seem to make the market imperfect. One of the most glaring barriers is the significant gulf remaining between the knowledge of judges and lawyers about the type of mediation available in these kinds of conflicts and the work of environmental conflict resolution experts. One solution is to provide more education and training for law students, lawyers, repeat parties, and court personnel about the growing field of public policy mediation. The idea of a multifaceted dispute resolution center has grown in the past 25 years into reality for many courts. In another 25 years, perhaps a number of courts will supplement the current ADR options with an option of early conflict assessment in public policy disputes so that parties can determine, with their lawyers, whether it is worthwhile exploring a mediated solution.

Note

I am grateful to Anessa Hart, Russa Kittredge, Kevin Mekler, and Sara Pirk for their excellent research assistance, to the judges and sponsors of this pilot for their cooperation, and to Gary and Anne Marie Galton for their generous financial support for this research. Thanks to Peter Adler, Elaine Hallmark, Kirk Emerson, and Dorothy Nelson for their comments on an earlier version of this chapter.

13

Dispute Resolution at the U.S. Environmental Protection Agency

ROSEMARY O'LEARY AND SUSAN SUMMERS RAINES

In 2000, the U.S. Environmental Protection Agency (EPA) announced plans to increase the use of alternative dispute resolution (ADR) techniques and practices across all agency programs.[1] ADR in this context means the variety of approaches that allow parties to meet face to face to reach a mutually acceptable resolution of the issues in a dispute or potentially controversial situation.[2] ADR is often viewed as intervention between conflicting parties or viewpoints to promote reconciliation, settlement, compromise, or understanding.[3] This includes mere assistance from a neutral third party to the negotiation process.[4] Such assistance can be directed toward settling disputes arising out of past events, or it can be directed toward establishing rules to govern future conduct.[5]

For the purposes of this research, we focus on ADR as a negotiation tool in which third-party neutral facilitators and mediators (herein referred to as "neutrals") are called upon to aid parties' attempts at finding resolutions to disputes related to enforcement activities at EPA. EPA's efforts in this area began in earnest in 1981. Box 13-1 provides a historical overview of ADR efforts at EPA.

This chapter is an analysis of the strengths and weaknesses of the program, as well as 10 lessons learned from EPA's experiences that can be used to improve the ADR programs at EPA and other federal and state agencies. The data for this effort were collected from archival records, government statistics, and interviews with four key stakeholder groups: EPA attorneys, potentially responsible parties (PRPs—defendants in EPA enforcement actions),[6] EPA's ADR specialists, and professional third-party neutrals. We hope that

---

Box 13-1. Historical Overview of Alternative Dispute Resolution at the U.S. Environmental Protection Agency

Date   *Action*
1981   EPA issued a policy on involving the public in agency decisions.
1983   EPA established the Regulatory Negotiation Project (RNP).
1985   The EPA Office of Enforcement piloted the use of ADR in enforcement activities.
1987   EPA issued guidance on the use of ADR in enforcement cases, establishing the review of all enforcement actions for the potential use of ADR processes.
       EPA expanded the RNP.
1988   EPA issued a contract for up to $1 million in third-party neutral (facilitator, convener, mediator) services.
1996   The EPA Office of Solid Waste and Emergency Response began to use ADR to resolve hazardous waste disputes.
1996   The EPA Office of Civil Rights piloted the use of ADR to resolve equal employment opportunity complaints.
1998   EPA appointed a dispute resolution specialist.
1999   EPA established a Conflict Prevention and Resolution Center in the agency's ADR Law Office.
1999   EPA issued a contract for up to $41 million in third-party neutral services.
2000   EPA issued an interim statement of policy announcing the drafting of a final ADR policy.

---

these findings will be of assistance to EPA as it plans to expand its ADR efforts and that it will also provide useful lessons for other agencies and organizations considering similar programs.

The Literature

The essence of ADR is face-to-face meetings of stakeholders to reach a consensus decision that best satisfies their interests. Based on the extant literature, O'Leary and others have identified five principal elements of ADR in the environment and public policy arena: (a) the parties agree to participate in the process, (b) the parties or their representatives directly participate, (c) a third-party neutral helps the parties reach agreement but has no authority to impose a solution, (d) the parties must be able to agree on the outcome, and (e) any participant may withdraw and seek a resolution elsewhere.[7]

The literature is ripe with normative pleas to increase the role of the lay public and interested stakeholders in the resolution of environmental disputes. One author, for example, argues that such participation in the resolution of water conflicts in the western United States is a fundamental tenet of our democratic government.[8] Other literature focuses on problems that might be more amicably and more efficiently resolved through ADR. For instance, Whitman argues that the use of ADR techniques could greatly improve the management of hazardous waste cleanups.[9] A study of intergovernmental conflict stemming from state law regulating solid waste in North Carolina concludes that state and local governments may be able to positively resolve such disputes by adopting a problem-solving stance and searching for win–win results.[10] EPA's Office of Site Remediation states in one of its publications that there are several benefits of ADR in its environmental enforcement actions: lower transaction costs, a focus on problem solving (as opposed to positioning), the generation of settlement options that are more likely to be tailored to stakeholders' needs, and the saving of time.[11]

Describing ADR as a more effective problem-solving or policymaking method than alternatives such as litigation or traditional rulemaking procedures, under certain circumstances, is a common theme. Examples of analyses that do not include enforcement ADR at EPA are DeHaven-Smith and Wodraska,[12] who examined consensus building in integrated resources planning within the Metropolitan Water District of Southern California; Kerwin and Langbein,[13] who analyzed negotiated rulemaking at EPA; Fiorino,[14] who looked at regulatory negotiation as a policy process at EPA; Blackburn,[15] who examined environmental mediation as an alternative to litigation; and Perritt,[16] who analyzed the use of ADR techniques in negotiated rulemaking. Public administration scholars have also examined generic conflict resolution techniques.[17] Thus, whereas the literature has generally advocated ADR as a public management response to the problem of environmental conflict, broad studies assessing the lessons from these programs are scarce.

The literature on EPA's use of ADR in enforcement actions is sparse. Peterson[18] tracked and evaluated the early use of ADR in EPA's Region 5 office and identified eight factors, listed in order of importance, used to explore the mediation potential of a Superfund case:[19] EPA's willingness to litigate, identification of issues suited to mediation, timing considerations, the nature of the parties in the dispute, the number of parties, participation by nonparties, the amount in dispute, and the ability of the parties to share mediation costs.

In 1990, Abbott[20] issued a somewhat cynical prognosis for EPA's ADR program as applied to Superfund cases. Abbott documented several cases where EPA successfully used ADR during the enforcement process but found an overall reluctance on the part of EPA officials to use the ADR process, as well as a fundamental distrust of settlement through ADR by PRPs. Public issues, she concluded in part, were more likely to be resolved among the private parties themselves, without EPA.

Focusing on steering committees formed by PRPs at EPA enforcement sites, Charla and Parry summed up the pros and cons of using ADR in Superfund cases in the early 1990s as follows: "When properly utilized, a number of ADR techniques provide good results at sites, including equitable allocations of liability, competent development of facts, facilitation and mediation services, and savings of time and transaction costs. Negatives can be high expenses, protracted delays, work product of questionable quality, and failure to accomplish outcomes intended ...".[21]

By 1995, Hyatt reported that ADR had become "virtually the norm [among private-party PRPs] at [EPA] multiparty Superfund sites for resolving contribution claims."[22] Consistent with our findings, before they negotiate with EPA, PRPs frequently use the services of a mediator or neutral cost allocator to help determine the percentage of the total settlement that each party should pay as part of any eventual settlement.

Our research provides a valuable opportunity to test some assertions from the existing ADR literature, while also providing helpful practical lessons to EPA and to other public ADR programs. The first and most common perception drawn from the literature is that ADR saves time and money and is preferable to litigation. To test this assertion we asked parties why they chose ADR, how satisfied they were with ADR compared to litigation or settlement through other means, and whether they would use ADR again.

Second, we set out to test the proposition that ADR provides an opportunity for the disputing parties to improve their working relationships.[23] This is probably one of the most important potential benefits for regulatory agencies. Better working relationships may translate into decreased litigation costs, improved compliance rates, a problem-solving (not a blame-casting) mindset among all parties, and an improved public image for the regulated entity. To test the claim that ADR can improve working relationships between parties, we asked respondents, through both scaled and open-ended questions, to discuss the effect that ADR had, if any, on their relationships.

Third, we set out to test the findings of Abbott, mentioned above. If members of the regulated community (e.g., PRPs) harbor doubts and skepticism about ADR, it is unlikely that such programs will be used often and with high rates of settlement. It is not enough to build programs. If private parties believe that the programs have built-in biases, they will avoid them or seek to undermine them. Interviews with private parties and their attorneys led us to question whether the findings of Abbott still held true almost 10 years later.

Research Design

To obtain an up-to-date and comprehensive picture of EPA's ADR activities as applied to enforcement, we conducted a four-part evaluation of these activities, gathering data between 1998 and 2000. Funded by the William and Flora Hewlett Foundation, this effort used in-depth telephone interviews, government statistics,[24] and archival records.[25] The authors, aided by master's degree and law students from Indiana University,[26] conducted the interviews. Each interview lasted between 25 minutes and two hours, with the average interview taking approximately 45 minutes. Interviewers recorded responses through handwritten notes, including both open-ended responses and responses based on 5-point Likert scales. Immediately after each interview, the responses were entered into a Microsoft Access database.[27]

The four groups interviewed were the following:[28]

— EPA's regional ADR specialists (18 out of 20, or 90% of the entire population)[29];
— PRPs to primarily Superfund cases, or their attorneys (a stratified random sample of 25, representing small and large firms from across the United States)[30];
— third-party neutrals used to convene, facilitate, or mediate the cases (22, for a response rate of 69%)[31]; and
— agency enforcement attorneys who had participated in agency enforcement ADR processes (61 out of a population of 78, for a response rate of 78%).[32]

The overall goals of this project included the following:

— evaluating the use of ADR in enforcement cases at EPA, particularly in Superfund cases;
— examining the sources of both obstacles and assistance to ADR efforts at EPA;
— suggesting ways in which EPA might improve its ADR programs; and
— drawing lessons from EPA's experiences that may be helpful to other agencies or organizations.

Interviewers spoke with EPA attorneys from 9 of EPA's 10 regions, not including Region 9 (San Francisco), which has been relatively inactive with respect to the use of ADR. Similarly, interviewers spoke with PRPs and their attorneys from around the country and all of the EPA regions, with the exception of Region 8 (Denver) and Region 9. Again, Regions 8 and 9 did not turn up in the relatively small sample of PRPs due to the smaller number of ADR cases from these regions.[33]

Research Findings

Levels of Satisfaction among EPA Attorneys and Potentially Responsible Parties

There is a high level of satisfaction among those who have participated in EPA ADR processes in enforcement actions. Tables 13-1, 13-2, and 13-3 compare the satisfaction of the EPA enforcement attorneys interviewed with the PRPs interviewed in three general areas: satisfaction with the ADR experience, satisfaction with the mediator, and satisfaction with the outcome. Although the averaged responses to all the questions are in the "very satisfied" to "somewhat satisfied" range, a few interesting similarities and differences emerge.

Overall, both EPA and private-party attorneys were fairly satisfied with the ADR process used by EPA. Both groups felt satisfied with their abilities to participate in the process and with the fairness of the process. This is an important finding because previous research showed that concerns about fairness are likely to influence the willingness of parties to use ADR programs.[34] Our findings from both the closed-ended and open-ended comments of PRP attorneys contrast with those of Abbott.[35] Rather than being reticent to participate, the PRP attorneys we spoke with repeatedly voiced enthusiasm for the use of mediation and other forms of ADR. In fact, they recurrently expressed their desire to see ADR used more frequently by EPA, both inside and outside of the enforcement arena.

It is somewhat surprising that the two groups do not exhibit larger differences in regard to their satisfaction with process fairness. Based on other

TABLE 13-1. Satisfaction of EPA Enforcement Attorneys and Potentially Responsible Parties: Part I—The ADR Process

Satisfaction with the ADR experience	EPA attorneys	PRPs
Information received about the process	1.66	2.05
Ability to present side of the dispute	1.43	1.23
Amount participated in the ADR process	1.37	1.39
Control over the ADR process	2.31	2.17
Examination of technical and scientific issues	2.17	1.75
Fairness of the ADR process	1.48	1.23

Note: 1 = very satisfied, 2 = somewhat satisfied, 3 = neutral, 4 = somewhat dissatisfied, 5 = very dissatisfied.

comments PRPs made in interviews, we conclude that they are comparing the fairness of the ADR process to that of litigation under the Superfund law. Under Superfund law, a person or company that contributed even a small percentage of the pollutants to a site can be held liable for all remediation costs. This law gives EPA a "big stick" and is often seen as unfair by PRPs. PRPs and their attorneys view mediation and other forms of ADR as fairer than litigation because it often increases the range of settlement options and results in a higher level of understanding regarding the technical and legal constraints faced by all stakeholders.

Though both groups are highly satisfied with most aspects of the ADR process, their average satisfaction levels are slightly lower in regard to the amount of control they exert on the ADR process (2.31 out of 5 for EPA attorneys and 2.17 for PRPs). Previous studies have shown that attorneys are often reluctant to try ADR, fearing that it may reduce the amount of control they have over their cases. Concerns over control, however, are common in

TABLE 13-2. Satisfaction of EPA Enforcement Attorneys and Potentially Responsible Parties: Part II—The Mediator

Satisfaction with the mediator	EPA attorneys	PRPs
Mediator's preparedness	1.46	1.60
Respect the mediator showed toward attorney	1.27	1.36
Mediator's knowledge of the dispute's substance	1.75	1.56
Mediator's impartiality	1.43	1.24
Mediator's skill at opening up new options	1.91	1.68
Mediator's skill at aiding parties to find a resolution	1.77	1.58
Mediator's fairness	1.48	1.32
Overall satisfaction with the mediator	1.57	1.46

Note: 1 = very satisfied, 2 = somewhat satisfied, 3 = neutral, 4 = somewhat dissatisfied, 5 = very dissatisfied.

TABLE 13-3. Satisfaction of EPA Enforcement Attorneys and Potentially Responsible Parties: Part III—ADR's Outcome

Satisfaction with the outcome	EPA attorneys	PRPs
Speed of resolution	2.36	1.92
Outcome compared to previous expectations	1.82	1.95
Control over outcome	2.15	2.30
Impact on long-term relationships between parties	2.13	1.44
Resolution's durability	2.20	2.00
Overall outcome	1.77	2.04

Note: 1 = very satisfied, 2 = somewhat satisfied, 3 = neutral, 4 = somewhat dissatisfied, 5 = very dissatisfied.

both ADR and litigation processes. One attorney, for example, remarked that "throwing a case before a judge" represents the ultimate loss of control, whereas mediation actually increases the amount of control attorneys and their clients have because resolution only occurs through consensus.

Although both groups agree that ADR increases the opportunity to discuss scientific and technical challenges that are often not adequately addressed in litigation, EPA attorneys were slightly less satisfied with this aspect than were PRPs (2.17 and 1.75). Because environmental disputes often involve highly technical scientific issues, the opportunity to examine and debate the scientific evidence in greater detail may be one of ADR's biggest advantages.[36]

Additionally, potentially responsible parties are slightly less satisfied (2.05) than EPA enforcement attorneys (1.66) with the amount of information about ADR that they were given before mediation. One possible explanation is the "home court advantage" of EPA attorneys when participating in EPA-sponsored enforcement ADR activities. In open-ended interviews, several EPA attorneys commented on how helpful the high-quality, dedicated ADR staff at headquarters and in the regional offices had been. Though it is not EPA's duty to educate private attorneys in the use of ADR, in the future it may be advisable for EPA to provide the relevant PRPs and their attorneys with more information about how ADR works at EPA. It is likely that this advice extends to other agencies involved in ADR as well.

Last, PRPs were more satisfied with the opportunity to present their client's side of the dispute (1.23) than were EPA attorneys (1.43). In open-ended questions, the majority of PRPs interviewed explained that the ability to tell their side of the story was one of the primary advantages of ADR in enforcement actions. If ADR encourages the parties to share their views and to adopt problem-solving attitudes, it is likely that it will improve the working relationships between them. Although this finding is important to numerous other agencies that must work with members of the regulated community on a day-to-day basis, it is especially important when Superfund sites are concerned because remediation of these sites often takes years to complete and EPA often works with the same attorneys and companies repeatedly on a number of environmental matters.

EPA attorneys were slightly more critical of mediators (1.57) than were PRPs (1.46). It is important to note that these scores reflect the mean average response, masking enormous variation, to be discussed later. Whereas both groups were generally satisfied with the quality of the mediators, the most common concerns expressed by EPA enforcement attorneys dealt with quality control issues, noting that some mediators were young, inexperienced, and uninformed about the law and science involved. It must be remembered, however, that all of these scores still reflect average satisfaction levels that fall between "very" and "somewhat" satisfied. The fact that both groups' scores are remarkably similar is good news; it means that the "neutrals" are generally viewed as neutral. If one group of attorneys expressed significantly more or less satisfaction with the mediators it could be a warning that the mediators or other neutrals are not adequately maintaining their neutrality.

The largest differences in satisfaction occur in reference to the outcome. Potentially responsible parties are less satisfied with the outcome overall (2.04) when compared with EPA enforcement attorneys (1.77). This is not surprising, given the fact that the outcome for PRPs usually means paying thousands, if not millions, of dollars. In response to an open-ended question, several EPA attorneys commented that they appreciated the flexibility made possible by EPA enforcement ADR processes in crafting solutions. On the other hand, PRPs are more content with the speed with which the dispute was resolved (1.92) than are EPA enforcement attorneys (2.36). This makes sense because PRPs and their attorneys may feel more pressure to reduce "billable hours" and settle the dispute as quickly as possible.

One of the most striking findings of this study concerns the usefulness of ADR in improving the working relationships between the parties. The PRPs are significantly more satisfied than EPA attorneys with the positive impact of the ADR process on the long-term relationship of the parties (1.44, compared to 2.13 for EPA attorneys). Because some members of the regulated community are likely to view EPA regulation as bothersome at best and as harassment at worst, it follows that the members of the regulated community and their representatives may stand the most to gain from improved relationships. Because both sides often work together for years, even decades, on some Superfund sites, the potential benefits of improved and more cooperative working relationships should not be underestimated. This finding is also quite important for other public agencies hoping to improve their working relationships with regulated community members while improving their public images.

Third-party neutrals and the EPA's regional ADR specialists correctly predicted that both groups of attorneys would generally be satisfied with the process and outcome of enforcement ADR at EPA and with the opportunities that ADR provides to improve the parties' understanding of each other and of the scientific and technical issues at stake (Table 13-4). These findings are important reasons for increased use of ADR because they support the arguments in the literature highlighted earlier, that agreements reached through ADR tend to be more flexible and are based on a deeper knowledge of the parties and the environmental problem than is often the case in litigation.

In sum, all four groups of participants interviewed were positive about EPA's use of alternative dispute resolution and were generally satisfied with the use of ADR in enforcement actions. Many believe, however, that EPA's ADR programs could get even better.

TABLE 13-4. Responses from Third-Party Neutrals and EPA's Regional ADR Specialists (percentage)

Responses	Neutrals strongly agree or agree	Specialists strongly agree or agree
Participants are satisfied with the ADR process	78	71
Participants are satisfied with the outcome of ADR	70	71
The ADR process opens a wider range of options than litigation	96	88
The ADR process yields a better understanding of parties' interests	87	88
ADR at the EPA allows for a deeper discussion of scientific and technical issues	83	—[a]

[a]This question was not asked of the ADR specialists.

Ten Lessons Learned

From the answers to closed-ended as well as open-ended questions, we have gleaned 10 lessons that should prove helpful to EPA as it plans to expand its ADR programs and services. At the same time, the lessons learned are applicable to other governmental organizations using, or contemplating using, ADR. Taken as a whole, they provide insights not found in the current literature.

Lesson 1: Pay attention to the concerns and comments of third-party neutral mediators and facilitators. Some of the third-party neutrals we spoke with expressed strong support for the ADR processes used but frustration with EPA itself. This frustration is most evident in their answers to open-ended questions. A few repeated themes emerged from the comments received.

First, within EPA there are deeply entrenched cultural norms for managing disputes with members of the regulated community. (See the quotes that follow in Box 13-2.) Some of these norms and habits may inhibit the spirit of cooperative problem solving required for mediation and other forms of ADR to be successful.

Second, EPA often signals its unwillingness to fully participate in ADR processes by sending its representatives to the bargaining table without the settlement authority necessary to make ADR resolutions possible. Both the neutrals and the PRPs repeatedly mentioned problems with authority to settle issues.

A third point, one that is also related to the organizational culture of EPA, is a lack of support for ADR from middle-level managers at EPA. Thirty-five percent of the neutral professionals "disagreed" or "strongly disagreed" that

Box 13-2. Insights from Neutral Professionals

Observations about the Conflict Culture

— "EPA often talks the talk, but doesn't walk the walk. EPA has a double standard. They expect understanding from other parties but fail to view themselves as a party required to be understanding of others."

— "I can't tell you how many times I've heard from EPA managers that ADR is great and creative and wonderful but not for this case."

— "EPA sees itself as the most important party to the negotiation."

— "The EPA culture does not embrace ADR."

— "There is a lack of shared values within the EPA about the worth of ADR."

— "The enforcement people at the EPA think ADR is unnecessary."

— "The ADR process does not have support from the powers that be."

Observations about Settlement Authority Issues

— "The key EPA decisionmakers are rarely at the table."

— "When the EPA would send key decisionmakers, they had no motivation to settle."

— "No EPA managers and supervisors were present, so I had to call EPA to check on whether the agreements we spent hours to craft would work. The answer was no."

— "Cases with both EPA and the Department of Justice [DOJ] are nightmares. The DOJ is totally inflexible, right up to the form upon which agreements may be written. I've spent hours trying to get key DOJ decisionmakers on the phone to sign off on an agreement."

Lack of Support from Middle Managers

— "Top management at the EPA supports ADR, while middle management is a mixed bag."

— "There may be a lack of incentives for EPA managers to use ADR."

— "There are major problems getting EPA management to be responsive."

— "The EPA has a strong tradition of being cautious, particularly in policy development. Issues must go up and down a chain of command, which makes the process lengthy."

Other Important Observations

— "The EPA would often lose interest in a case once it started."

— "EPA staff commit to mediation before understanding what it means."

— "The EPA doesn't know how to handle conflicts when the federal government is a PRP."

— "EPA does not evaluate whether ADR is appropriate."

— "The EPA tries to dictate who comes to the table."

ADR processes have the support of the majority of EPA managers. Only 26% of the same group "agreed" or "strongly agreed" with the same statement. This is particularly important because the middle-level managers, those who work with case attorneys on a day-to-day basis, most directly affect the extent to which ADR programs are implemented. In the open-ended questions, a clear feeling of concern emerges.

Box 13-2 overviews some of the more frequent comments received from neutral professionals in response to open-ended questions concerning their views of the EPA ADR program. Yet even with these concerns in mind, 74% "disagreed" or "strongly disagreed" with the statement, "The ADR processes utilized by the EPA are having little impact in settling environmental disputes."

Lesson 2: There is a need for consistent quality among mediators. The previous discussion of mediator quality revealed overall high levels of satisfaction with the mediators, but this level of satisfaction was based on a mean average of the responses, thereby masking some troubling variation in mediator quality. PRPs commonly complained that some mediators were not adequately knowledgeable about the substance of their disputes. Several pointed to young and inexperienced mediators who could not handle aggressive attorneys with strong personalities. Others commented that EPA should invest in the best mediators, even if the cost is somewhat higher. A majority (65%) of the PRPs said that their willingness to participate in future ADR efforts hinged on the choice of the neutral.

Similarly, EPA enforcement attorneys uniformly said that they want mediators who are strong, prepared, and informed, with consistent skill levels. The future success of ADR in enforcement cases at EPA partially hinges on the ability to find consistently high-quality mediators. The vast majority of the people we spoke with stated that the mediators or other neutrals were not evaluated at the end of the mediations. Allowing participants to evaluate the mediators' strengths and weaknesses would allow the mediators to improve their skills, while also providing a source of institutional memory to avoid rehiring unsatisfactory mediators.

The majority of attorneys in our study stated that they would use ADR again if it was appropriate for a given case. Most stated that they believed that ADR saved time and money, with some respondents using the phrase "ADR reduces transaction costs." However, this willingness to use ADR was also tempered by concerns over mediator quality and the ability of the parties to choose their own mediators. The act of working together to successfully choose a mediator for their case can provide both the sense of control and quality assurance necessary for parties to feel confident in the use of ADR. This is an important lesson for EPA, as well as for other agencies and organizations using ADR.

Lesson 3: Stronger educational efforts are needed within EPA to educate managers about the basics of ADR. As mentioned earlier, mediators reported strong support from top EPA management but far from adequate support from EPA middle management. Lack of uniform support from managers results in great disparities in ADR usage among the 10 EPA regions and lends weight to the argument that guidelines should be developed by which all cases would be evaluated to determine if they are suitable for ADR. Guidelines may be especially important for those regions where ADR is rarely used.

Based on interview responses, education across all EPA boundaries is needed. The majority of those interviewed expressed a need for and interest in further ADR training. To their credit, almost all of the EPA attorneys we talked with had received some ADR training. Only 14% stated that they had not received any ADR training. In contrast, 67% of PRPs stated that they had not been trained in the use of ADR. Because the respondents in this study had taken part in one or more ADR processes, this group was more likely to have been trained than were those attorneys who had never tried ADR.

For ADR to be a success at EPA, stronger educational efforts are needed, especially those focusing on attorneys who have never used ADR and on middle-level EPA managers. The mid-level managers at EPA have the authority to approve or deny the use of ADR, to suggest its use to other EPA attorneys, and to approve or reject settlement plans crafted through ADR. EPA is not responsible for training PRPs or their attorneys, but they should nonetheless be aware of the lower levels of training of most members in this group. To encourage the full and successful participation of PRPs and their attorneys, EPA might consider increasing its efforts to inform and educate PRPs when ADR is proposed.

Because variation among EPA regions is great, EPA should consider studying those regions in which ADR is having the most success to better understand what other regions can do to improve their use of ADR. Success stories need to be advertised. According to those we interviewed, the most successful use of ADR in EPA regional offices often stems from the commitment of certain enthusiastic and skillful key staff persons, rather than an overall institutionalized evaluation of when and where to use ADR. This point is supported by the 1997–1998 ADR Status Report from EPA (published in 1999), which shows that about one-third of the EPA regions are responsible for the vast majority of cases in which ADR is used, while another third rarely use it at all.[37] Whereas many cases settle without the assistance of outside neutrals, the regional variation clearly indicates room for increased use of ADR in some, if not all, regions.

There are, however, a number of barriers to increased use of ADR. For example, some lawyers asked, "If I can win, why mediate?" This is a good point because EPA "wins" most of the cases it takes all the way through litigation. However, the time and resources necessary to resolve cases in this way, and the damage to working relationships, as well as to EPA's public image due to the adversarial nature of this process, may make the option of ADR more attractive in many cases.

Some attorneys reported a perception that suggesting the use of ADR signals that one's case is weak. Still others reported fearing a loss of control over their case once in the ADR process. Some attorneys are also concerned that asking for a mediator will be interpreted as a sign that they need help as a lawyer or as a negotiator. Much of this reluctance stems from traditional law school education, in which attorneys are taught that to represent their clients' interests zealously, they must act in an adversarial fashion. Mediation and other forms of ADR call for cooperative attempts at problem solving that may be new and unfamiliar to attorneys trained in traditional adversarial methods, despite their knowledge of negotiation. Here again, training in interest-based negotiation may prove helpful in widening the range of problem-solving behaviors used by attorneys in negotiations.[38]

Lesson 4: A roster of easily accessible mediators, not paid for exclusively by EPA, should be used. In 1988, EPA issued a contract authorizing up to $1 million in neutral services over three years. Another contract, issued in 1999, has a ceiling of more than $41 million. Whereas the intent of these EPA contracts is to provide dispute resolution services from outside the agency, the perception of the intent differs. First, many of the EPA enforcement attorneys and PRPs interviewed thought that the mediator in their cases was selected by EPA, rather than by a neutral entity (e.g., 30% of EPA attorneys stated that EPA headquarters chose the mediator in their cases).

Second, 58% of the PRPs stated that the mediator's neutrality is best ensured when all parties, including EPA, share the costs of the mediator. Typically, these respondents stated that sharing the costs of mediation shows a commitment to the process and constitutes a sign of good faith. Twenty-five percent of PRPs, generally those from smaller companies, preferred that EPA pay all of the costs of mediation. Another 21% felt that it does not matter who pays for the mediation because the cost of mediation is generally small in comparison to overall litigation and settlement costs.

In an effort to provide the services of a neutral party in cases where funds are unavailable, EPA has occasionally used "in-house" neutrals. These in-house neutrals are EPA staff, trained in dispute resolution, with no direct connection to the particular case. Eighty-three percent of the PRPs stated that in-house neutrals would not be acceptable to them. Concerns included skepticism about EPA staff's "credibility," "neutrality," and "confidentiality." One respondent stated that an in-house neutral would be "laughed out of the room." Still another noted that EPA staff are already overwhelmed and would not be able to devote the time necessary to the mediation. In defense of in-house neutrals, three of the respondents who participated in mediations that used an in-house neutral stated that their initial skepticism was overcome by the mediator's skill and demonstrated neutrality. Additionally, two respondents stated that in-house neutrals would be "better than nothing." If in-house neutrals are to be routinely used, it will be important for EPA to develop safeguards to help ensure confidentiality and to insulate the mediator from internal pressure, biases, and retaliation. This lesson should apply equally to other agencies and organizations considering the use of inside versus outside neutrals.

Third, a number of problems concerning the mediator contracting process became evident. Even the third-party neutrals disliked the contracted arrangement. Mediators hired pursuant to the first contract indicated a feeling that they were considered "EPA consultants," not neutral professionals. Several mediators reported that EPA tried to dictate who was at the table. Finally, EPA enforcement attorneys reported that the contracting process to obtain a mediator was slow, bureaucratic, and cumbersome. The red tape involved in orchestrating mediation may make ADR an undesirable option, according to some respondents.[39]

Rather than contract for these services, a neutral roster of easily accessible mediators, not paid for exclusively by EPA, should be used. The U.S. Institute for Environmental Conflict Resolution, a relatively new federal agency located in Tucson, Arizona, has developed such a roster. This development may help streamline the process of selecting neutrals, but it is too soon to evaluate the effect of this effort at this time.

Lesson 5: Assistance is needed to help nonprofits, community groups, and de minimus PRPs[40] participate in ADR efforts. As mentioned earlier, the majority of PRPs interviewed wanted more ADR in enforcement actions. They generally see ADR as a way to get EPA's attention and to tell their side of the story. They also perceive the process as fairer and faster than traditional Superfund litigation. Furthermore, PRPs stated that they want to be trained in ADR to minimize power differences at the table due to ADR knowledge differentials. When nonprofits, community groups, and de minimus PRPs participate in EPA enforcement ADR efforts, they express a need for outside technical support that is not needed by the large corporate and governmental parties in the dispute. Although current EPA grant programs for communities and nonprofit groups are commendable, they are insufficient. Additional funds are needed to assist these groups in participating fully and equally in EPA enforcement ADR programs.[41] Increased participation by these groups may improve the outcome of the ADR process, while also lending them legitimacy.

Lesson 6: An established referral and screening mechanism is needed for determining whether ADR is appropriate. In 1987, EPA issued "Guidance on the Use of Alternative Dispute Resolution in EPA Enforcement Cases," establishing, in part, "the review of all enforcement actions for the potential use of ADR processes."[42] Yet in the year 2000, such a comprehensive review was nonexistent. Thirteen out of eighteen ADR specialists, for example, indicated that the review process in their region was ad hoc, informal, and dependent upon the ADR views of the regional counsel or the persuasive abilities of the ADR specialists. This could, in part, account for the enormous discrepancy among regions concerning the percentage of cases that use ADR.[43]

EPA attorneys expressed a concern about the inadequate screening of cases, which particularly is problematic when cases that are not "ripe" go to ADR. Furthermore, some cases simply are not amenable to environmental ADR (e.g., those cases in which one or more parties seek to set a judicial precedent). Finally, the majority of enforcement ADR cases are Superfund cases. If a comprehensive review of all enforcement actions for the potential application of ADR truly were implemented, a more balanced array of ADR enforcement cases might emerge. The time for an institutionalized review process has come.

Lesson 7: EPA needs to evaluate its ADR efforts continually. Half of the EPA attorneys interviewed knew of no evaluations of enforcement ADR efforts at EPA. Most said that they had no opportunity to evaluate the process or the neutral professional used in their cases. In addition, both PRPs and ADR specialists expressed a desire to evaluate EPA's enforcement ADR process and program. As long as evaluative efforts are absent or their findings remain unpublicized, a great deal of useful information is not being captured. Greater evaluation efforts can only strengthen ADR efforts at EPA and would provide an important lesson for other public organizations using or contemplating using ADR.

Lesson 8: EPA needs to take advantage of the growing demand by PRPs for ADR. Almost all the PRPs expressed strong support for ADR processes, but their views of EPA's actions in the area of enforcement ADR were split. Concerning ADR processes in enforcement actions, there is a perception among PRPs and their attorneys that ADR saves money in transaction costs and resolves disputes more quickly than does litigation. There also is a perception that they "get a better deal" through ADR. PRPs uniformly reported feeling that they have more control over their cases when they use ADR. Finally, PRPs reported that ADR helped them to control their risks and gave them a chance to educate EPA. As long as the mediators are strong and skillful, most would use ADR again. This is an interesting finding because it appears that the supply of ADR services is not keeping pace with the demand for ADR services among members of the regulated community.

Views were mixed concerning EPA's performance and treatment of PRPs during the ADR process. Approximately one-half of the PRPs stated that EPA was not helpful in setting up or assisting the ADR process. Some PRPs reported that EPA agreed to use ADR only after repeated prodding by PRPs. Some private-party attorneys said that EPA encouraged the parties to use ADR but did nothing more. Others said that EPA was less flexible during ADR than it should have been. Still others felt that EPA treated them poorly during the ADR process.

However, the other half of the PRPs was effusive in its praise for EPA. For example, one PRP attorney said, "David Batson [senior EPA ADR specialist] and his crew are very skillful at bringing parties together and getting ADR started." Yet another praised David Batson's shop for providing a conflict assessment and seed money to get the ADR effort started. Others indicated that if it were not for EPA, their ADR negotiations would never have happened.

Lesson 9: An evaluation of ADR efforts initiated by administrative law judges is needed. Although this research effort did not set out to evaluate the relatively new ADR program offered through the Office of the Administrative Law Judge (OALJ), a number of EPA attorneys had strong feelings about this program and urged us to draw attention to some potential problems. As we only received information about this program from 12 EPA attorneys, our sample size is too small to allow firm conclusions to be drawn, but it is large enough to suggest further evaluative work concerning this program.

Since 1996, the number of EPA cases using ADR has skyrocketed, going from approximately 18 cases in 1995 to 116 cases in 1998. This increase is largely due to the creation of an ADR program through the OALJ. According to the 1997–1998 EPA ADR status report, this program "has now become the ADR process EPA enforcement personnel participate in most frequently."[44]

According to EPA attorneys, in this program an administrative law judge (ALJ) acts as a mediator to try to encourage settlement by the parties. Mediations are generally conducted over the phone, with the judge speaking to the parties separately or all together. Rather than being confined to Superfund cases, any case filed with the OALJ may be a candidate for mediation. If the case does not settle, a different ALJ is appointed to hear the case. Approximately 77% of these cases result in settlement, whereas non-ALJ mediations have a 79% settlement rate.[45]

Whereas the majority of respondents stated that they were either somewhat or very satisfied with this program, a number of concerns were mentioned repeatedly. For instance, some attorneys commented that the ALJs often conduct the mediations without properly preparing by learning about the case history and issues. Others mentioned that it is difficult for ALJs to "switch hats," changing from their typical roles as authoritative decisionmakers to mediators involved in consensus decisionmaking. Others felt pressured to settle out of a belief that the ALJ may become biased against them in future interactions.

Overall, the OALJ ADR program is likely to widen and improve the dispute resolution options available to both EPA and private-party attorneys. However, a comprehensive evaluation of this program is needed to ensure that the aforementioned concerns are addressed.

Lesson 10: For ADR to be successful, it must be part of the dominant culture at EPA. Despite the fact that EPA's Interim Statement of Policy on Alternative Dispute Resolution states that "it is the policy of the Environmental Protection Agency to work to prevent disputes and to use ADR techniques where appropriate ...,"[46] ADR is not part of the day-to-day business of EPA. Rather, it is the exception to the rule. As one mediator put it, it is as if EPA has a split personality: David Batson's [senior ADR specialist at EPA] shop is promoting ADR, while many EPA attorneys are fighting it. From our interviews, it appears that a small percentage of EPA attorneys have incorporated ADR into their dispute resolution repertoires, but most others have not yet tried it. As a result, the decision to use or not use ADR is based more on an individual attorney's familiarity with the ADR process rather than with the needs of the particular case. Also, as mentioned earlier, support from middle-level managers is inconsistent at best, nonexistent at worst.

Conclusions

In sum, taking into consideration the generally high satisfaction levels of the participants and their perceptions that ADR saves time and money while contributing to better working relationships, it is time for EPA to show a stronger internal commitment to ADR. Furthermore, there is a need for a comprehensive agency policy that will assess and encourage the use of ADR across the agency's programs and regions, which is why recent pronouncements about the increased use of ADR at EPA give us hope.

ADR, however, is not a panacea. Every case does not need a mediator to be settled, and not all cases are amenable to ADR. Furthermore, although it may be argued that ADR should not be institutionalized in a regulatory agency whose job is to enforce environmental laws, there will always be a portion of cases for which ADR is appropriate. At the very least, ADR should be institutionalized for that percentage of appropriate cases. At the present time, the program depends on those who have the personal interest in ADR and the personality to persuade others to use it. Based on the comments received by program participants, it is clear that ADR at EPA has tremendous untapped potential.

Notes

The authors thank the U.S. Environmental Protection Agency, particularly David Batson, Deb Dalton, and Lee Scharf, for their assistance with this research. The authors also thank the Hewlett Foundation for funding this research and three anonymous reviewers for comments on a previous draft. Finally, the authors thank Lisa Bingham and the staff of the Indiana Conflict Resolution Institute for their support.

1. ADRWorld.com (accessed January 27, 2000); *Federal Register,* vol. 65, no. 49 (March 13, 2000).

2. Gail Bingham, *Resolving Environmental Disputes: A Decade of Experience* (Washington, DC: Conservation Foundation, 1986).

3. John McCrory, "Environmental Mediation—Another Piece for the Puzzle," *Vermont Law Review,* vol. 6 (Spring 1981), pp. 49–84.

4. Gail Bingham, Frederick R. Anderson, R. Gaull Silberman, F. Henry Habicht, David F. Zoll, and Richard H. Mays, "Applying Alternative Dispute Resolution to Government Litigation and Enforcement Cases," *Administrative Law Review,* vol. 1 (Fall 1987), pp. 527–551.

5. Melvin Aron Eisenberg, "Private Ordering through Negotiation: Dispute-Settlement and Rulemaking," *Harvard Law Review,* vol. 89 (February 1976), pp. 637–681.

6. Attorneys for potentially responsible parties often attend mediations or other processes on behalf of their clients. Therefore, when the PRPs themselves were unable to answer our questions about the ADR process, the neutral, and the outcome, we often interviewed their attorneys instead.

7. Rosemary O'Leary, Robert F. Durant, Daniel J. Fiorino, and Paul Weiland, *Managing for the Environment: Understanding the Legal, Organizational, and Policy Challenges,* Nonprofit & Public Management Series (San Francisco: Jossey-Bass, 1999).

8. Tom Waller, "Knowledge, Power, and Environmental Policy: Expertise, the Lay Public, and Water Management in the Western United States," *Environmental Professional,* vol. 17 (1995), p. 153.

9. Bradford F. Whitman, "Alternative Dispute Resolution Needed for Superfund Remedies," *BNA Environmental Reporter,* vol. 17 (1993), p. 1533.

10. Stephen Jenks, "County Compliance with North Carolina's Solid Waste Mandate: A Conflict-Based Model," *Publius,* vol. 17 (1994), p. 35.

11. Environmental Protection Agency, "Use of Alternative Dispute Resolution in Enforcement Actions," *BNA Environmental Reporter,* vol. 26 (May 1995), pp. 301–304.

12. Lance DeHaven-Smith and John R. Wodraska, "Consensus Building for Integrated Resources Planning," *Public Administration Review,* vol. 56 (1996), p. 367.

13. Cornelius Kerwin and Laura Langbein, *An Evaluation of Negotiated Rulemaking at the Environmental Protection Agency, Phase I* (Washington, DC: Administrative Conference of the United States, 1995).

14. Daniel J. Fiorino, "Regulatory Negotiation as a Policy Process," *Public Administration Review,* vol. 48 (1988), p. 764.

15. J. Walton Blackburn, "Environmental Mediation as an Alternative to Litigation," *Policy Studies Journal,* vol. 16 (1988), p. 562.

16. Henry Perritt, "Negotiated Rulemaking in Practice," *Journal of Policy Analysis and Management,* vol. 5 (1986), p. 482.

17. See, e.g., Zhiyong Lan, "A Conflict Resolution Approach to Public Administration," *Public Administration Review,* vol. 57 (1997), p. 27.

18. Lynn Peterson, "The Promise of Mediated Settlements of Environmental Disputes: The Experience of EPA Region V," *Columbia Journal of Environmental Law,* vol. 17 (1992), pp. 327–380.

19. Superfund cases are hazardous waste cleanup cases filed pursuant to the Comprehensive Environmental Response, Compensation, and Liability Act (CERCLA).

20. Heidi Wilson Abbot, "The Role of Alternative Dispute Resolution in Superfund Enforcement," *William and Mary Journal of Environmental Law*, vol. 15, no. 1 (1990), pp. 47–64.

21. Leonard Charla and Gregory Parry, "Mediation Services: Successes and Failures of Site Specific Alternative Dispute Resolution," *Villanova Environmental Law Journal*, vol. 2 (1991), pp. 89–97.

22. William Hyatt, "Taming the Environmental Litigation Tiger," *Journal of Environmental Regulation*, vol. 5, no. 1 (1995), pp. 91–98.

23. Robert A. Baruch Bush and Joseph P. Folger, *The Promise of Mediation: Responding to Conflict through Empowerment and Recognition* (San Francisco: Jossey-Bass, 1994).

24. The primary source of statistics was the U.S. EPA Enforcement ADR Program's "Status Report on the Use of Alternative Dispute Resolution in Environmental Protection Agency Enforcement and Site-Related Actions" (December 1999).

25. The primary sources of archival records were U.S. EPA Office of Site Remediation records and Lexis consent decree files.

26. These students worked through the Indiana Conflict Resolution Institute and were earning M.P.A., M.S.E.S. (environmental science), and J.D. degrees.

27. All participants were guaranteed anonymity, and Indiana University's Human Subjects Committee reviewed and approved each of the four interview protocols before their use.

28. Information about age and race was not gathered, as the existing literature does not indicate that these traits influence satisfaction with ADR processes, mediators, or outcomes.

29. This group was composed of an equal number of men and women and represented all of EPA's regions.

30. The names of companies and their attorneys were derived from EPA consent decrees. The companies in the sample represent all regions within the United States and firms of varying sizes. In all but two cases the attorneys, and not the firm owners or managers, directly took part in the ADR processes. For this reason, most of the interviewees for this group of stakeholders consisted of PRP attorneys. The authors interviewed two company owners who had directly participated in ADR processes.

31. The EPA sent us a list of 45 third-party neutrals. From this list, 7 stated that they had never served as a neutral on an EPA case, 5 could not be located due to a change of address, 3 declined to participate, and 7 could not be reached. Of the 22 interviewed, 15 were men and 7 were women.

32. Men and women were equally represented among the EPA attorneys interviewed, but the sample of PRP attorneys was exclusively male. Nationally, women are underrepresented among corporate lawyers. We do not believe that satisfaction with the ADR process, the mediator, and the outcome vary based on gender, but we cannot reach any firm conclusions based on our sample.

33. For a detailed report on the number of ADR cases listed by EPA region, see the December 1999 "Status Report on the Use of Alternative Dispute Resolution in Environmental Protection Agency Enforcement and Site-Related Actions," published by the EPA.

34. See Richard A. Posthuma, James B. Dworkin, and Maris Stella Swift, "Arbitrator Acceptability: Does Justice Matter?" *Industrial Relations*, vol. 39, no. 2 (April 2000), pp. 313–335.

35. Abbot, "The Role of Alternative Dispute Resolution in Superfund Enforcement."

36. See Nancy J. Manring, "Reconciling Science and Politics in Forest Service Decisionmaking: New Tools for Public Administration," *American Review of Public Administration*, vol. 23, no. 4 (1993), p. 343. Also see Abbott, "The Role of Alternative Dispute Resolution in Superfund Enforcement."

37. U.S. Environmental Protection Agency, "Status Report on the Use of Alternative Dispute Resolution in Environmental Protection Agency Enforcement and Site-Related Actions."

38. See Roger Fisher, William Ury, and Bruce Patton, *Getting to Yes: Negotiating Agreement without Giving In* (Penguin Books, 1981).

39. Current managers in EPA's enforcement ADR program office report that these contracting difficulties have been addressed and that the turnaround time for hiring a mediator is two weeks as of April 15, 2000.

40. The term "de minimus PRPs" generally refers to parties who contributed only a small percentage of the total pollution at a Superfund site.

41. The U.S. Institute for Environmental Conflict Resolution announced that it will provide some matching funds to assist nonprofits, community groups, and de minimus PRPs.

42. U.S. Environmental Protection Agency, "Guidance on the Use of Alternative Dispute Resolution in EPA Enforcement Cases," (1987), p. 1.

43. See U.S. Environmental Protection Agency, "Status Report on the Use of Alternative Dispute Resolution in Environmental Protection Agency Enforcement and Site-Related Actions."

44. Ibid.

45. Ibid.

46. *Federal Register*, vol. 65, no. 49.

PART V

Downstream Environmental Conflict Resolution and Outcome Measures

There is much discussion in the public policy and public management literature today about the difference between measuring outputs and measuring outcomes. Measuring outputs connotes bureaucratic "bean counting." For example, how many water treatment plants did we build? How many environmental managers did we train? How many polluters did we sue? Although there is nothing inherently wrong in tracking outputs—these numbers can be quite useful in answering certain questions—doing so often fails to address the most crucial public policy questions. Did the environment improve because of our programs? Did we protect human health in any documentable way? How many deaths did we prevent?

The outcome measurement movement has reached ECR in a way that is both appropriate and challenging. Fiscal stringency and increased customer demands have ECR managers and ECR participants looking for meaningful ways to measure outcomes. The most striking common threads running through the chapters in this section are the need to focus on outcome measures and the challenges in developing a measurement strategy that makes sense.

In Chapter 14, Mette Brogden provides a concrete example of a successful initial effort to assess the environmental outcomes of an ECR process. Then, from six rich case studies and data derived from focus group discussions, Brodgen presents a useful evaluation checklist that can be applied to virtually any ECR case. Recommendations for future research are highlighted.

Bonnie G. Colby analyzes the challenges of evaluating ECR programs and policies through an economic lens. In Chapter 15, Colby presents 10 criteria for judging the efficacy of ECR outcomes: positive net benefits; well-defined, measurable objectives; cost-effective implementation; financial feasibility; fair distribution of costs among parties; flexibility; incentive compatibility; improved problem-solving capacity; enhanced social capital; and clear documentation protocols. Lessons learned are examined. A negotiated agreement checklist is presented.

The importance of delivering ECR programs in a way that engenders public trust and confidence cannot be overestimated. ECR programs that are both efficient and effective are a must. Part V offers insights into how to develop efficient and effective programs.

14

The Assessment of Environmental Outcomes

Mette Brogden

W hat are the environmental outcomes of environmental conflict resolution (ECR) processes? To make wise decisions about investment of time and financial resources, funders, legislators, other political leaders, potential participants, and the general public need a means to assess the environmental effects of ECR processes compared to other methods of resolving conflicts. Assessment methods must be both cost-efficient and time-efficient, or they will not be undertaken.

For a variety of reasons, assessment of the environmental outcomes of ECR processes seems an elusive goal. This chapter explores why and then examines data from a national policy dialogue on state conservation agreements (SCAs), during which focus group participants representing a wide variety of interests considered how to create an effective tool for proactive conservation of species and ecological systems. The resulting draft SCA proposal included guidelines for assessing the potential and actual environmental outcomes of multiparty conservation agreements at the time of signing and during implementation, with implications for both prospective and retrospective evaluation. The guidelines, data from six case studies, and additional information produced by the focus groups are used here as a basis to construct a prospective evaluation checklist that can be applied to ECR cases more generally. The chapter concludes with (a) a discussion of how prospective evaluation instruments may also be used to evaluate ECR programs and to aggregate data from streams of cases and (b) recommendations for future research.

Why Don't We Know the Environmental Outcomes of ECR Processes?

The field of environmental and public policy conflict resolution is relatively new. Its practitioners have needed time to grow, accumulate experience, develop ideas about best practices, and learn from trial and error. Research is likewise an iterative process. Several factors complicate the task of evaluating environmental outcomes of individual cases, ECR programs, and the field as a whole.

Heterogeneity of the Field

Practitioners define the field of ECR quite broadly. It encompasses a variety of approaches that enable stakeholders to meet face to face and reach mutually acceptable solutions to environmental conflicts and planning issues.[1] These approaches typically use a neutral third party whose role may vary from formal mediation of lengthy and complex public policy disputes to facilitation of a single meeting around an issue or decision. Conceivably, an environmental issue could involve only two parties, but usually, environmental disputes and planning involve multiple parties, as well as a broad public interest.

ECR processes are directed toward resolving both conflicts about the establishment of rules or plans for future conduct and conflicts that arise from past events, such as allocation of responsibility for hazardous waste site cleanups or violation of laws.[2] These conflicts are identified within the field as "upstream" and "downstream" disputes, respectively. This chapter is primarily concerned with ECR processes that address the upstream category and leaves unexplored the question of measuring environmental outcomes of ECR processes directed at the downstream category, but I acknowledge its importance for future inquiry.

Within the upstream category, significant heterogeneity exists and complicates the question of how to determine environmental outcomes. Processes target different natural systems and systemic levels, from local land-use plans to management of regional watersheds or ecological systems that span several counties, states, or countries, to (conceivably) global systems such as climate. Even local ecological processes are far more complex than we currently understand, but local planning processes that seek to develop and implement management plans may be the most directly and easily accessible for measurement of environmental outcomes. However, ECR processes also range from determining how to allocate a resource in a local geographic area to broad-scale policy-development processes. Many steps occur on the way to decisionmaking, and each step may be undertaken with neutral facilitation and ECR methods. Study circles, consensus-building policy dialogues, mediated agreements, ad hoc citizen processes, and temporary commissions established by law and charged with development of recommendations that must then survive political implementation are some examples of processes that engage interested parties in achieving mutually acceptable decisions. All of these, however, are at least one step removed from the specific actions that manipulate the natural environment, thus vastly complicating the establishment of causative links between processes and their outcomes.

Our research would benefit from the development of a typology of efforts, matched to a set of methodologies for tracking environmental outcomes. For example, in assessing environmental outcomes of collaborative policy-development processes, we need further understanding of how these processes relate to existing pluralistic decisionmaking processes that set many of the rules of human engagement with the environment. That understanding will require in-depth case studies that follow the implementation of agreements and decisions. More descriptive research is needed to develop better theory, which in turn helps to refine research questions that can test theory and evaluate outcomes.

Finally, the very nature of upstream disputes requires the use of both *prospective* and *retrospective* assessment methods. That is, because upstream disputes occur during the development of plans or policies that guide future conduct, we need ways to evaluate the potential for the agreement or plan to affect the environment in the desired way at the time of agreement signing, as well as the actual environmental outcomes during and after its implementation.

Methodological Problems

Even if we had already agreed upon useful categories of ECR efforts within each category, the ECR field would face difficult challenges in establishing causality between its processes and environmental outcomes. Use of rigorous experimental or quasiexperimental design or natural experiments is compromised by a lack of baseline data about pretreatment environmental conditions to compare to posttreatment measurements. Also, differing human and natural contexts, selection biases, and a host of other confounding variables challenge comparability of cases. Political and perceptual paradigms shift under participants' feet during complex processes that may take years to complete. Likewise, the natural environment is changing, and other external factors affect it.

We lack adequate data to compare many ECR processes with environmental outcomes that would have occurred without the process; alternative legal and policy-development processes may not collect environmental outcome data. When they do, they likely suffer from the same methodological issues in measuring their outcomes.

As the field of ECR has developed, so have environmental practices and understanding. Managers use "adaptive management" in trial-and-error learning and to handle naturally occurring chaotic fluctuations in ecological systems. The term "adaptive management" is of relatively recent origin, first used in the 1970s. It is difficult to tell whether and how many ECR processes have developed systematic protocols for adaptive management, have implemented monitoring, and are using results to adjust management. At the same time, lack of data to prove causality does not mean that causal relations do not exist. Many land managers, using only pencils and their own senses, have managed to produce desired environmental outcomes on ranches and farms for many years, adjusting their management through trial and error without calling it "adaptive management."[3]

We deal with complex issues. We need to shift our thinking from linear, mechanistic views of environmental processes to the more organic paradigm outlined in the science of complexity. Our familiar experimental designs were developed within the mechanistic paradigm of Newtonian science, with inputs and outputs observed through narrow disciplinary lenses that also circumscribe the questions that can be asked. Less familiar to ECR practitioners and observers are new research methods, such as agent-based modeling, which handle continuous adaptation, learning organisms and organizations, and the distributed intelligence of complex adaptive systems.[4] A new, "generative" social science has resulted from the marriage of game theory and evolution and seeks to explain the self-organizing nature of systems. It may assist our evaluation research because it accomplishes both prospective and retrospective research. However, for the moment, its specialized computer-modeling methods are outside the expertise of most ECR practitioners, observers, sponsors, and clients.

Even without these technically intensive research methods, ECR programs and practitioners have yet to benefit widely from in-depth, theory-informed case studies that could help establish validity of claims about environmental outcomes.[5] Policymakers must regularly proceed in the absence of information that achieves the 95% validity standard accepted by science. Evaluative standards are necessarily lower, though that fact should not excuse a continual quest for more knowledge and valid outcome measures for our processes. However, we must recognize where we are and what our needs are and develop systematic assessments of qualitative data in addition to descriptive research. As we accumulate case data, quantitative methods become useful, although we need to resist a persistent tendency to hold quantitative studies in higher esteem than qualitative ones. This is especially true given the practice nature of this work,[6] which parallels the fields of law and medicine, where execution of practice within cases remains context-dependent and details of specific cases are important in the construction of remedy.

Practitioners can easily complete participant-observation research as part of their reflective practice; indeed, many practitioner-authored reports of cases would certainly fall into this category. Policy dialogues may be particularly well suited to leveraging participant-observation data when they engage focus groups of experienced participants in reporting and reflecting on their experiences and in predicting how they might respond behaviorally within changed policy environments. The case presented in this chapter illustrates this kind of leveraging.

The Nature of Environmental Outcomes

Recent views of the environment—first chaos theory and now complexity theory—see the environment as continuously changing, affected by naturally occurring chaotic fluctuations and changes in behavior occurring simultaneously at many different levels within the system. Environmental outcomes are therefore moving targets.

Systems theory originally took a mechanistic view of systems as bounded, homeostatic units. Its most recent version—complexity theory—has a more organic view of systems as dynamic and learning, characterized by a nonstop flow of information within the system and by individual actors making continual adjustments in their choices and behaviors.

"Emergence" is a key concept that explains the self-organizing character of systems. This term describes a systemic pattern or structure that emerges from local interactions or decisions of individuals but is not available to or producible by any single individual or single interaction. The aforementioned agent-based modeling "observes" systemic emergence by running a population of diverse agents through many rounds of decisions and actions within a varied environment and can be used to predict emergent outcomes or to assess factors that have led to emergent outcomes. Emergent patterns are unplanned, unexpected, and may be undesired. Real-world examples of unwanted environmental outcomes are urban sprawl, global warming, groundwater depletion, and large-scale imperilment of species. The last problem served as the impetus for the case presented below.

Agent-based computer modeling "observes" systemic emergence by running a population of diverse agents through many simulated rounds of decisions and actions within a varied "landscape" and according to a fixed set of decision rules. It can be used to predict emergent outcomes or to assess factors that have led to emergent outcomes so as to identify changes in agent behavior or decision rules that could "tip" the system to a new emergent outcome.

It may be that many environmental outcomes of ECR will need to be understood as something between a process and an endpoint, much as sustainability is best described as a process goal. We have yet to develop language and concepts that pinpoint such outcomes effectively; the problem is similar to our difficulty in defining "health." We know "unhealthy" when we see it. We have developed indicators that point to relative health. However, it is difficult to define "health" as an endpoint except to think in terms of properly functioning process qualities, such as "resilience" and "resistance," concepts that capture the ability of systems to recover from or resist outside perturbations. We are still developing the necessary theoretical constructs that can characterize what we seek.

Case Study: A National Policy Dialogue on State Conservation Agreements

This case is an example of an upstream policy dialogue process to develop a policy tool proposal that, upon completion, had to move into a political implementation phase and be implemented on the ground to achieve actual environmental outcomes. I will be following the case through both its political implementation and its on-the-ground implementation because the case offers the opportunity to study the articulation of an ECR process with existing national and state pluralistic political processes.

Here we use the findings from the dialogue itself as research data gathered from eight focus groups. The case is relevant because it parallels the problem that the field of ECR faces in determining the environmental outcomes of its processes. We think we have a better mousetrap, yet we have little empirical data about our environmental outcomes to support this assertion. However, potential participants and funders have to make decisions about the value of our mousetrap anyway. Planners face a similar problem. How do they know that one plan is superior to another before its implementation? Upstream cases require both prospective and retrospective evaluation methods, and that is exactly what participants in the SCA dialogue developed as they tackled their problem. The case thus has important implications for the evaluation of upstream ECR cases more generally.

Description of the SCA Dialogue

From November 2000 to July 2001, the International Association of Fish and Wildlife Agencies (IAFWA) sponsored a national policy dialogue to develop a proactive tool for conservation of species and ecological systems. During eight workshop sessions held across the United States, participants grappled with the question of measuring environmental outcomes of multistakeholder partnership approaches to conservation. Many of the participants represented federal or state agencies that either administer and enforce the Endangered Species Act or implement its protections for species in crisis. Other stakeholders included representatives of nongovernmental organizations (NGOs) concerned with conservation, industry, private-property interests, other state and federal agencies, local government officials, and staff from quasigovernmental organizations and state and national legislative bodies.

Participants brought expertise in plant and animal species biology, hydrology, policy formation and political implementation, natural resource production and extraction, business, agriculture, education, and management. Many were veterans of both successful and unsuccessful conservation efforts, where success is defined as achievement of conservation goals, political legitimacy for the effort, or both. More than 225 individuals participated in at least one of the eight workshops.

A planning team from the IAFWA's Threatened and Endangered Species Policy Committee managed the project and engaged a facilitation team that I headed. The IAFWA is a quasigovernmental organization that works on conservation and serves as a professional organization for personnel employed in state fish and wildlife agencies, as well as federal agencies with conservation responsibilities. Conservation NGOs are also members. A report detailing the impetus, method, and a synthesis of workshop results is available from IAFWA and forms the basis of reporting on the case below.[7] Although the dialogue did not set out to gather information about the evaluation of multistakeholder ECR processes, relevant information surfaced as participants tackled their task.

The Context and Impetus for the Dialogue

Perceiving an accelerated rate of species endangerment as related to human practices within ecological systems, the public in 1973 supported passage of national legislation to protect species threatened by or in danger of becoming extinct. The Endangered Species Act (ESA) is administered by the U.S. Fish and Wildlife Service (USFWS) and the National Marine Fisheries Service (NMFS), while other federal and state agencies are required to conserve threatened or endangered species inhabiting lands and using resources that they manage.

In the almost 30 years since its passage, the ESA has not been without controversy. It provides "emergency room" care for species in crisis through regulatory action, but it is cost-intensive and crisis-oriented, and it reduces management flexibility for government agencies as well as private interests. As species decline and the need to be listed grows more urgent, decisionmaking authority accrues to the federal level, and state and local governments retain significant implementation responsibilities.

Species in crisis have become the focus of conflicts over land use. Some groups see the ESA as a means to gain leverage in regulating land and resource use. Some private landowners who would be interested in protecting and enhancing habitat on their properties become reluctant to do so out of fear that such activities will encourage listed species to take up residence on their property and lead to limitations on its future use.

The ESA has enabled regulatory actions that have helped stop the precipitous decline of some species and recover others, but the overall trend toward imperilment of a staggering array of species continues. According to the Nature Conservancy's rankings on 44,359 of the 204,700 described plant and animal species in the United States, 15,224 were at risk or imperiled—more than one-third of those ranked. As of January 2001, 1,233 species were federally listed as threatened or endangered, 56 species had been proposed for listing, and 283 were candidates for listing.[8] It is becoming clear that to ensure conservation of many species, habitat conservation efforts by private landowners and industry are needed; public lands neither encompass enough acreage nor provide the habitat connectivity that some species require.

By the late 1990s, state and federal agencies were working to develop mechanisms for voluntary conservation that could complement the ESA's regulatory approach. The USFWS and NMFS developed ways to partner with private landowners through such mechanisms as safe harbor agreements, habitat conservation plans, candidate conservation agreements, and candidate conservation agreements with assurances. The latter two mechanisms begin species recovery efforts before listing decisions; the lengthy and resource-intensive listing process had forced the USFWS and NMFS to develop priorities for listing, and they needed a way to start helping candidate species that awaited listing decisions.

Meanwhile, several state wildlife agencies had initiated development of formal conservation agreements with a variety of private and government partners, also aimed at helping species that were in decline but not yet federally listed. When they successfully stabilize and recover species, proactive agreements help stakeholders retain management flexibility that federal listing greatly reduces. Stakeholders are therefore motivated to start developing effective conservation measures for species of concern much earlier.

Section 4(b)(1)(A) of the ESA requires the Secretary of Interior or the Secretary of Commerce to take into account any formal efforts by a state, foreign nation, or any of their political subdivisions to protect species when making listing decisions. However, judicial rulings have reversed several decisions not to list species that had been made on the basis of existing proactive conservation agreements,[9] dampening enthusiasm for such efforts and raising concern that any proactive efforts on behalf of declining species would draw the attention of potential petitioners (for federal listing) and litigants.

Faced with (a) a mandate to consider formalized conservation efforts by states, as stated in the ESA; (b) litigation that had successfully challenged decisions not to list on the basis of such existing agreements; and (c) an acute awareness that effective partnerships with private interests and local governments were critical to stemming the overall trend toward species imperilment, the USFWS and NMFS worked to clarify how they would assess conservation efforts in making listing decisions. They published a draft, "Policy for the Evaluation of Conservation Efforts" (PECE) in the *Federal Register*, in which they proposed criteria for evaluating "whether formalized conservation efforts contribute to making listing a species as threatened or endangered unnecessary."[10] The proposed policy indicated that the USFWS and NMFS would look for certainty that

— conservation efforts would be implemented, including documentation of when actions would be implemented, who would implement actions, and that signatories to the formal agreement have authority to implement actions and
— conservation efforts would be effective, as shown by (a) completion of a description of threats to species and how the agreement will address the threats; (b) establishment of performance measures; (c) a plan for monitoring compliance and effectiveness; and (d) delineation of adaptive management plans.

Their dilemma illustrates the classic problem decisionmakers face when they need to act without access to data and analyses that could withstand the 95% validity standards acceptable to scientists. The USFWS and NMFS faced the problem of assessing and predicting the likelihood that state-led conservation agreements would produce desirable environmental outcomes before implementation of the agreements was complete and before the outcomes of implementation were known. Their criteria are aimed, therefore, at accomplishing *prospective* evaluation.

Purpose of the SCA National Policy Dialogue

The SCA National Policy Dialogue was convened by the SCA planning team. Its purpose was to engage a wide variety of stakeholders in a dialogue about proactive approaches to conservation. The team hoped to build consensus about a prevention approach that would encourage private entities and government agencies to conserve declining species early and sufficiently enough "so that no one would feel compelled to try to list them."[11] This statement reflects the need to remedy the dampened enthusiasm for proactive approaches that resulted after courts later reversed decisions by the Services not to list some species based on existing proactive conservation agreements.

Project Design

Two national workshops engaged state and federal agency representatives in a review of conservation efforts to date. Presenters described and analyzed six conservation agreements developed in the 1990s, reviewed conservation tools developed by the Services, explained the proposed PECE policy, and reviewed case law concerning formal conservation agreements.[12] Participants used the findings from these presentations, as well as their own experiences, as a basis to outline the concept for a proactive conservation tool—a state conservation agreement (SCA). The participants thought that the tool should complement, not replace, existing tools developed by USFWS and NMFS, and should aim to conserve precandidate species,[13] enable partnerships across jurisdictional boundaries, encourage voluntary efforts, and focus efforts on restoring the health of ecological systems and communities. Their initial outline described a purpose, a list of potential participants in developing the agreement, the method to be used in developing the agreement, and a list of items the written agreement would include. Participants also saw a need to detail what would happen if the effort failed to conserve the species, if a chaotic fluctuation in natural systems or the species population occurred, or if an endangered species took up residence on property involved in implementation of the SCA.

The draft concept was taken to six regional workshops, where representatives of a wide variety of government and private interests did the following in this order:

— reported their experiences to date with conservation of species and ecological systems,
— listened to a case study of an existing conservation agreement,
— thought abstractly about the necessary elements of a proactive approach to conservation,
— reviewed and discussed the draft concept from the national meetings to which their ideas from the "necessary elements" conversation had been added, and
— worked in three smaller groups to imagine how they would create an SCA to conserve one of two species or an ecological system of concern in their region.

The results of the workshops were collated and reviewed with the SCA planning team. The team used the results as the basis to write a final draft of the SCA concept to be taken for review to the IAFWA annual meeting. A significant addition to the final draft was a checklist for evaluating the agreement at the time of signing—a category suggested in several of the regional workshops.

After the concept's adoption by the IAFWA, the SCA planning team planned to take the concept through a political implementation phase, where it would seek endorsement and funding from elected officials at both federal and state levels. Concurrently, they would invite participation in a process to prioritize species and systems to be targeted first for development of SCAs.

Facilitation Methodology

The workshop design accomplished de facto focus group research in the process of building consensus around an SCA tool proposal. Facilitators used a variety of methods to systematically elicit informed opinions from stakeholders experienced with conservation and species endangerment, including:

— round robin collection of ideas from every participant,
— free-flowing discussion about the draft concept,
— recording of comments and disagreements during development of a single text,
— anonymous written reflections about cases,
— observation and recording of sparingly facilitated SCA process enactments, and
— anonymous written evaluations of the workshop findings and process.

The SCA workshops produced a significant amount of data relevant to assessing environmental outcomes of multistakeholder conservation agreements. The six case studies of multistakeholder conservation agreements included outcome data. The workshop focus groups made recommendations about prospective evaluation and systemic indicators, and facilitators observed how the groups worked with these ideas as they tested the concept in scenarios. With these inputs, the SCA planning team included in the SCA proposal recommendations for both prospective and retrospective evaluation protocols.[14]

Findings Related to Assessment of Environmental Outcomes of ECR Processes

Below, general categories of findings are listed, followed by specific findings and a brief discussion of evidence.

Finding 1. Multistakeholder processes increase both scientific and individually held knowledge about the natural environment.

A. At the outset of multistakeholder processes to develop proactive conservation agreements, insufficient data exist about species, habitats, and the threats to species. Data need to be collected and analyzed while the agreement is under development.

SCA workshop presenters described six cases of proactive conservation agreements targeting barrens topminnow, Barton Springs salamander, blacktailed prairie dog, copperbelly watersnake, robust redhorse, and virgin spinedace. In the last five cases (all but barrens topminnow), conservation requirements of the species were poorly understood at the outset of the agreement-development process. Insufficient data existed about species biology and life cycle, population numbers, range, habitat requirements, and threats to the species. Development of the agreement required ongoing collection and assessment of data, and continuing data-collection needs were included as part of the agreement and its implementation plans.[15]

Workshop participants confirmed this finding. Speaking from their experience and expertise, participants discussed what information would be necessary for the agreement-development process. The resulting SCA concept encourages sponsors and stakeholders to assemble existing pertinent biological and social data at the outset of the process, including

— existing agreements,
— the regulatory structure and climate,
— relevant programs and plans,
— jurisdictional boundaries,
— habitat assessments,
— biological cycles,
— population levels and distribution, and
— historical data.

With this information assembled, participants and sponsors can then identify data gaps and make arrangements to fill these gaps as they develop and implement the agreement.

B. In the absence of multistakeholder agreement-development processes, the likelihood of achieving comprehensive and systematic data collection is less.

Without a formalized, multistakeholder process such as an SCA, it is difficult to secure funding for required research or to develop a comprehensive assessment of knowledge gaps and then conduct research to fill in those gaps. State wildlife agencies' budgets for conservation and management of nongame species have overwhelmingly been devoted to activities on behalf of federally listed species.[16] Stakeholders may have information that no one knows about because there is no place to report it, no reason to report it, no request to report it, and no benefit to reporting it. A formalized conservation effort such as an SCA can provide participants with some risk protection; without risk protection, some stakeholders may have motivation to conceal information for fear of increased regulation. One university-based biologist told of working to recover an endangered bat. He reported watching landowners "get a stricken look upon learning that [the bat] was located on their property. It's always a negative response, and you may come back and find the species gone. People are fearful."

The case studies demonstrated that starting a process such as that outlined in the SCA proposal makes it more likely that the funds necessary for research will be made available and that the research program will be more comprehensive. Agreement partners help procure and focus the resources necessary for comprehensive environmental research. Industry and conservation groups with research budgets provide monetary and staff contributions. Workshop participants believed that formalized efforts give legislators a rationale to increase funding for nongame species conservation when they have the political support of stakeholders. For example, the Utah legislature, when approached by its state wildlife agency, allocated funding that enabled development of the agreement for the virgin spinedace.

C. Knowledge of individual stakeholders also increases during the process. If additional knowledge affects human behavior, we may infer that environmental outcomes will result simply from the processes having occurred, with or without achievement of a signed agreement.[17]

Regional workshop participants considered how science and technical information would be developed during the process. After sometimes-intense discussions about how to handle the inevitable controversies over the science, participants concluded that all stakeholders involved in the development of the SCA need to be included in scientific discussions as they are occurring, even if a technical advisory team is used. As one SCA workshop participant suggested, "Do it in front of everyone, like full disclosure."[18] This process will result in increased stakeholder understanding if we assume that participants are listening to the discussion, a reasonable assumption because scientific findings will undoubtedly affect their interests.

D. Outreach to members of the wider public raises citizens' knowledge as well and encourages their voluntary participation in conservation activities for the target species or system.

Through outreach and education, people who are not signatories to the agreement can develop an interest in and contribute to on-the-ground implementation of conservation actions. Participants suggested such things as curriculum development for schoolchildren and development of a media campaign that first publicizes the problem and then celebrates success. They thought that both should begin early in the agreement-development process, long before the agreement is signed. Though this recommendation was made in all of the regional workshops, participants in the Midwest workshop were particularly emphatic about it, based on their experience with recovery efforts for the Karner blue butterfly, a federally listed species.

Finding 2. Achieving successful environmental outcomes requires a plan for accommodating new information about the environmental target, naturally occurring fluctuations, and poor outcomes of management regimes.

A. Processes develop hypotheses about human activities and environmental factors that are leading to a decline in the species.

An early step in the agreement-development process is the assessment of threats and a listing of conservation actions to address threats. Threats to the species may not be known at the outset of a process, requiring basic research. In the case of the robust redhorse, for example, the fish was thought to be in rapid decline because surveys for the fish turned up no juveniles. Additional research located the juveniles and added necessary information about the life cycle of the fish, thereby allowing researchers to develop better hypotheses about threats to the species.

B. Because environmental outcomes are often moving targets, development of goals and indicators is essential to test and revise hypotheses over time, using adaptive management.

Participants noted that conservation efforts require ongoing assessment in which a variety of indicators are tracked. SCAs need to include plans for accomplishing adjustments in management when indicators move.

Over time, indicators pick up emergent patterns at different levels and time scales of the system. Innes and Booher[19] have suggested development of broad-scale system performance indicators that inform stakeholders and the public about the overall "health" of the system and policy and performance indicators that provide managers with feedback about how management actions are working. Each level needs adaptive management protocols.

C. Participants need to agree on systemic indicators and define the magnitude of response that is required to consider the effort a success.

For SCAs targeting species, the systemic indicator is a population response of some magnitude. SCA workshop participants wanted to know an overall population threshold for decisions about whether species would need to be federally listed, as well as some indication of necessary distribution of the species or discrete populations of the species that would be required to maintain species in a nonimperiled status. Regulators responsible for enforcing the environmental protection laws that are backstopping these processes need to participate in the development of indicators and target goals.

The case studies showed that, in practice, these goals may be difficult to develop—much less measure—even though several case studies reported discovering more individuals and populations than were known at the outset of the process. Participants recognized the iterative nature of knowledge building. The magnitude of response that would define success may change as more becomes known.

Workshop groups testing the SCA concept on ecological systems mentioned a variety of other systemic indicators, including population levels of keystone species, diversity of plant and animal communities, number of roads, area of land covered by impervious surfaces, and level of nutrient loading in rivers.

D. Management efforts must be monitored during implementation so that they can be assessed.

Workshop participants suggested development of monitoring programs at two levels: local volunteer programs and state-level oversight programs.

They recommended that on-the-ground monitoring programs should be simple and feasible to implement, using participating landowners and other volunteers. "It could be taking a picture at the same spot once a year, or [an indicator] that a landowner could wander out after supper occasionally and check."[20] They also noted that volunteer monitors need a place to report their findings. One Midwest workshop participant described a successful volunteer monitoring program:

> The Wisconsin Sandhill Crane Count is an event that sees 3,000 citizens out before dawn on a Saturday to do the count. It's not people who you'd expect to see. This is citizen science, and they see themselves making a contribution. The sandhill crane is coming back in Wisconsin. It is now a harbinger of spring.

A state-level oversight monitoring program would be needed for more complicated monitoring activities that track implementation and results of conservation actions and to provide an overall data collection point where volunteers can report their data, so that the information becomes part of the adaptive management cycle. An assigned team needs to be responsible for making adjustments to management and conservation strategies in the face of a downward movement in systemic health indicators.

E. The SCA concept document lists a means to test for adequacy of management plans for achieving environmental goals before an agreement is signed using the criterion "sufficient certainty that the SCA will be effective in conserving the target species or ecological systems."

The SCA Planning Team listed the following criterion indicators (quoted from the document):

— Is the SCA reasonably likely to meet the conservation goals and objectives for the species or ecological systems?
— Is the SCA based on the best available science?
— Does the SCA sufficiently describe adaptive management strategies that are reasonable, logical, and straightforward and that can be implemented without confusion or controversy? At minimum, the adaptive manage-

ment framework should include appropriate protocols for administrative and management processes and for monitoring and measuring (evaluating) progress through objective (preferably quantitative) standards or benchmarks.

—Do the SCA and its adaptive management protocols adequately address how conflicts among signatories or partners will be discussed and resolved?

Finding 3. Successful environmental outcomes depend on implementation of the agreement.

A. Political will and effective social process at the outset of an effort as well as at the time of signing are necessary to ensure that implementation will take place.

One of the interesting aspects of this dialogue was the inclusion of both expert biologists and natural scientists with political specialists, as well as the degree to which many agency participants exhibited both political and biophysical science skills. This enabled them to think *as a group* through the complex of human activities that affect the viability of species and ecological systems by understanding human roles and activities—social, political, and economic—as integral parts of the ecological systems.

No one would question the idea that politics affects the environment; the innovation here, however, is that politics was recognized as an ecological factor and planned for as part of the management of ecological systems. Participants with expertise in biology were adamant that such expertise is not enough—they had tried to manage on the basis of technical expertise and intervention alone, but found that it did not work because they had not "gotten the 'social' and the 'political' [aspects] right."[21]

This is an extremely important finding for ECR practitioners and evaluators, who have noticed their own tendency to focus on the social outcomes of ECR processes. It provides a compelling argument that such attention to social outcomes and the field's focus on the importance of stakeholder inclusivity are warranted. Our tendency to consider social outcomes as separate from environmental outcomes reflects a reification dating from the rise of disciplinary approaches in the modern period and against which historians of science have written in the postmodern era.[22]

B. The SCA proposal includes the following criterion for the prospective evaluation of the agreement at the time of signing: there is "sufficient certainty that the SCA can and will be implemented."

The SCA planning team listed the following criterion indicators (quoted from the document):

—Do the participants recognize the time required to accomplish the goals and objectives, and have they assessed the long-term feasibility of the management approach?

— Does the SCA provide enough flexibility for landowners to manage their lands [enabling them to apply their experience, knowledge, creativity, and analytical thinking]?

— Can other stakeholders interested in implementation participate easily and effectively?

— Does the SCA address the fact that some of the stakeholders involved will inevitably change in the long term, as new partners emerge and others are replaced?

— Is the SCA operationally and economically practical?

— Does the SCA meet and comply with existing laws and regulations?

— Do the signatories have the legal or decisionmaking authority necessary to implement the provisions of the SCA?

C. Both the process and its implementation should be fun.

Participants recommended beer, barbecue, bumper cars, and donuts.

Applying Findings from the SCA Workshops to Assessing Outcomes of ECR Cases

ECR processes certainly address more than species, but they always address issues located in ecological systems, making the findings from the SCA dialogue pertinent to the problem of assessing environmental outcomes of ECR cases. For example, the increased knowledge about environmental processes and biology resulting from SCAs is likely also true of ECR processes in which scientific and local environmental knowledge has been part of discussions.

Generalized from the findings of the SCA workshops and the resulting SCA concept document, the prospective evaluation checklist in Box 14-1 can be applied to upstream ECR cases. The checklist is designed to assess potential environmental outcomes of cases. Some individual items may not be applicable to specific cases. The checklist should always be completed with the participation of stakeholders.

Discussion

This chapter has dealt primarily with evaluation of upstream ECR cases that involve on-the-ground management planning and implementation. We also need to document the environmental outcomes of the ECR field as a whole and of ECR programs that are funded and operated at the state and federal levels. The prospective checklist can help meet both needs.

Within the field as a whole, the checklist offers a means of standardizing data collected and reported across cases to aggregate and analyze data on

Box 14-1. Checklist for Prospective Evaluation of Environmental Outcomes

Increased Overall Knowledge about the Environment
— Existing data and knowledge on the environmental target were assembled at the outset of the case.
— Information about ___ existing agreements, ___ regulatory structure and climate, ___ relevant program and plans, and ___ jurisdictions were collected at the outset of the case.
— Knowledge gaps were identified.
— Research was undertaken to fill knowledge gaps. (List what has been completed and what remains to be completed during agreement implementation.)
— All groups with knowledge of or stake in the environmental target have been represented in the agreement-development process.
— Stakeholders thoroughly educated each other about their interests and the factors they consider in making individual decisions about the environmental target.
— A process for resolving disputes over science was agreed to early in discussions and adhered to during the process.

Increased Knowledge Held by Individual Stakeholders
— Scientific discussions occurred in front of stakeholders.
— If there was a technical committee, controversies and conclusions were fully disclosed to stakeholders during regular meetings.

Threats, Goals, and Management Actions
— Threats to the environmental target or system were identified and assessed.
— Actions that would address threats were identified.
— Environmental goals and objectives for the process were developed and included in the written agreement.
— Indicators of broad-scale systemic performance were developed during the process.
— Environmental regulators were involved in the development of broad-scale indicators to ensure that regulatory sideboards were addressed adequately.
— Indicators for feedback on management were developed during the process as part of an adaptive management program.
— An on-the-ground monitoring program for participating landowners and volunteers was developed.

— An oversight monitoring program was developed for complex monitoring operations and tracking of implementation.

— The oversight program provides a centralized data collection point to which volunteers can report their monitoring data.

— The written agreement lists who is accountable for accomplishing what actions, by when, and with what funding.

— Someone is assigned to monitor implementation of actions.

Outreach to the Wider Public Resulting in Increased General Knowledge about the Environment

— An ongoing communication strategy was developed early in the process.

— The strategy has been implemented throughout the process.

— Media have publicized the problem under consideration by the group.

— Media have publicized the scientific research completed by the group.

— Media have helped to celebrate successes during the discussions.

— Curriculum based on increased knowledge about the environment has been developed and provided to educators.

— A plan to continue outreach during implementation exists, with assigned responsibilities.

Adequacy of Management Plans in Achieving Environmental Goals at the Time of Agreement Signing

— The agreement is reasonably likely to meet the environmental goals and objectives.

— The agreement is based on the best available science.

— The agreement sufficiently describes adaptive management strategies that are reasonable, logical, and straightforward, and these can be implemented without undue confusion or controversy.

— The adaptive management program framework includes, at minimum, appropriate protocols for administrative and management processes, protocols for monitoring and measuring indicators, and plans for who or how the group will make decisions to adjust management based on monitoring results.

— The agreement and its adaptive management protocols adequately address how conflicts among signatories will be discussed and resolved.

continued on next page

Box 14-1. Checklist for Prospective Evaluation of Environmental Outcomes (continued)

Certainty of Agreement Implementation

— Participants recognize the time required to accomplish goals and objectives.
— Participants have assessed the long-term feasibility of the management approach.
— The agreement-development process enjoyed full support of political stakeholders who can help guarantee resources and other support necessary for implementation.
— The agreement provides enough flexibility for local managers to manage, enabling them to apply their experience, knowledge, creativity, and analytical thinking.
— Other interested members of the public can participate in implementation of management actions easily and effectively.
— The agreement addresses the fact that some of the stakeholders involved will inevitably change in the long term, as new partners emerge and others are replaced.
— The agreement is operationally and economically practical.
— The agreement complies with existing laws and regulations.
— Signatories to the agreement have the legal and decisionmaking authority necessary to implement its provisions.
— The agreement-development process has been fun.
— The agreement-implementation process provides opportunities for participants to have fun with implementation.

---

streams of cases. Aggregate outcomes can then be compared to results obtained through other kinds of dispute resolution or planning processes or in the absence of any process.

With respect to ECR programs, staff members (and subcontractors) have an understandable reluctance to assume responsibility for environmental outcomes of processes in which they serve a neutral role.[23] They have no stake in issues, no control over individual stakeholder behavior, and no responsibility for implementation. If it were otherwise, their neutrality could be compromised for future efforts. However, using an instrument such as the prospective evaluation checklist before a case is completed would provide a fair means to evaluate a program's environmental outcomes overall because it essentially outlines best practices without specifying substantive outcomes. For example, increased knowledge about the environment constitutes a successful outcome even if a process does not reach an agreement that is to be implemented. When a process does reach an agreement, it is reasonable to say that its quality will be higher and that implementation is more likely if the checklist items have been accomplished. Neutral programs and practitioners can assume responsibility for ensuring that the items on such a checklist have been completed, and if not, to state the circumstances and rationale for why they haven't. The latter information, collected and analyzed over time and across programs, provides important information for increasing efficacy in the field.

We need to recognize an important difference in responsibility for environmental outcomes between the ECR field as a whole and ECR programs. The field developed in part because of stakeholder dissatisfaction with existing means of settling environmental disputes. It should offer better outcomes for stakeholders and for the environment than what existed before. Accumulating actual environmental outcome data on streams of ECR cases is critical to establishing the overall record of the field with respect to the environment.

Whereas responsibility for implementation and monitoring of an agreement and its outcomes resides with stakeholders and not with programs, it would be simple for programs to collect data for *retrospective* evaluation of case outcomes as long as one of the checklist items has been accomplished: the delineation of a collection point for monitoring data. Programs can request that monitoring data be routinely reported back to them, perhaps on a yearly basis, so that they can be included in the case file. The number of complex ECR cases programs handle is not high enough that this would require a costly investment of staff time, and such a practice would assist outside researchers who are trying to assess streams of cases from the field as a whole.

Conclusions

This chapter reviewed some of the reasons that we do not know the environmental outcomes of ECR processes. The heterogeneity of ECR processes presents a major challenge to answering this question. Developing a typology of ECR efforts with corresponding methods for assessing their environmental outcomes is a necessary preliminary step.

One category of ECR cases is composed of upstream efforts that engage interested stakeholders in developing management plans for specific locales or aspects of the environment. This chapter reported findings from a national policy dialogue on SCAs that provided a basis for creating a prospective evaluation checklist to use with this category of ECR cases. This checklist could facilitate the evaluation of ECR programs and the field as a whole because it provides a set of best practices for evaluation of cases while enabling collection of standardized data across cases. Finally, the chapter outlined a strategy for streamlining research on actual outcomes of such upstream cases that reach agreement and have moved to implementation.

Upstream policy dialogues such as the SCA dialogue itself make up another category of ECR cases. Environmental outcomes of this category of cases are actually second- or even third-order effects of cases, vastly complicating their evaluation. The results of most multistakeholder policy dialogues require effective dissemination into pluralistic decisionmaking processes if they are to be implemented. However, pluralistic lawmaking processes characteristic of Congress and state legislatures operate under a rationality that is very different from collaborative processes.[24] We need descriptive research and in-depth case studies that examine how the results of ECR policy dialogues are implemented politically so that we can develop a theory of practice concerning the effective articulation of collaborative and pluralistic processes and a corresponding set of best practices. We can work, too, from the opposite direction by mining such fields as epidemiology (which studies how a disease moves through a population), network theory in sociology (which explores connections and communications across nodes), and political ecology (which explores how macro- and micro-level activities and structures influence and affect each other) for methodologies they use to answer analogous questions about articulation and dissemination.

Finally, we need to develop ways to predict policy outcomes under consideration using modeling that accounts for patterns emerging at the level higher than the individual actor. Methods coming from the science of complexity (e.g., agent-based modeling) may assist decisionmaking and provide a means for such prospective evaluation. Where political implementation of the results of ECR processes is successful and on-the-ground implementation has been in place for a time, we can use agent-based modeling retrospectively for assessing these second- or third-order environmental outcomes.[25] However, this research goal is a long way off, likely expensive, and probably not necessary to meet more immediate evaluation needs of the field.

Good descriptive research that can develop and refine valid theories of practice, coupled with an accumulation of standardized data across similar cases, would go a long way toward helping us know the environmental outcomes of our field. We need to develop case databases that collect prospective evaluation data; track indicators developed during ECR processes; and report on the use of adaptive management, including whether adjustments to management have occurred in response to indicator movement.

Notes

1. Gail Bingham, *Resolving Environmental Disputes: A Decade of Experience* (Washington, DC: Conservation Foundation, 1986).

2. Rosemary O'Leary, "Environmental Mediation and Public Managers: What Do We Know and How Do We Know It?" (Indiana Conflict Resolution Institute Research Paper: Environmental Mediation, 1997).

3. Daniel Robinett, "Ecological Sustainability of Ranching in the Southwest" (unpublished manuscript prepared for the Arizona Common Ground Roundtable's Sustainable Ranching Study Committee, 2001).

4. Joshua Epstein and Robert Axtell, *Growing Artificial Societies: Social Science from the Bottom Up* (Washington, DC: Brookings Institution, 1996).

5. O'Leary, "Environmental Mediation and Public Managers."

6. Stephen Edelston Toulmin, *Cosmopolis: The Hidden Agenda of Modernity* (New York: Free Press, 1990). In his book, Toulmin argues persuasively for the need for research that is contextualized, evaluates practice, and accommodates and accounts for diversity, based on his research of the historical conditions that gave rise to the Cartesian search for universal laws. He says that the search for certainty and universal laws at that time was driven by the extreme political instability associated with and advanced by wars over religion. People were looking for ways to find ultimate truths that could resolve the seemingly unresolvable. In so doing, Descartes and his contemporaries derailed the rich exploration of diversity, context, and variability that had been underway as Europe emerged from the Middle Ages.

7. Mette Brogden, "Toward a Theory of Practice for an Applied Political Ecology," paper presented to the Society for Applied Anthropology Annual Meeting, San Francisco, CA, March 24, 2000.

8. Figures were reported by Mark Bosch, U.S. Department of Agriculture Forest Service; Nancy Gloman, USFWS; and Margaret Lorenz, NMFS.

9. Reported by Holly Wheeler, an attorney from the solicitor's office at the U.S. Department of Interior, when she reviewed case law regarding conservation agreements at the first two SCA workshops.

10. *Federal Register* (June 13, 2000, p. 37102). When this paper was written, the proposed PECE policy had not yet been formally adopted.

11. Terry Johnson, "How We Arrived Here and Where We Are Going," in M. Brogden, ed., *Digest of the Midwest Regional Workshop of State Conservation Agreements* (Washington, DC: International Association of Fish and Wildlife Agencies, 2001), pp. 4–6.

12. Summaries of cases, case law, and existing conservation tools are provided in Mette Brogden, *State Conservation Agreements: Report of a National Policy Dialogue Sponsored by the International Association of Fish and Wildlife Agencies* (Washington, DC: International Association of Fish and Wildlife Agencies, 2001). Summaries of presentations are included in the appendices of individual workshop digests, available from me.

13. For the USFWS, "candidate species" are species that are candidates for listing but are of lower priority than other species in more immediate danger. For the NMFS, "candidate species" are species about which they need more information before they can determine the need for listing.

14. See Brogden, *State Conservation Agreements.*

15. The barrens topminnow and Barton Springs salamander are species that were highly specific to tiny locales at the outset of the agreement-development process, making data collection easier and understanding of conservation parameters and

management needs significantly less challenging. Conservation needs of the barrens topminnow were largely known at the outset of developing formalized conservation agreements. Indeed, because the USFWS knew what the species requirements were, it determined that a conservation agreement approach would be the best strategy to achieve conservation, given that the two known wild populations occurred solely on private land. Few species, however, have such simple conservation needs. I am indebted to Lee Andrews, USFWS, for information on this case.

16. Johnson, "How We Arrived Here and Where We Are Going."

17. The truth of this statement is testable, but it would require basic research that is beyond the usual purview of evaluators or applied researchers. It should, however, become part of any program of research for the ECR field as a whole.

18. *Digest of the Mid-South SCA Workshop* (2001).

19. Judith Innes and David Booher, "Indicators for Sustainable Communities: A Strategy Building on Complexity Theory and Distributed Intelligence," (working paper 99–04, Berkeley: Institute of Urban and Regional Development, University of California, 1999).

20. Digest of the Mid-South SCA Workshop.

21. Digest of the Mid-South SCA Workshop.

22. See, e.g., Stephen Edelston Toulmin, *The Return to Cosmology: Postmodern Science and the Theology of Nature* (Berkeley: University of California Press, 1982). See also Toulmin, *Cosmopolis;* Mitchell Waldrop, *Complexity: The Emerging Science at the Edge of Order and Chaos* (Simon and Schuster, 1992).

23. Juliana Birkhoff and Mette Brogden, "Home Rule," paper presented at the Syracuse University–Indiana University Greenberg House Conference, "Evaluating Environmental Conflict Resolution," March 2001.

24. See Brogden, "Toward a Theory of Practice for an Applied Political Ecology."

25. Jeffrey S. Dean, George J. Gumerman, Joshua M. Epstein, Robert Axtell, Alan C. Swedlund, Miles T. Parker, and Steven McCarroll, "Understanding Anasazi Culture Change through Agent-Based Modeling," in T.A. Kohler and G.J. Gumerman, eds., *Dynamics of Human and Primate Societies: Agent-Based Modeling of Social and Spatial Processes,* SFI Studies in the Sciences of Complexity (New York: Oxford University Press, 1990).

15

Economic Characteristics of Successful Outcomes

BONNIE G. COLBY

I dentifying those factors that contribute to successful conflict resolution is a particularly pressing concern. Multiple environmental and public policy (E&PP) disputes are in the process of litigation, negotiation, legislative consideration, or administrative rulemaking in any given week in the United States. In courtrooms, agency hearings, and corporate conference rooms, multiple stakeholders and regulators (along with their respective attorneys and technical experts) square off and expend considerable time and money.

Although some disputes are resolved expeditiously, many involve protracted and bitter struggles, culminating in ambiguous and hard-to-implement court rulings or in administrative rulings that will be contested by dissatisfied stakeholders. Meanwhile, key policy questions are left in limbo, creating further costs and uncertainties. If a "resolution" is achieved, it may neglect to identify financial mechanisms to pay for implementation, to allocate costs clearly, or to anticipate future contingencies. Once these gaps become apparent, stakeholders are faced with another round of potential conflict.

Resolution of E&PP conflicts represents a substantial investment by the public, private, and nonprofit sector. The purpose of this chapter is to suggest strategies to provide useful economic analysis of E&PP dispute resolution and to summarize implications for public dispute resolution programs and policies. The chapter begins by identifying economic criteria and concepts relevant to evaluating dispute resolution efforts and reports preliminary results from a pilot study that applied these criteria to western U.S. water conflicts. Though the chapter frequently refers to water conflicts, the criteria can be applied to a diverse range of E&PP conflicts—from adjacent landowners disputing over pesticide drift, to environmental groups suing the U.S. Environmental Protection Agency (EPA) for failure to enforce water quality standards, to nations arguing over how best to reduce ozone depletion under the Montreal Protocol. The chapter proposes new ways in which economic expertise can be useful and concludes with a descriptive list of economic and financial characteristics that contribute to successful E&PP dispute resolution outcomes.

Economic evaluation of public expenditures can be traced to the early 1800s in the United States, when the Secretary of the Treasury requested comparison of costs and benefits for water projects.[1] In the almost two centuries since then, economic analysis has become commonplace for major federal projects and for new regulations; it is now mandatory for many types of federal actions. Given the emphasis on evaluation of public expenditures in the United States, the time is ripe to consider how best to evaluate the financial and economic aspects of E&PP dispute resolution.

Policymakers and the public seek accountability for the manner in which environmental conflicts are resolved. Public agencies are often stakeholders in conflicts, public resources are expended in addressing conflicts, and issues of public interest—such as air and water quality, endangered species, and management of public lands—are frequently the subjects of disputes. Public officials and taxpayers want to know how much money, time, and other resources were expended and whether the costs incurred were justified by positive results of the dispute resolution process.

Economic and financial evaluation of E&PP dispute resolution outcomes serves a number of purposes. First, it directs the use of public money and agency staff toward more effective dispute resolution. Economic analysis of past efforts also generates information to help stakeholders and mediators learn what strategies are most likely to be effective in resolving current disputes. In addition, thoughtful evaluation suggests changes in public policies and institutions to facilitate more efficient resolution of conflicts.

Economic Criteria for Evaluation of Environmental and Public Policy Dispute Resolution

E&PP dispute resolution literature has focused on achieving agreements, designing processes perceived as fair and inclusive, and building better relationships among stakeholders, with little direct emphasis on economic factors. Numerous studies of individual cases have analyzed negotiation processes and mediation techniques. However, comparison across cases and particularly across dispute resolution methods (litigation versus mediation) is relatively rare, as is critical analysis of the quality of the outcome achieved.[2]

The criteria presented below were developed as part of a larger framework for evaluating "success" in resolving environmental disputes.[3] The framework was developed to compare litigation, administrative remedies, legislation, and negotiated agreements as dispute resolution mechanisms, measuring performance against 27 criteria (see Box 15-1).

The existing literature on resolving environmental conflicts focuses primarily on achieving an outcome (I), process quality (II), and relationships among the parties (V). There is relatively little emphasis on outcome quality (III) and relationship of the parties to the outcome (IV). Only a subset of the original 27 criteria is discussed here, and some of these criteria have been modified for the purposes of this chapter.[4]

Dispute resolution involves both processes (such as litigation or multiparty negotiations) and outcomes. "Outcomes" include court rulings, legislation, public agency administrative actions, and negotiated agreements. Typically, a complex case includes several phases with successive processes and outcomes. For instance, a dissatisfied stakeholder may initiate litigation, prompting negotiations, which eventually result in legislative action to appropriate public money for implementation of a negotiated solution. Consequently, any evaluation must specify which phases of the case (processes and outcomes) are being analyzed. In this chapter, the focus is on economic and financial characteristics of the dispute resolution outcome. The larger d'Estrée and Colby[5] evaluation framework addresses many additional elements of the dispute resolution process and outcome.

A pilot evaluation of environmental disputes involving water in the western United States began in 1998, with case researchers instructed to rely only on publicly available sources to evaluate each case based on 27 criteria. The reliance on accessible sources was intended to test whether dispute resolution processes and outcomes could be evaluated in a low-cost and nonintrusive manner, without the use of stakeholder interviews. This pilot project indicates that economic criteria can be difficult to assess without obtaining additional information directly from stakeholders. However, for environmental conflicts involving National Environmental Policy Act (NEPA) processes, economic data are available from environmental impact reports, assessments, and statements. For conflicts involving expenditures of federal dollars, the Office of Management and Budget, the General Accounting Office, the Department of Interior's Office of Policy Analysis, and the Congressional Budget Office sometimes perform economic or financial assessments.

Access to economic data varied substantially across the cases. In principle, expenditures by public agencies are public information. However, agencies rarely compile cost data in a systematic manner on a case-by-case basis. Expenditures by corporations and nonprofit organizations to participate in a particular E&PP dispute resolution process and to implement outcomes are not public information and are available at the discretion of the organization.

Box 15-1. Effective Environmental Conflict Resolution Criteria Categories

I. Outcome reached
    Unanimity or consensus
    Verifiable terms
    Public acknowledgement of outcome
    Ratification

II. Process quality
    Procedurally just
    Procedurally accessible and inclusive
    Reasonable process costs

III. Outcome quality
    Cost-effective implementation
    Perceived economic efficiency
    Financial feasibility and sustainability
    Cultural sustainability and community self-determination
    Environmental sustainability
    Clarity of outcome
    Feasibility and realism (legal, political, and scientific)
    Public acceptability

IV. Relationship of parties to outcome
    Outcome satisfaction and fairness as assessed by parties
    Compliance with outcome over time
    Flexibility
    Stability and durability

V. Relationship between parties (relationship quality)
    Reduction in conflict and hostility
    Improved relations
    Cognitive and affective shift
    Ability to resolve subsequent disputes
    Transformation

VI. Social capital
    Enhanced citizen capacity to draw on collective potential resources
    Increased community capacity for environmental and policy
        decisionmaking
    Social system transformation

Source: Tamra Pearson d'Estrée and Bonnie G. Colby, *Guidebook for Analyzing Success in Environmental Conflict Resolution,* ICAR report 3 (Fairfax, VA: Institute for Conflict Analysis and Resolution, 2000).

To illustrate how certain criteria can be applied, summaries of preliminary comparisons between litigated cases and cases resolved through negotiated agreements follow. These comparisons are based on eight cases in the d'Estrée and Colby pilot analysis, my earlier research on a dozen tribal water cases,[6] and my research on a dozen additional cases reported in National Research Council[7] and the Western Water Policy Review Advisory Committee Report.[8] Whereas this body of cases cannot be construed as a representative random sample of any particular type of dispute, it does represent an impressive cross section of western U.S. water conflicts and is useful for a pioneering attempt to compare and evaluate dispute resolution mechanisms and outcomes.

The 10 criteria discussed below examine economic and financial aspects of implementing an outcome (such as a court ruling or a negotiated agreement) in a particular dispute. Though presented here as criteria upon which public E&PP conflict resolution programs focus, these also are characteristics of successfully implemented outcomes. Successful cases are the foundations for successful dispute resolution programs. Economic factors should be examined across parallel sets of cases that were resolved using different processes to learn whether one mechanism, such as litigation, tends to be more costly than other approaches, such as negotiated agreements. Comparable data from similar cases resolved using different mechanisms is limited. However, over time, with a sufficiently large number of carefully documented cases, it will be possible to more thoroughly compare the costs of litigation, mediation, administrative actions, legislative remedies, and other means of resolving E&PP disputes.[9] The following criteria are discussed in this chapter as economic characteristics of successful dispute resolution outcomes:

— positive net benefits;
— well-defined and measurable objectives;
— cost-effective implementation;
— financial feasibility;
— fair distribution of costs among parties;
— flexibility;
— incentive compatibility;
— improved problem-solving capacity;
— enhanced social capital; and
— clear documentation protocols.

The *positive net benefits* criterion asks whether the dispute resolution outcome creates for the party's net benefits (benefits minus costs) that would not have been available otherwise.[10] Trades among the negotiating parties are the basis for producing net gains. Such bargaining can involve many different types of assets—trades of pollution control improvements for cost-sharing money, or lending political support for a new regulation in exchange for compromise on how quickly the new regulation will be phased in. In the case of voluntary, negotiated agreements, this criterion is almost always satisfied. If the agreement fails to provide improvements for those who sign on, compared to their BATNAs (best alternative to a negotiated agreement), they would decline to bind themselves to the agreement.[11] Litigated outcomes, and other outcomes that do not involve voluntary consent of the parties, are unlikely to satisfy this criterion.

Net benefits may arise from avoiding the costs of prolonged litigation, from improved natural resource management, from cleaner air or water, and from better sharing of information and technology among the parties. Many analytic challenges arise in documenting and quantifying the various types of benefits that arise from resolving a dispute. These are discussed in a later section of the chapter.

Well-defined and measurable objectives are essential to a successful, implementable outcome. The parties may be able to agree, in principle, to improve air quality or to reduce regulatory costs. However, the devil is in the details— in this case the precise and measurable objectives that the parties are willing to commit to. Agreements often emphasize broad, vague goals. These may represent important breakthroughs, but they are easier to agree upon than specific actions because their vagueness imposes no costs or threats. However, they are not implementable without specificity as to the degree of improvement or change required, computed from a mutually agreeable baseline and measured using particular techniques at specific locations and time intervals. Details of this type are frequently absent in both court decrees and negotiated agreements, leading to problems with implementation. Many agreements hailed as successful when initially achieved flounder during implementation.

The next two criteria apply specifically to the implementation plan developed to achieve objectives. The implementation plan must balance specificity and flexibility. For instance, a plan to reduce pollution emissions can either instruct power plants to install specific pollution control equipment or give plant managers discretion to choose their compliance methods and require submission of an annual compliance plan.

Cost-effective implementation also applies to the implementation plan and includes a number of subcomponents. When the outcome of an E&PP dispute resolution process sets specific goals (e.g., a 10% increase in summer stream flows for fish, or a 20% reduction in SO_2 emissions by power plants), this criterion asks whether the goal is being achieved in a cost-effective manner. Implementation costs are likely to be higher under court rulings because courts are not required to consider costs as an element in crafting their rulings.[12] Rather, courts are focused on rights and consistency with the existing body of law. In contrast, legislative mandates, administrative actions, and agreements negotiated among stakeholders are more likely to carefully weigh costs to limit financial burdens on taxpayers, firms, and property owners, and because the parties negotiating the agreement will bear some of the costs themselves.[13] To date, only incomplete data are available on implementation costs for a cross section of cases, so the rationales offered to explain cost differences and similarities cannot be verified empirically. Implementation cost data are incomplete for several reasons: costs incurred by private parties are not public information; costs incurred by public agencies are not compiled systematically; and implementation takes many years for some cases, delaying the examination of complete implementation costs.

Cost comparisons are confounded by the fact that stakeholders receive different "products" for the money they invest in different processes. In litigation, the ideal payoff for environmental advocates is a ruling that both favors their position in the particular case at hand and sets a favorable precedent for future disputes. In market transactions, the payoff is acquisition of the land and water needed to resolve a specific environmental problem. Investments in different dispute resolution strategies provide differing types of results, and this makes comparisons complex.

Evaluation of cost-effectiveness needs to consider not only costs to the direct participants in the dispute resolution process (litigants and negotiants), but also costs to public agencies, courts, taxpayers, and more dispersed interests, such as ratepayers, recreationists, and property owners in the affected region. Some costs to consider when evaluating cost-effectiveness include direct monetary outlays, contributions of staff time and natural resources (such as water or electric power), and costs of raising money for implementation. Transaction costs are a subset of the overall costs involved in implementing a dispute resolution outcome and are particularly easy to overlook.

Transaction costs are "information, contracting, and enforcement" costs and include verifying legal rights (such as land ownership) and regulatory requirements, gathering data on compliance, assessing and collecting penalties, and monitoring the condition of the natural resources that are the subject of the agreement. Transaction costs have a powerful (but largely unquantified) influence on dispute resolution. They may arise from public policies that require stakeholders to follow specific procedures in the quest to solve a problem. Examples include court formalities to undertake litigation, mandatory state agency review of proposed water transactions, and the various NEPA procedures for major federal actions. The effects of such policy-induced transaction costs have been shown to influence disputes over water transfers,[14] and analysis of water and environmental cases suggests that they are a significant influence on the strategies that stakeholders employ in resolving environmental conflicts.

Stakeholders in E&PP dispute resolution also possess, to varying degrees, the ability to impose delays and direct costs on one another. This ability is an important component of their bargaining power. When an environmental group threatens to sue a regulatory agency or a resource user (with accompanying high transaction costs), and the threat is perceived as credible, the other stakeholders have a powerful incentive to compromise. Possible actions include litigating to force the designation of critical habitat for a federally listed endangered species, or forcing an agency to regulate nonpoint source discharges to comply with water quality standards. Administrative processes can also be used to impose transaction costs on others. Filing an objection to another parties' proposed water transfer could delay approval for many months and require costly administrative hearings. Consequently, a threat to object could cause other parties to accommodate the potential objector's concerns.[15]

Financial feasibility is concerned with the mechanisms adopted to cover implementation costs. Different stakeholders have widely differing abilities to levy taxes, issue bonds, raise investment capital, and charge user fees. Moreover, these various financial instruments, with different costs per dollar made available for implementation, so relying on one instrument over another, could save money. Financial feasibility examines the ability of different stakeholders to cover their share of costs, identifies unfunded mandates (increases in agency responsibilities without commensurate increases in staff and budget), scrutinizes unrealistic reliance on federal money not yet appropriated, and examines the costs of the overall plan to raise money needed for implementation, looking for potential cost savings.

Negotiated agreements sometimes defer costs and hard choices about resource management to the future, transferring the burden from those currently involved in the dispute to future decisionmakers and disputants. Financial mechanisms that shift repayment of loans and other costs to future periods may later prove infeasible—particularly if predictions of the future ability to pay are based upon unrealistic assumptions about future growth in tax base or continued access to public subsidies (such as low-cost water and electric power).

Litigated outcomes assign costs to parties but generally do not address financial mechanisms, leaving the parties to figure out how to meet their obligations. Negotiated agreements tend to be more specific in identifying mechanisms to cover implementation costs. However, among the negotiated agreements evaluated, there were numerous instances of optimistic reliance on federal dollars not yet secured and of failures to specify how increased agency responsibilities would be funded.

The next two criteria address distribution of costs and benefits. *Fair distribution of costs among parties* is problematic by definition because perceptions of fairness vary among stakeholders. Nevertheless, fairness appears consistently on stakeholders' and policymakers' lists of desirable E&PP dispute resolution characteristics.[16] The distribution of costs among parties is an issue distinct from cost-effectiveness and controlling the overall costs of a dispute resolution outcome. Ideally, the stakeholders first will identify the most cost-effective plan to accomplish the objectives they have agreed upon. Then they can use cost-sharing principles and compensation packages to even out discrepancies between parties who gain a lot from the outcome and parties who gain very little or face high costs of implementation.

There are many possible cost-sharing principles, each with differing implications for different stakeholders. One principle, for instance, is sharing costs in proportion to the benefits received from an agreement ("beneficiaries pay"). Another possible principle is sharing costs in proportion to past damages ("polluter pays"), or bearing costs proportional to one's financial assets ("deep pockets pay"). There may be opportunities to assess dispersed interests (who may not be direct parties to an agreement) for some implementation costs—recreationists, for instance, who will benefit from environmental restoration could pay higher access fees. Two common principles for sharing pollution control costs are equal proportional emissions reduction and costs prorated based on tax base or customer base among utilities or local governments. Common principles in western water conflicts include no diminishment of existing water rights, federal government pays for environmental improvements, and Native American tribes receive compensation for past wrongs to develop water for reservations.

There are clear cost distribution differences between negotiated and litigated dispute resolution strategies. In a voluntary process (such as multiparty negotiations), the primary burden is on those parties seeking change. They must call the relevant stakeholders together and initiate a bargaining process. They must offer sufficiently attractive financial (or other) inducements to persuade holders of rights to sell or lease their land and water, or to donate conservation easements and consent to changes in dam operations and land management.[17] In a litigation framework, the burden of initiating litigation also falls on the advocates for change. However, once the legal process is set in motion and affected parties begin to take it seriously, they, too, must spend money on attorneys, experts, and court costs. The cost burden is spread among the stakeholders, providing impetus to settle the problem. As noted earlier, the "fairness" of different cost distributions depends on the benefits different parties obtain from resolution, and perceptions of fairness vary among stakeholders.

Flexibility is another criterion with economic implications. Flexibility refers to the ability of the dispute resolution outcome to withstand changing conditions and unexpected events. Ideally, outcomes will be responsive to natural contingencies (drought or a disease outbreak affecting a particular species), as well as to political and economic contingencies, such as a change in federal administration or a recession.[18] Negotiated agreements, legislative solutions, and agency actions have begun to specifically provide for adaptive management, altering resource management in response to resource needs, changing conditions (e.g., flood or drought), and biological conditions (e.g., pest infestation).[19] Adaptive management, in the past 10 years, has come to be viewed as essential to the success of environmental problem-solving efforts.

Court rulings do not contemplate changing natural resource and public policy needs over time, unless key legal issues raised by the parties require them to do so. Consequently, rulings are unlikely to explicitly provide for flexibility. Litigation may be reactivated if court-ordered actions prove unsatisfactory as conditions change over time or are unsuited for unusual conditions such as drought. Improved flexibility is a hallmark of negotiated transactions, and rightly so, when compared to litigation. However, the cases examined suggest that simple market transactions (such as purchases of water rights) may not be sufficiently sophisticated to meet environmental needs. Consider, for instance, restoration of a fishery or an aquatic ecosystem. These require flows that vary seasonally and mimic the natural hydrograph.[20] A market acquisition of water rights can improve base streamflow conditions, but more complex, flexible arrangements with upstream dam operators and irrigation districts are needed to manage rivers in ways that approach pre-dam conditions in flow levels, water temperature, and flood magnitude and frequency. Examples of such arrangements include dry-year options and contingent water leases. Several such arrangements have been negotiated in the western United States to address water conflicts.[21]

Incentive compatibility means that the dispute resolution outcome (e.g., a negotiated agreement, court ruling, or legislation) generates signals that assist, rather than obstruct, successful implementation of the outcome. Two specific elements of incentive compatibility are incentives to comply with the terms of the agreement and incentives for more efficient resource use and conservation.

Outcomes that specifically incorporate economic incentives for compliance will involve lower monitoring and enforcement costs than those that rely on command-and-control directives or that fail to consider incentives at all. Compliance incentives can come in the form of rewards or penalties. Ideally, the dispute resolution outcome will specify consequences for violations and allocate money for monitoring and enforcement. For instance, settlements of Native American water claims sometimes specify a penalty to be paid by the federal government if water is not developed and delivered to the reservation by the date promised.[22] High penalties are not effective in inducing compliance, however, if there is little monitoring and therefore little probability of actually being caught and fined.[23] For instance, disputes to reduce overfishing require patrolling of fishing waters and monitoring of fish landing ports. Farmers' agreements to use best management practices need on-farm inspection for monitoring.

With respect to incentives for better resource management, some dispute resolution outcomes specifically provide for market transactions—tradable water rights or discharge permits. Market transactions create incentives by providing a known market price for the resource being traded (e.g., water rights). That price shows resource users that water has value beyond their own immediate use of it. Irrigators, for example, will realize that on-farm water conservation may enable them to sell or lease the water no longer needed for irrigation, and this opportunity provides an incentive for more efficient water use.[24] Water prices set by an administrative agency can be compatible with resolving a dispute over scarce water, if they are structured to encourage reduced consumptive use.[25] However, subsidized water prices, or failure to link water bills to water use, do not provide incentives compatible with the goal of leaving more water in streams to satisfy environmental needs in disputes for which in-stream flows are a critical issue. If wastewater dischargers can buy and sell transferable discharge permits, they have an incentive to be more cost-effective.[26]

Litigated outcomes seem to give less consideration to economic incentives than other dispute resolution outcomes. Courts generally do not have jurisdiction to alter resource prices, and court rulings seldom are incentive compatible. Judicial processes do send incentive signals, deterring violation of established environmental policies that would bring the violator before the courts, with attendant costs and uncertainties.

Analysis of a cross section of western water disputes indicates that court rulings, agency rulings, legislation, and negotiated agreements do not routinely consider incentive compatibility. The potential benefits of more emphasis on economic incentives in E&PP dispute resolution are large.

Another criterion with economic implications is *improved problem-solving capacity*. The stakeholders engaged in E&PP conflicts often must address multiple resource problems over a period of years.[27] For instance, this year's conflict may concern providing in-stream water for endangered fish recovery, but in the next few years the same stakeholders might confront a drought or a water quality problem. Consequently, ability of the stakeholders to work together effectively can be an important asset. Negotiated agreements provide some clear advantages over litigation processes because they engage stakeholders in identifying strategies to accomplish restoration, debating their merits, allocating the cost burden, and building consensus for a particular approach.[28] The process gives stakeholders experience in teamwork, which can make solving the group's next problem much easier. In contrast, litigation encourages an adversarial approach among the parties rather than a problem-solving stance.[29] After stakeholders solve a conflict through multiparty negotiations, one would expect subsequent disputes to be solved with lower cost problem solving than would occur after resolving an earlier conflict through litigation. At this time, adequate longitudinal data on subsequent phases of related disputes are not available, so this hypothesis cannot be tested empirically.

Enhanced social capital is a broad criterion that encompasses improved problem-solving capacity along with many other elements.[30] Social capital includes the potential benefits, advantages, and preferential treatment resulting from one person or group's sympathy and sense of obligation toward another person or group. Social capital encompasses mutual obligations and expectations of others' behavior. It affects economic and political transactions by altering the terms of agreements reached and the costs of achieving those agreements and monitoring and enforcing their implementation. Like economic assets, social capital requires investment to maintain its value. The economic benefits of possessing social capital serve as an incentive to refrain from behaviors that reduce one's social capital, such as violating the terms of an agreement or failing to follow through on a commitment. Enhancing social capital can reduce transaction costs (such as attorneys' fees) and improve the productivity of other inputs (such as managers' time spent in negotiations).[31]

Empirical research on the role of social capital in business transactions and government activities documents the economic importance of social capital and provides some insights on the magnitude of its role in reducing costs, addressing uncertainties, and enhancing economic performance.[32] The role of trust and reputation has also been shown to be an important factor in bargaining situations.

Analysis of dispute resolution cases makes it evident that stakeholders' social capital affects their ability to obtain an outcome that addresses their interests. Moreover, groups of stakeholders who encounter one another over multiple rounds of problems in managing a natural resource (such as a river basin or regional air quality) clearly can build up trust (one element of social capital), as they deal with each other over time. This enhanced trust can, in turn, lead to easier, lower cost problem solving. The stock of social capital held by a stakeholder can also be a partial substitute for other means to obtain a desired outcome, such as wealth, strong legal arguments, or political connections.

Clear documentation protocols are essential to ongoing implementation of a dispute resolution outcome and to providing data for improved future understanding of dispute resolution processes and outcomes. As many chapters in this volume emphasize, systematic collection of dispute resolution data will further our collective learning. Ideally, reporting requirements would be specified up front in the negotiated agreement (or court ruling). Parties could be asked to report monthly on implementation expenses and on the progress of specific implementation activities, with a central repository established to compile and track such information. Also, regular reports on changes in resource pricing and resulting changes in use of the resource (known to economists as "own-price elasticity of demand") would be useful for future efforts to adjust resource use and promote conservation through changes in prices. Data on specific cases provide the foundation for evaluating public programs that manage dispute cases.

To summarize, these 10 criteria (along with many others described in d'Estrée and Colby[33] were developed for a pilot effort to compare across dispute resolution cases, knowing that economic analysis would necessarily be rudimentary due to limited data. Useful observations have emerged from applying these criteria, and a checklist for evaluating the economic and financial aspects of negotiated agreements is provided in the appendix of this chapter.

Economic Principles for Evaluation of Dispute Resolution Outcomes

A number of economic principles and concepts are important in economic and financial evaluation of outcomes.

Defining an Appropriate Baseline from Which To Evaluate Dispute Resolution Outcomes

The baseline is a crucial concept in evaluating a dispute resolution outcome (the court ruling, negotiated agreement, or legislation). The analytic goal is to identify effects specifically caused by the outcome being evaluated. However, also likely to occur are other changes that are not attributable to the specific outcome but that affect the same region and natural resources. For instance, a water district may vote to increase water rates for reasons unrelated to the outcome under scrutiny. The increased water rate and the outcome will affect water use patterns simultaneously.

To differentiate between the effects of the outcome and changes occurring due to other factors, it is necessary to define a baseline. The baseline consists of those conditions that would exist in the absence of the outcome being analyzed. Effects due to the outcome are those that would *not* have occurred without that outcome. This is the "with and without" principle—attributing to the dispute resolution outcome only those effects that would not have occurred anyway.[34] For instance, if a fish species is declining and the E&PP dispute resolution outcome provides for fish recovery 100,000 acre feet per year of additional water that would not otherwise have been available for this purpose, then improvements in fishery conditions linked to the new 100,000 acre feet can be properly described as benefits of the outcome. Now suppose that during the time period being studied, unusually favorable rainfall in one year brings another 200,000 acre feet for the fish. The outcome should not receive credit for these additional fishery benefits.

It is essential to separate the economic effects of general regional changes in resource use and management from changes that are properly attributable to the outcome being evaluated. However, judgement calls are necessary to isolate the effects of a specific outcome from the other events that affect the resources and the region involved in a specific E&PP dispute resolution case.

Quantification of the costs and benefits of a particular E&PP dispute resolution process and outcome relies upon defining the appropriate counterfactual. What does one conjecture would have occurred in the absence of the actual outcome? Would the environmental or policy problem have continued without any resolution? Would a different E&PP dispute resolution process and outcome have occurred? In cases involving negotiated agreements, litigation seems to be a natural baseline for comparison because nonlitigation alternatives evolved in response to dissatisfaction with expense, delays, and uncertainties of litigation.[35] First, litigation and other processes are not mutually exclusive options. Most litigation is settled by negotiation rather than a full trial. (More than 90% of civil cases settle before the trial, concludes Bingham.[36]) Sometimes, negotiations or legislative solutions commence after litigation has begun. Both litigation and negotiation require collection and analysis of technical information, preparing positions, and analyzing trade-offs among different outcomes. Moreover, litigation often provides the incentives necessary for negotiated agreements to be achieved. In addition, different processes generate different "products," so comparison of costs alone would neglect potentially large differences in benefits generated by negotiation versus litigation for different parties. Litigation may not be an option for the case at hand if administrative processes have not yet been exhausted.[37] In these instances, an administrative hearing and ruling might constitute the most realistic baseline. Still, one must conjecture regarding the length, cost, and outcome of the hearing.

To date, there is a dearth of similar, comparable cases that have been addressed by different processes and were carefully documented so that rigorous comparisons can be made. Data that are available on environmental litigation suggests that it is more protracted than civil litigation in general.[38]

Difficulties Agreeing on a Baseline

Careful attention to the baseline is not merely an essential analytic principle. Disagreement about the baseline (what will occur in the absence of an agreement) actually forms the core of many environmental conflicts. Different stakeholders have differing (and often contradictory) views of their baseline rights. For instance, tribal governments and non–Native American water users rely on differing legal doctrines to buttress their claims to exactly the same water sources.[39] A court ruling that allocates to a tribe water customarily used by non–Native Americans leaves the non–Native American water users claiming a large welfare loss, whereas the tribe argues that the water was never theirs (the non–Native Americans) to lose and no compensation is due to the non–Native Americans.

Another contemporary example of conflict rooted in differing baselines is the disagreement over who should bear the costs of environmental restoration in the California Bay Delta. Environmentalists argue that, because the degradation of the Bay–Delta ecosystem is intimately linked to decades of massive water diversions, water users should primarily bear restoration costs. A variant of the "polluter pays" principle, they argue that those who caused the problem should clean it up. Although there is an element of fairness in this argument, large water users respond that they were playing by the rules of the era in developing and diverting water over the decades of the 1900s. To retroactively penalize them now for behavior that was fully encouraged by earlier laws and policies would be unjust. In fact, water users argue, tax dollars and environmental agencies should pay for restoration because new social values and concerns are demanding it. This is a variant of the "beneficiary pays" principle. Environmental advocates respond that such a principle ignores more than 100 years of environmentally damaging water development activities—much of them taxpayer funded. The debate continues.

There is no simple strategy that stakeholders can use to reach agreement on the baseline from which progress in achieving implementation will be measured or on the baseline for determining cost shares or shares of access to the benefits of a negotiated agreement. In some instances, the case evaluator might need to select a baseline for evaluating progress toward implementing a dispute resolution outcome, even though the disputants themselves have not agreed upon a baseline.

Well-Defined Accounting Stance

Careful definition of accounting stance is another crucial analytic decision in evaluating dispute resolution outcomes. The accounting stance determines how widely (across time, layers of parties, and geographically) to count costs, benefits, and other effects.

Here is an example of an explicit statement regarding accounting stance for a hypothetical case:

> The conflict over water allocation on the Middle River has been ongoing since the early 1900s. In this evaluation, we examine the period 1960 to the present. The primary parties are farmers, cities, and anglers. Boaters are also affected by the conflict, but were not key players in the process. Consequently, there is little information on effects on boating, and we do not include boater effects in our detailed analysis. The conflict affects river management in several downstream states, but the primary effects

are in Nebraska. We do not assess effects in downstream states because there is little information, and these effects are peripheral to the process and outcome we are analyzing, which occurred within Nebraska. We count costs and benefits to federal taxpayers, but examination of all other economic effects is limited to Nebraska.

As the example illustrates, any evaluation of a dispute resolution outcome must clearly state the time period and geographical area the analysis covers and the range of parties it considers. There could be legitimate reasons for excluding some time periods, regions, and parties (i.e., they are not central to the case, or limited information is available), and these reasons must be explained. The issue of whether to focus on local, regional, or national costs and benefits is one commonly encountered in benefit–cost analysis. However, selecting a time period to cover is uniquely complicated for E&PP dispute resolution. Many disputes continue over decades in varying forms with no clear beginnings or endings. The case analyses on which this chapter is based described the history of each conflict in its entirety (the longest case extended over 90 years). However, a discrete time interval and outcome were selected for detailed analysis—a particular court ruling or negotiated agreement that represented a turning point in the case.

Benefits and Costs over Time

With respect to accounting for benefits and costs over time, several interesting issues arise in economic evaluation of environmental conflict resolution. Economists have well-defined techniques, using net present values and discounting procedures, for comparing benefits and costs occurring at different points in time. However, one must estimate the longevity of benefits that arise from resolving a particular phase of a dispute. For instance, achieving a collaborative negotiated agreement may produce smoother, more cost-effective intergovernmental working relationships. Are these benefits assumed to grow over time or to decay, to remain robust in the face of new conflicts or to dissipate? Litigation outcomes face the converse issue. Does the immediate animosity (and inclination to impose transaction costs on other parties) abate over time or grow?

Potential New Contributions by Economists to Conflict Resolution

I have highlighted limitations in rigorous economic analysis of dispute resolution processes, but economists can also make immediate contributions, both to specific, ongoing dispute resolution efforts and to furthering our collective ability to provide more well-informed advice to policymakers, dispute resolution programs, and stakeholders in the future.[40]

Economists have long played a role in dispute resolution as consultants to different stakeholders and as public agency analysts, but the time is ripe for a new role. Many E&PP disputes involve professional facilitators or mediators. Substantial benefits might come from economists serving as neutral advisors to E&PP dispute resolution processes—working in concert with the process mediator or facilitator.

Acrimonious stakeholders often find themselves beginning to solve problems together as they examine potential solution scenarios. An economic advisor retained by the whole group to model various scenarios and to estimate their costs, benefits, and financial implications would be valuable. When someone in stakeholder negotiations proposes a solution, the first question (after, "Will it actually solve the problem it is intended to address?") is, "How much will it cost?" followed by, "How will it be paid for?" An economist, ready to estimate costs over time and the financial implications of cost-sharing alternatives, would greatly facilitate this phase of negotiations. Without such a neutral advisor, the stakeholders go back to their own consultants, who make cost estimates based on differing assumptions tailored to their client's interests.

In addition to estimating costs and financial implications, a neutral economic advisor could provide many other services to a group of stakeholders (see the checklist in the appendix). These services include guiding the language of proposed agreements to include incentive-compatible resource pricing and appropriately structured compliance incentives and penalties. The issue of compensation for parties who will have reduced access to resources (or bear increased costs) inevitably arises in E&PP disputes. The neutral economist can suggest levels of compensation based on documentable welfare losses or increased costs, bringing some measure of balance to a typically bitter debate about whether compensation is appropriate and how much should be paid. In general, there are alternative mechanisms to pay for implementation costs—issuing bonds, user fees, and tax increases of various types. The advisor can compare the advantages and disadvantages of financial mechanisms available to the parties. A neutral advisor can be on the alert for externalities imposed upon third parties who are not involved in the negotiations, so that the stakeholder group can proactively mitigate third-party concerns and avert political opposition to the proposed solution. The economist can also note the taxpayer and property owner burdens associated with proposed solutions, so that opposition on these fronts can be anticipated.

A group may identify goals essential to achieving an agreement, such as 25% increases in summer stream flows for endangered fish, or an additional 100,000 acre feet ensured for a city during drought. Once a goal is clearly articulated, economists can present and compare different means to achieve that goal. They can compare, for instance, agricultural and urban conservation practices as methods of reducing consumptive use and increasing stream flows, or water leasing from farmers and revised operating criteria for upstream dams as ways to improve urban supply reliability. This type of comparison, provided in the midst of negotiations, will highlight the most cost-effective means of addressing the conflict and might help reduce the incidence of extravagant solutions to resource problems.

In addition to assisting E&PP dispute resolution efforts that are underway, economists may be able to contribute to conflict prevention and to ex post facto evaluation of recently completed E&PP dispute resolution processes. "Common sense" economic principles are often overlooked, despite their potential for reducing the public costs of conflict.

First, prices charged for use of public resources (timber from national forests, water from public projects, electricity from federal dams, forage on Bureau of Land Management lands, recreation in the national parks) should reflect their scarcity value and any externalities (third-party effects) associated with their extraction and use. Economically rational resource pricing has the potential to reduce conflicts associated with subsidies and with resource degradation. Second, resource problems should be addressed by altering resource use incentives when possible, rather than by (generally more expensive) engineering and technical solutions. For instance, the construction of the Yuma Desalting Plant in the 1980s, as a solution to an international dispute over salinity in the Colorado River, violated all notions of cost-effectiveness. Farmlands contributing much of the salt loadings could have been bought out and retired for a fraction of the cost of constructing the plant.[41] Similar examples abound, and any economist working on applied policy problems has his or her own favorite anecdotes of costly solutions to problems that could have been mitigated by better use of economic incentives.

Conclusions

There is some reason to be hopeful that more comprehensive case-level data will be available to evaluate the economic aspects of E&PP dispute resolution in the future. The U.S. Institute for Environmental Conflict Resolution, established by Congress in 1998 to assist in resolving disputes involving federal agencies and federal interests, is establishing a standardized framework for documenting all cases under its auspices. In addition, most U.S. states now have conflict resolution programs housed within the state court system or in the administrative branch of state government. These programs have the potential to provide more comprehensive data on E&PP cases than have been available in the past.

Public dispute resolution programs and policies can benefit from becoming more cognizant of economic and financial aspects of negotiated agreements, such as the items listed in the checklist in the appendix. Dispute resolution professionals can enhance their abilities to serve their clientele by learning more about the role of economic incentives in structuring agreements, allocating implementation costs, designing compliance mechanisms, and raising funds for implementing agreements. Economists—even those who specialize in public policy and environmental matters—are not well acquainted with the field of dispute resolution. There is need for outreach and mutual exchange in both directions.

Environmental conflicts and their resolution consume public, nongovernmental organization (NGO), and private-sector resources. There has been relatively little economic analysis of E&PP dispute resolution outcomes. Whereas it is desirable to evaluate the economic and financial aspects of dispute resolution outcomes, a pilot effort to do so found notable data limitations. Several analytic challenges must also be faced to evaluate the economic and financial implications of a specific case or to compare across cases. Despite these challenges, economists can fruitfully contribute on many fronts: advising specific, ongoing dispute resolution efforts; helping public dispute resolution programs and policymakers to more fully integrate economic aspects of dispute resolution into their activities; and establishing documentation protocols that will allow more complete economic assessments in the future. Possibly because E&PP dispute resolution has received relatively little attention from economists, the marginal gains from further research and professional involvement are high.

Appendix. Negotiated Agreement Checklist: Economic and Financial Considerations

Note: All examples below are illustrative and are not intended to describe actual cases.

1. Does the agreement specify well-defined baselines and measurement protocols for each objective it identifies?

   Example: "Irrigators will reduce their water use by half," vs. "Irrigators will reduce their water diversions to 150,000 acre feet per year (50% of their 1990 diversions), as measured weekly at Irrigation District Pumping Station 1A with records to be compiled by the Bureau of Reclamation."

   Example: "Power plants will improve their pollution control efforts," vs. "The 20 power plants located in EPA Region X will reduce their overall SO_2 emission levels to 50% of baseline 1995 emissions, as measured and recorded by the EPA using round-the-clock monitoring devices, to be installed at each plant at the expense of the power plant owners.

2. Does the agreement include the principle of achieving goals and raising funds to cover implementation costs in a cost-effective manner, with a commitment to use cost-sharing agreements and transfer payments to make equity adjustments?

Example: The irrigation district will borrow money under a federal program at low interest rates, and other parties agree to make regular payments to the district to assist with repaying the loan, which provides funds for implementation programs that benefit other stakeholders.

Example: A conservation organization has donor funds to immediately purchase an environmentally valuable property, without incurring debt to do so. Other stakeholders agree to repay the purchase price over time, recognizing the benefits of the acquisition in implementing a negotiated agreement.

3. Are the actions that different parties must take to achieve the goals of the agreement clearly specified? Agreements can be very specific regarding technologies to be used and management practices to be followed, or the parties may be left free to evaluate and use whatever means for achieving the goals seem best to them. Specific requirements make it easier to verify compliance but may limit parties' flexibility to adapt to changing conditions and to explore new opportunities to accomplish objectives in a more effective manner.

Example: "The irrigation district must implement the five best management practices for water conservation detailed in *Bureau of Reclamation Manual 1996B*, with full implementation to be certified by the Bureau no later than December 31, 2001."

Example: "Power plant operators must evaluate the comparative costs and effectiveness of installing scrubbers (or other technologies) versus trading discharge permits among plant locations and present an annual compliance plan to cost-effectively achieve compliance for the following year."

4. Are the mechanisms selected the most cost-effective means of achieving goals? Incentive-based mechanisms generally are more cost-effective and flexible than mandating a specific technology or management practice. If costs change or new technologies become available, then the specific actions that seemed most desirable when the agreement was drafted may become outmoded.

Example: "The irrigation district commits to alter its water rate structure to promote water conservation and to establish trading mechanisms for water permits to promote more efficient water management and to comply with the annual cutback in water use mandated in the negotiated agreement."

5. Are the types of costs involved in implementation fully identified, including direct monetary outlays, contributions of staff time and other resources, costs of borrowing and raising money, and transaction costs?

6. Are the costs to all affected parties considered—those stakeholders at the table, public agencies, taxpayers, and dispersed interests not at the table who may be affected?

7. Are cost-sharing principles and compensation packages well defined? Keep in mind that costs include not only monetary outlays, but also reduced access to natural resources.

   Example: "Historical fish catch quotas will be cut 20% during the next five years to allow fish stocks to improve. Job retraining will be provided to all persons employed on fishing boats during the past three years, along with an expanded unemployment compensation program. Economic development grants will be provided to towns in which 10% or more of the employed population has been employed in the fishing industry over the past three years. The federal government will provide economic development grants and unemployment compensation. The state of Alaska will provide job retraining."

8. Are the costs assigned to various parties realistic, in terms of their ability to pay and the financial mechanisms available to each of them? Are loan repayment assumptions based on realistic projections about such issues as economic growth, future costs of inputs (cheap water or electric power), and access to subsidies?

9. Are the financial instruments necessary to raise implementation money specified, and is the overall mix of financial instruments cost-effective, given the powers of the various parties to levy taxes, issue bonds, and otherwise raise money?

   Example: "The municipality will issue bonds, the public utility will obtain a low-interest public loan, the water agency will increase its tax on groundwater use, and the NGO will provide foundation money to begin the most urgent habitat restoration projects."

10. Do cost sharing agreements provide a contingency fund (as with construction projects) for unanticipated costs, or specify the share of such unanticipated costs to be paid by the parties?

11. Are sanctions for noncompliance specified, including deadlines, performance benchmarks, and assessment of penalties?

12. Has an implementation team with authority to monitor compliance, impose sanctions, and evaluate progress toward achieving the specified goals been designated?

13. Is an adaptive management team and process in place to respond to unexpected changes in natural, political, or economic conditions relevant to implementing the agreement?

14. Does the agreement incorporate incentives to help achieve goals, such as directing parties to set resource prices and user fees in a manner consistent with the objectives of the agreement?

    Examples: Alter water rate structures to encourage water conservation; charge visitors higher entry fees during peak-use seasons and days at national parks to ease congestion problems.

    Note: An agreement that requires parties to undertake specific expenditures and alter pricing structures can assist them in getting approval from regulatory authorities (if needed) for new expenditures and changes in pricing structures. Such regulatory authorities could include state corporation commissions or water boards, for instance.

15. Are new decisionmaking structures and resource management tools that have proved effective and acceptable to the parties been "institutionalized" in the agreement (so that the parties return to these venues for problem solving and they become the new way of doing business)?

16. Have formal policy changes to support the agreement been achieved (or begun), authorizing such things as legislation or public appropriations?

17. Are documentation protocols such as monthly reporting of staff time, travel, professional services, and other financial expenditures associated with implementation incorporated into agreements and compiled by a central recorder? Documentation should also include changes in resource pricing and subsequent changes in resource use, so that price elasticity can be more accurately estimated and used in the future. Reports on related disputes and how they were handled would also be useful to observe how problem-solving capacity changes over time.

Notes

This research has been supported by grants from the Udall Center for Studies in Public Policy and by the National Science Foundation–U.S. Environmental Protection Agency Water and Watersheds Grant Program. The author appreciates input from Tamra Pearson d'Estrée and from participants in the March 2000 workshop organized by Rosemary O'Leary and Lisa Bingham.

1. Nick Hanley and Clive Spash, *Cost Benefit Analysis and the Environment* (Brookfield, VT: E. Elgar Publishers, 1993).

2. Kirk Emerson, "A Critique of Environmental Dispute Resolution Research," Conflict Analysis and Resolution Working Group Seminar, University of Arizona, April 1996. For a comprehensive review of the literature on defining success in environmental conflict resolution, see Tamra P. d'Estrée, Connie A. Beck, and Bonnie G. Colby, *Criteria for Evaluating Successful Environmental Conflict Resolution* (unpublished manuscript, 1999).

3. Tamra Pearson d'Estrée and Bonnie Colby, *Guidebook for Analyzing Success in Environmental Conflict Resolution Cases* (Fairfax, VA: Institute for Conflict Analysis and Resolution, George Mason University, 2000).

4. The full body of criteria that were developed, along with a research instrument for collecting the data needed for evaluation, is reported in d'Estrée and Colby, *Guidebook for Analyzing Success in Environmental Conflict Resolution Cases*. Some criteria have been added, and others have been reworked for this chapter.

5. d'Estrée and Colby, *Guidebook for Analyzing Success in Environmental Conflict Resolution Cases*.

6. Elizabeth Checchio and Bonnie Colby, *Indian Water Rights* (Tucson: University of Arizona, 1992).

7. National Research Council, *Water Transfer in the West* (Washington, DC: National Academy Press, 1992).

8. Western Water Policy Review Advisory Commission, *Water in the West: Challenge for the Next Century,* Final Report (Washington, DC, June 1998).

9. The framework in d'Estrée and Colby (*Guidebook for Analyzing Success in Environmental Conflict Resolution Cases*) is intended to provide a standardized method of documenting complex disputes so that multiple cases can be readily compared and evaluated.

10. In d'Estrée and Colby (*Guidebook for Analyzing Success in Environmental Conflict Resolution Cases*), this is called "perceived economic efficiency." This concept of weighing benefits and costs is central to the "mutual gains" negotiation framework described in Roger Fisher, William Ury, and Bruce Patton's *Getting to Yes, 2nd ed.* (Penguin Books, 1991) and applied to E&PP disputes in Lawrence Susskind, Sarah McKearnan, and Jennifer Thomas-Larmer, eds., *The Consensus Building Handbook: A Comprehensive Guide to Reaching Agreement* (Thousand Oaks, CA: Sage Publications, 1999) (see pp. 236–239, 273–276). It is sometimes called "creating value," or converting zero-sum negotiations to positive-sum negotiations.

11. A broader version of this criterion could inquire whether the outcome provides net gains to the larger community and society, beyond the immediate signatories.

12. David L. Horowitz, *The Courts and Social Policy* (Washington, DC: Brookings Institution, 1977), pp. 28–39.

13. Ibid.

14. Bonnie G. Colby, "Transaction Costs and Efficiency in Western Water Allocation," *American Journal of Agricultural Economics,* vol. 72 (1992), pp. 1184–1192.

15. Ibid., note 11.

16. The larger evaluation framework in d'Estrée and Colby (*Guidebook for Analyzing Success in Environmental Conflict Resolution Cases*) contains additional criteria that address other aspects of fairness.

17. National Research Council, *Water Transfer in the West,* pp. 27–36.

18. See Peter Adler, R. Barret, M. Bean, Juliana Birkhoff, Connie Ozawa, and E. Rudin, "Managing Scientific and Technical Information in Environmental Cases: Practices and Principles for Mediators and Facilitators" for a thoughtful discussion of uncertainty in environmental dispute resolution, available at www.ecr.gov (accessed January 17, 2003).

19. Western Water Policy Review Advisory Commission, *Water in the West,* pp. 3-40 to 3-44.

20. Western Water Policy Review Advisory Commission, *Water in the West,* pp. 3-58 to 3-60.

21. Bonnie G. Colby and J. Pratt, "Innovative Strategies To Acquire Water for Environmental Needs," draft manuscript, University of Arizona (2001).

22. Ibid, note 5.

23. See Lawrence Susskind, Mieke van der Wansem, and Armand Ciccarelli, *Mediating Land Use Disputes: Pros and Cons* (Cambridge, MA: Lincoln Institute of Land Policy, 2000) for a discussion of economic sanctions in enforcing environmental laws.

24. Terry L. Anderson and Pamela Snyder, *Water Markets: Priming the Invisible Pump* (Washington, DC: Cato Institute, 1997), pp. 47–66.

25. Ibid.

26. Bonnie G. Colby, "Negotiated Transactions as a Conflict Resolution Mechanism," in Michael Rosegrant, Ariel Dinar, and William K. Easter, eds., *Markets for Water—Potential and Performance* (Kluwer Academic Publishers, 1998), pp. 72–79.

27. Ibid., note 23.

28. Western Water Policy Review Advisory Commission, *Water in the West,* pp. 3-40 to 3-44.

29. Ibid.

30. See d'Estrée and Colby, *Guidebook for Analyzing Success in Environmental Conflict Resolution Cases,* for a more complete discussion of social capital and dispute resolution. The focus in this chapter is limited to economic aspects of social capital.

31. Douglas North, "Institutions and Transaction Cost Theory of Exchange," in J. Alt and K. Shepsle, eds., *Perspectives on Political Economy* (Cambridge, U.K.: Cambridge University Press, 1990), pp. 182–194; Paul Wilson, "Social Capital, Trust and the Agribusiness of Economics," *Journal of Agricultural and Resource Economics,* vol. 25 (2000), pp. 1–13.

32. Steven Knack and Paul Keefer, "Does Social Capital Have an Economic Payoff?" *Quarterly Journal of Economics,* vol. 112, no. 4 (November 1997), pp. 1251–1288.

33. d'Estrée and Colby, *Guidebook for Analyzing Success in Environmental Conflict Resolution Cases.*

34. John Loomis, *Integrated Public Lands Management* (Columbia University Press, 1993).

35. Gail Bingham, *Resolving Environmental Disputes* (Washington, DC: Conservation Foundation, 1986).

36. Ibid.

37. Ibid.

38. Ibid.

39. Ibid.

40. See Adler and others, "Managing Scientific and Technical Information in Environmental Cases" for a thoughtful discussion of contributions by scientists to environmental dispute resolution.

41. Edward Marston, "Reworking the Colorado River Basin," in *Western Water Made Simple* (New Paonia, CO: High Country News, 1987), pp. 199–210.

PART VI

Conclusion

16

Fulfilling the Promise of Environmental Conflict Resolution

LISA B. BINGHAM, DAVID FAIRMAN,
DANIEL J. FIORINO, AND ROSEMARY O'LEARY

The contributors to this volume have created a rich reference with a wide variety of theoretical and practical perspectives, methods, research questions, measures, indicators, and empirical findings on environmental conflict resolution (ECR). Despite the wealth of information provided, the contributors agree that more empirical work on ECR is needed. Whereas the contributors use different terms to refer to their work—evaluation, program evaluation, evaluation research, applied research, field research, or just plain research—they nevertheless reach a broad interdisciplinary consensus about the usefulness of this variety of perspectives. This volume covers a wide swath of a diverse ECR landscape and provides much food for thought.[1] Nevertheless, the current underdeveloped state of theory is a challenge for ECR research and evaluation.

Part I provided an overview of the field of ECR and reviewed the literature, including an assessment of the challenges involved in ECR research and evaluation. It outlined the ECR processes, including assisted negotiation, policy dialogues, facilitation, mediation, fact finding, and arbitration. It arrayed the processes along points on the environmental conflict continuum, which ranges from upstream conflicts (in policymaking, rulemaking, and planning), to midstream conflicts (in administrative permitting and enforcement), to downstream conflicts (in environmental litigation).

In Chapter 2, Juliana E. Birkhoff and Kem Lowry provided a synthesis of claims and criticisms in the research literature, arguing that insufficient attention is paid to helping mediators and facilitators learn how to improve. Claims for ECR include individual satisfaction and met needs; individual empowerment and capacity building; personal transformation; and relationship-, social-, and ecological-level outcomes and benefits. ECR is criticized for delegitimizing conflict; producing lowest common denominator outcomes; including stakeholders with unequal abilities or undefined roles; addressing national policy issues through local dialogue; excluding or disempowering urban environmental groups and national or local majorities; circumventing public authorities; and co-opting environmental advocates. This discussion in Part I framed the challenges of ECR research and evaluation that followed.

Parts II, III, IV, and V of this volume illustrated the variety of ways that researchers and evaluators have approached research on ECR. Research has focused on upstream ECR use in public participation, policymaking, and rulemaking; on midstream ECR use, getting inside the black box of a given environmental conflict; on the outcomes of ECR; and finally on ECR program design and evaluation. Table 16-1 categorizes the contributors' efforts across both dimensions of ECR technique (assisted negotiation, facilitation, mediation, and ECR programs) and ECR context (upstream in public participation, policymaking, or rulemaking; midstream in permitting or enforcement; and downstream in litigation). Of course, many of the chapters cross these boundaries.

The contributors also suggest a wide variety of criteria for assessing and evaluating ECR. Table 16-2 summarizes these criteria, also organized across the dimensions of ECR techniques and contexts. Again, many of the contributors make suggestions that could fit into multiple cells of the table.

Some evaluation criteria apply all along the ECR stream, but there are clusters unique to the stage of conflict at which ECR is used. For example, our contributors suggest a number of criteria related to improved participation in the democratic process when ECR is used upstream in policymaking. For facilitation and mediation upstream, they suggest that we measure whether the public and stakeholders experience enhanced information exchange and communication, and they suggest that ECR will improve decision quality, public legitimacy, public outreach, and representation of diverse constituencies and viewpoints. In contrast, our contributors suggest that ECR mediation further downstream be evaluated more in terms of outcomes such as public benefits, contribution to policy objectives, cost-effectiveness, fairness, and improvement in ongoing relationships and problem-solving capacities. The criteria suggested for evaluating conflict resolution programs differ substantially from criteria for evaluating ECR in a given case. They tend to focus on the best practices in service delivery and program administration. Across all contributions, researchers tend to embed the criteria for judging an ECR process or program in its context. However, throughout this volume, the contributors make it clear that whereas much useful research on ECR exists,[2] much more empirical work must be done.

TABLE 16-1. The Contributors, Arrayed on the ECR Continuum

Process or context	ECR continuum		
	Upstream[a]	*Midstream[b]*	*Downstream[c]*
Facilitation assisted negotiation	Beierle and Cayford Brogden Coglianese Sabatier Stephens	Birkhoff and Lowry Kaufman and Gray	
Mediation	Birkhoff and Lowry Brogden	Gail Bingham Birkhoff and Lowry Caton Campbell Colby d'Estrée	Berry, Stiftel, and Dedkorkut Kloppenberg
Conflict Resolution Programs	Emerson and Carlson Rowe	Emerson and Carlson Rowe	Berry, Stiftel, and Dedkorkut Emerson and Carlson Kloppenberg O'Leary and Raines Rowe

[a]Public participation, policymaking, and rulemaking.
[b]Permitting and enforcement.
[c]Environmental litigation.

In this concluding chapter, we identify several core research challenges in answering questions about the value of ECR and suggest a model of continuous learning about when and where collaborative ECR is likely to be most useful. We then outline a future research program that could assist environmental policymakers and stakeholders, ECR practitioners, and academic researchers in deepening our understanding of ECR. We conclude with a call for a much more systematic and integrated effort to gather data on both traditional and collaborative ECR. With more systematic data collection and analysis, we can dramatically increase both opportunities for evaluation research and the potential impact of that research on the theory and practice of environmental conflict resolution.

Comparing "Collaborative" Environmental Conflict Resolution to "Traditional" Alternatives

The contributors' collective effort yields important questions for future research. First, policymakers would like to know whether "collaborative" processes of public dispute resolution are better than the next-best alternative available to stakeholders (e.g., administrative decisions, litigation, or legisla-

TABLE 16-2. Evaluation Criteria, Arrayed on the ECR Continuum

Process or context	ECR continuum		
	Upstream[a]	*Midstream[b]*	*Downstream[c]*
Facilitation assisted negotiation	Incorporating public values Improving decision quality Resolving conflict Building trust Educating the public Socioeconomic representativeness Consultation, outreach with wider public Diversity of views represented Integration of concerns Information exchange Mutual learning Effectiveness, efficiency, and equity Cost avoidance Project or decision acceptability as legitimate Mutual respect	Participant communication frames	
Mediation	Increased overall knowledge about environment Increased individual stakeholder knowledge Identifying threats, goals, and management actions Wider public outreach Adequacy of plan to achieve goals Certainty of agreement implementation	Positive net benefits Measurable objectives Cost-effective implementation Financial feasibility Fair distribution of costs among parties Flexibility Incentive compatibility Improved problem-solving capacity	

Conflict resolution programs	Relation of program, cases, and practice	Enhanced social capital	Participant procedural justice
	Use of best practices	Clear documentation protocols	Mediator effectiveness
	Training	Reduction in conflict and hostility	Comparative satisfaction of different categories of disputants
	Meeting facilitation	Improved relations	
	Conflict assessment	Cognitive and affective shift	
		Ability to resolve subsequent disputes	
		Transformation	
		Relation of program activities to program theory	Reducing or narrowing issues
		Durable agreements	Case management outcomes
		Comprehensive or complete agreements	Docket improvement
		Mediator and party capacity improved	Referrals, voluntary use rates
		Government decisionmaking improved	
		Right parties	
		Best practices in screening	
		Sound process design	
		Non-ADR used when appropriate	
		ADR systems design	
		Conflict assessment	
		Roster management	
		ADR use	
		Benefits and incentive structures	
		Barriers and obstacles	

[a]Public participation, policymaking, and rulemaking.
[b]Permitting and enforcement.
[c]Environmental litigation.

tion). In responding, researchers need to define ECR alternatives and measure relative benefits with care.

The counterfactual question, "what would have happened if the stakeholders had used a noncollaborative alternative process?" presents conceptual and methodological challenges. For disputes in settings with well-institutionalized and frequently used procedures for dispute resolution (e.g., administrative hearings), we may be able to come up with a set of process and outcome measures that allow fairly straightforward comparison between collaborative and noncollaborative approaches (e.g., see Chapter 12). For disputes in arenas where there are not clear or obvious alternative resolution procedures, the counterfactual question becomes much more difficult to answer with confidence. Participants' own views on the question, "what would you have done to resolve this conflict if you had not used an ECR process?" are arguably the best starting point for defining and assessing the counterfactual alternative. Methodologically, the more congruent the participants' views of the most likely alternative and its most likely outcome, the greater should be the evaluator's confidence in using that alternative as the counterfactual.

A second question is when and how we should apply terms like "success" and "failure" to collaborative dispute resolution processes. Consider, for example, that a collaborative process fails to produce full agreement, but does significantly narrow the range of disagreement and significantly improves relationships among participants. Is the process a success, a failure, neither, or both? A recent survey of environmental attorneys found that when ECR successfully resolved a dispute, a majority of attorneys who participated in the ECR effort reported several positive benefits. At the same time, the survey found that when ECR did *not* successfully resolve a dispute, attorneys nonetheless reported positive benefits.[3]

Public agency managers, foundation funders, and many scholars and practitioners may want simple, definitive answers to questions like, "how did we do?" Yet group interaction tends to produce many kinds of outcomes, from the relationship impacts categorized by d'Estrée in Chapter 6 to the multitiered institutional and environmental effects of the watershed partnerships assessed by Leach and Sabatier in Chapter 8. The responsible evaluator's goal should be to identify and assess the multiple types of costs and benefits associated with a particular case or set of cases. Evaluators should only make broad statements about success and failure when the outcomes of all the most important measures are pointing in the same direction.

If we want to make accurate comparisons between ECR and alternative means of conflict resolution, how much should our comparative evaluation depend on the intractability of the underlying conflicts? For example, should we give greater weight to modest process or outcome benefits when the challenge is finding a place to put radioactive waste than we do when it is finding a place to put a new health clinic? If we think that it makes sense to consider the "degree of difficulty" in a particular conflict, then we may need some measures of intractability, as Caton Campbell suggests in Chapter 5. One way to make a rough assessment of intractability would be to survey stakeholders before the initiation of a collaborative process. We could measure intractability by the degree to which participants agree with the statements that (a) they have reached a stalemate and (b) they see no way out. The greater the degree of intractability, the greater the credit we should give to any collaborative process (or traditional process) that produces significant gains in understanding, relationships, or problem solving.

Compounding the challenge of comparative evaluation is the fact that that there is a kind of "anti-innovation bias" in the way that we evaluate new ideas when they are applied to public policy. The burden of proof is usually placed on the innovator to prove conclusively that the change produces superior results to the existing way of doing business. This bias has several consequences. First, it takes time to get these things right, so having to overcome a great burden of proof may be the death of many promising ECR innovations. A new procedure or policy will inevitably have flaws in its design or early implementation that will lead people to compare it unfavorably to the existing approach. Critics of change often want to draw conclusions too quickly, before there has been time to test and work through the early wrinkles.

Conversely, there is often no comparable expectation that the existing approach will have to stand up to critical scrutiny. Although federal rulemaking has been criticized on multiple grounds for being costly, slow, adversarial, and so on, the burden of proof always appears to fall on the advocates of a new approach, not the defenders of the old one. This second form of bias has immediate and critical consequences. Many efforts at change are dropped on the argument that there are too few resources to invest in experimental or pilot approaches. If an innovation does not show immediate and measurable payoffs, there is pressure to discontinue it. This leads to the odd result in which the innovator is asked to justify the use of perhaps 1% of the organization's resources by proving its superiority over the use of the other 99% of the resources, which face no such burden of proof.

Evaluation as a Learning Process

Given the many methodological challenges involved in answering the basic question, "is collaborative ECR better than other approaches to resolving environmental conflict?" it seems clear that evaluation of ECR should be seen as a matter of continuous learning over time rather than as a definitive thumbs up or down on the merits of an ECR process itself.[4] To be sure, many researchers and most practitioners assume that consensus-based decision-making achieved with the support of professionals in dispute resolution will be superior to adversarial, legalistic decision process in many circumstances. Once we accept that premise, as most people with experience in the field probably do, the role of evaluation then becomes one of learning when, where, and how ECR may be used most successfully.[5]

This changes the emphasis on evaluations slightly. The primary question is less about the merits of ECR and more about the design and application of ECR to different settings and conflicts. Regulatory negotiation illustrates this point. There are regulatory issues for which negotiation is clearly inappropriate. These include decisions in which the rule is precedent-setting or has effects well beyond the range of interests that may be represented at the negotiating table, or the relationships are so adversarial that any constructive dialogue that could lead to an agreement is unlikely. The purpose of evaluation should not be to tell us whether regulatory negotiation is superior to conventional rule-making; the answer to that question is "it depends." The purpose of evaluation should be to tell us when, how, and in what form ECR should be applied.[6]

Similarly, evaluation may tell us a great deal about the limitations of consensus-based processes such as ECR. In the 1990s, for example, the U.S. Environmental Protection Agency (EPA) and many other government organizations became so uncritically fond of consensus processes (e.g., those in which the agreement of all the parties involved is a condition for making a decision) that they were used inappropriately. As the EPA and others learned in such programs as the Common Sense Initiative and Project XL, we cannot lock a policy system into a process that requires the consent of all the affected or interested parties every time we try to adapt to the need for change.

Drawing from these observations about regulatory negotiation, the role of evaluation should be as part of an extended, systematic, learning process about ECR. Learning through detailed case studies as well as through quantitative evaluation methodologies is important. To be useful to practitioners, however, evaluations should extract practical lessons on how best to design ECR processes and should advise in what circumstances (that is, the nature of the issues, relationships among parties, and the context of the dispute) alternative designs should be used, if indeed they should be used at all. It is also important, at times, for evaluators to work with ECR practitioners in the field, to provide immediate feedback on their process, roles, and strategies. This in-the-field approach should not replace the more traditional evaluator role but should supplement it.[7]

A Future Research Program: From Satisfaction, Single Cases, and Snapshots to Dynamic Systems and Environmental Impacts

There is a consensus among the contributors to this book that the field of ECR must move beyond settlement and satisfaction as the principal measures of effectiveness. In making this move, ECR would join the rest of the alternative dispute resolution (ADR) field. Across subject matter contexts, researchers and evaluators are working on contingency theories to broaden and deepen our understanding of what these processes can and cannot deliver and under what circumstances. We must move from experimental research on negotiating dyads to field research on aggregate use; we must look at ECR from a systemic perspective (i.e., looking not only at individual ECR cases, but also looking both at the ECR process as a system and at ECR cases within a larger system). We must adapt our methods to capture the time-extended nature of ECR, and we must begin to examine ECR in terms of its environmental effects.

From the Dyad to the Aggregate

A first insight concerns experimental research on negotiation. Much of the research examines the interpersonal dyad: two people. It also tends to examine negotiated decisions that are binary: agree or disagree, cooperate or defect. This is true of both experimental social psychology research on negotiation and game theoretic research from the disciplines of economics and political science.[8] Although it is important, this type of research has some drawbacks. First, interpersonal relationships do not exist in a vacuum. Increasingly, researchers recognize that the dyad must be seen and studied within its organizational, cultural, and historical context.[9]

Second, and perhaps more importantly, ECR processes consistently involve groups of people, not dyads. Researchers have begun to model the aggregate, that is, they are examining group cooperative behavior.[10] Finally, the outcomes of ECR are not binary decisions; they are complex, multifaceted agreements that address many issues. These agreements do not cover a simple purchase and sale between parties who will not have future dealings; they instead provide a framework for carrying out a plan of action over time that affects the environment and thus members of the general public not at the table. All of these factors contribute to making ECR difficult to research in ways that we can quantify. For ECR research, we must look at the aggregate.

To move to the aggregate means that researchers must not simply ask whether a given party was satisfied with and derived personal value from participating in an ECR process. They should instead attempt to measure how that party, by virtue of the ECR experience, does or does not behave differently in handling conflict with others. Do people who have experienced consensus-based processes form more cooperative groups? Do they participate more in future policymaking either individually or as part of a group? Are they empowered or exhausted by their experience with consensus processes?

Research from political science is especially valuable for addressing these questions. For example, game theory research suggests that cooperators will conquer a population of competitors in iterated prisoner dilemma games using genetic algorithm modeling.[11] In addition, researchers are examining the Truth and Reconciliation Commission in South Africa and the genocide trials in Rwanda to determine the effectiveness of conflict resolution institutions for healing the wounds of large-scale racial and ethnic conflict.[12] They are looking for evidence that those who witnessed testimonials to truth and scenes of reconciliation have different beliefs about the racial divide in their society and indicate changes in their own behavior and attempt to determine whether these institutions are effective means to promote reconciliation. These are only a few of the many research initiatives in other ADR subfields that could prove to be valuable to the field of ECR.

Understanding ECR as Complex Systems

A second and related insight evolves from the recognition that relatively little empirical work on dispute resolution examines the systemic level.[13] Complexity theory is influencing research across many disciplines and provides insight on how researchers might understand ECR, both as a complex system in itself and operating within a larger societal system for managing conflict. Complexity theory suggests that individual actors behaving in accordance with simple decision rules may, through the aggregation of their individual acts, give rise to a self-organizing complex system.[14]

One area where researchers have tried to examine dispute resolution at the systemic level is in the context of labor and employment dispute processes, including studies of grievance procedures and voice mechanisms in organizations in the fields of industrial relations, human resource management, and organizational behavior.[15] These disciplines attempt to measure the effect of aggregate participation in the dispute resolution process on productivity and efficiency. Within the organizational context, researchers have examined the function of dispute resolution programs and compared different dispute system designs. For example, the U.S. Postal Service (USPS) has worked with the Indiana Conflict Resolution Institute since the inception of its mediation program for employment disputes to collect comprehensive, national data on the system using a number of different variables and indicators. This permitted a controlled comparison of two different dispute system designs, one using inside neutral USPS employee mediators and the other using outside independent contractor mediators. Although there was selection bias intended to send the easier cases to mediation in the inside neutral program, while almost all cases were permitted in the subsequent outside neutral program, the outside neutral program produced superior results.[16] A subsequent study examined the effect of the program on national patterns of formal discrimination complaint filing and found a statistically significant drop in formal complaints correlated with implementation of the mediation program.[17]

These studies were only possible because the organization maintained comprehensive, aggregate data not simply on rates of settlement in the mediation cases, but on other aspects of the system for handling various kinds of conflict within the organization. This allowed researchers to examine not only what happened to the immediate disputants involved in ADR, but also to look at the pattern of outcomes produced by all the disputants' uses of various dispute processing mechanisms within a single organization.

This type of aggregate and systemic research is needed in the field of ECR. Some of the contributors to this volume suggest ways for evaluating ECR and public policy dispute resolution programs in this manner. They suggest that for these evaluations to be comparable to the organizational-level literature, they would have to examine how the ECR or public policy program affects the organization of which it is a part, as well as the cumulative effect on the government as a whole. This examination is theoretically possible, but such an effort requires baseline data on disputing through civil justice and agency administrative systems that are simply not available at present.

Outside the programmatic context, it is even more difficult to do research analogous to the organizational-level literature in the field of ECR. The cases are fewer, bigger, and more time-extended, and they involve many more participants than is typical in an ADR case. Moreover, they generally occur outside of any single organizational context and often involve participants who are appearing as representatives of organizations or unincorporated associations of stakeholders with shared interests.

Notwithstanding these complicating factors, our contributors' discussions suggest one possible method for conducting analogous research: to view a single ECR case as a system in itself. One reason this construct might assist in analysis is that ECR processes often involve multiple interim determinations and agreements, for example in response to ongoing monitoring for levels of a contaminant in water or air. These might be viewed as individual disputes. The ECR process functions over time as a system for resolving these interim disputes. Instead of treating the case as the unit of analysis, our contributors have suggested taking numerous measurements of diverse indicators across participants and at multiple points in time, both before and after participants reach a framework settlement or agreement. This would permit us to develop much better descriptions of how the ECR case-as-system develops over time. Even if we think of ECR as a dispute system, there remains the question of context. If ECR is not a case but a system, what is the organization for which it constitutes the dispute system design and how would we capture data to evaluate it as a system within that context?

There are other systemic-level approaches from other branches of ADR that may prove to be inspiring to ECR research. For example, one scholar has proposed a systemic approach to identity-based conflicts.[18] The argument begins with the individual actor operating to reduce anxiety by identifying with beliefs and values. The author then examines social identity and its attendant stereotyping. In-groups tend to institutionalize status differences between in- and out-groups and in doing so render institutions carriers of beliefs that may include biases. Institutions may become elements of the dispute if they are part of the mechanisms that society uses to create or enforce fairness through access to opportunities or resources or if they are mechanisms for society to manage conflict, such as through the justice system. When institutions become the focus of disputes, social identity may deepen and tempt out-groups toward violence. When in-groups respond with repression, a feedback loop is created and the system comes into existence. This systemic approach to examining racial or ethnic conflict may also apply to other identity-based conflict, for example, water resource allocation disputes involving multiple cultural and economic groups.

Other scholars have recently begun to examine how individual litigants and lawyers organize into the more complex systems of class action and large-scale litigation.[19] This work focuses on mass tort claims and has relevance to mass environmental tort litigation. There is a complex mix of multiple actors, perceptions of information, calculations of costs and benefits, and political judgements that interact and give rise to this relatively small but growing subset of litigation. The phenomenon can best be understood from a systemic perspective. Systemic-level analysis is as important in this downstream use of ECR in litigation as it is in the midstream use involving disputes over the allocation of resources.

Traditionally, ECR researchers have used the case study method to examine relations among disputing groups at a given locale or site. This method allows for the identification of patterns across cases and lessons learned about ECR. Likewise, contributors to this volume examine how communication frames change over time and contributes to attitudes among participant groups in an environmental conflict. Whereas this work at the case level is valuable, much work remains to be done at the systemic level. Until such systemic research is available, we cannot adequately understand the dynamics of protracted environmental conflicts as complex systems.

Time-Extended Phenomena

A third insight involves the fact that ECR takes place over protracted periods of time. Practitioners have observed that disputants often think ADR ends with the settlement agreement, as if settlement were a fixed point at the end of a linear process. Instead, one researcher argues that settlement is part of a wavelike form.[20] The settlement occurs after the peak point of tension, just past the crest in a diminishing sine wave. After settlement, the level of tension falls to a low point. The parties have agreed on a structure for handling conflict in the future. Conflict will recur and tensions increase; the wave will recrest. However, the earlier agreement means that when conflict does recur, the parties will handle it differently. Moreover, tension will not rise to the same level of intensity. Thus, to examine ECR as time-extended phenomena, we must test the hypothesis that people who participate in ECR processes think or behave differently over time from those who do not.

This research agenda has been advanced in areas other than ECR. For example, some negotiation researchers have proposed a method for evaluating how collaborative negotiation training affects individuals and groups.[21] They emphasize the need to examine participants for evidence of change not just in perceptions immediately following training, but instead in behavior over time. Others argue that consensus building and other collaborative planning processes must be examined in light of how members of the group and subgroups learn to work together over time, after the process concludes, because there may be intangible products in the nature of social, intellectual, and political capital that are outcomes of value for evaluation.[22] Finally, some theorists have asserted that the policy sciences have as their goal helping the body politic clarify its values and options to achieve the maximum human dignity.[23] One aspiration for future ECR research is to build on these differing disciplinary traditions to find ways of measuring whether ECR does enhance these less tangible forms of capital and does contribute to maximum human dignity over time.[24]

Studying ECR as time-extended phenomena poses several difficulties, one of the most immediate being that it is difficult for a researcher to begin and to complete a study of an environmental dispute on a time scale commensurate with incentives for productivity in academia. However, this and other difficulties simply mean that we must adapt our research methods to the phenomena to learn from them. It also means that we may want to reexamine academic incentives with a view toward promoting better longitudinal ECR research.

Outcomes for the Environment

A final insight concerns measuring outcomes for the environment. Although the contributors to this volume suggest a rich variety of ways to determine whether ECR is more effective than traditional methods of handling environmental and public policy conflict, there is broad consensus that the field must examine outcomes more carefully. They suggest a variety of outcome measures carefully tailored to the process and context, yet there is a general reluctance to impose on ECR the burden of proving itself effective in advancing substantive environmental policy.

This is not the case in other conflict resolution contexts, where researchers attempt to relate participation in an ADR program to objective indicators of impact. For example, work on employment conflict resolution in the organizational context also looks at large-scale effects, including the effect of ADR systems on complaint filing rates, productivity, morale, and retention of employees. Court ADR evaluators have examined overall case processing and docket times, looking for differences in negotiated, mediated, arbitrated, and adjudicated cases. Some of these researchers examine a variable, often named "outcome," defined as the percentage of the claimant's original demand that the claimant ultimately settles upon through mediation or arbitration of the dispute.[25] There is an ongoing debate about whether this ratio ought to be equal for arbitrated and litigated outcomes. For example, if a hypothetical set of disputants in arbitration recovers on average 20% of what they claim, whereas a different hypothetical set of disputants in litigation recovers 60% on average, does this mean that arbitration is flawed or somehow unjust? Does it simply mean that there is selection bias because some disputants opt for arbitration while others opt to litigate? Does it mean the two sets of disputants and their claims differ? Are the disputants different in terms of personality characteristics? In some evaluations of court-annexed arbitration, the fact that the outcome ratio is the same for arbitrated and litigated cases is cited as evidence that the program is effective. All of these are efforts to relate participation in ADR to some "objective" measure of impact beyond individual case settlement and participant satisfaction.

Despite their reluctance, ECR researchers and evaluators must begin to consider outcomes in terms of protecting the environment. In principle, it is important as a matter of public policy and management to determine whether ADR methods are effective tools for implementing environmental policy. Other policy implementation tools must withstand substantive scrutiny; why should ECR evaluators limit their analysis to whether a given set of participants can reach agreement and are happy about it? This requires information not only about ECR but also about the consequences of using traditional mechanisms for policy development and enforcement.

For example, in the area of environmental enforcement, what does on-the-ground, in-the-water, or in-the-air testing reveal about the state of the environment and its ecosystems in environmental conflicts before and after ECR activities or in cases with and without ECR interventions? Does remediation happen faster? These questions move beyond assessments of ECR use in allocating costs after the fact. They go to the heart of ECR use to agree upon a remediation plan. How do cleanup activities conducted at sites that use ECR compare with those at sites that do not? What is the success of permitting using ECR compared to permitting under traditional processes? At what rate do different groups of permits generate litigation or enforcement conflict? These questions move beyond the mere fact of settlement to the quality of that settlement. They put ECR advocates' claims of superior resolutions to the test.

William Leach and Paul Sabatier in Chapter 8 take one step in this direction, examining perceptions of success in terms of environmental protection.

Ideally, researchers seeking to assess environmental impacts should also look beyond participant perceptions by gathering and analyzing "hard" data on environmental indicators whose improvement is one key goal of ECR.

To do so, they would need to collect the best available scientific and technical information about the state of the environmental media, resource, ecosystem, or species that is the subject of the environmental conflict.[26] However, scientific information itself is often the subject of conflict and negotiation in environmental conflicts.[27] Therefore, it may be difficult to gather "objective" scientific information to assess environmental outcomes. Even if it were possible, researchers would need to control for the many variables other than the outcomes of the conflict resolution process that may affect environmental quality. Nevertheless, researchers need to confront the question of environmental impact using the best available methods.

Researchers may not be able to draw meaningful conclusions about outcomes for the environment with data from a single case or site; however, by consistently collecting data at multiple times for a number of cases or sites, they could aggregate results and begin to look for patterns. ECR cases differ widely. Unlike small claims cases, they are often sui generis. However, it might be possible to examine them for patterns in the proportion of improvement in various scientific indicators over time. By what percent does a given contaminant decline in the water or soil? By what proportion of the ultimate goal does the environment improve or does a species population recover? Are these proportions similar over a given period of time for cases where ECR is and is not used?

Institutionalizing Data Collection on ECR

The fundamental obstacle to deepening our research and evaluation of ECR is the data gap. We cannot develop broad generalizations, controlling meaningfully for selection bias, without collecting detailed descriptive information about ECR and non-ECR cases. To do so, we need a great deal of data that are hard to collect. These problems can, however, be overcome by serious and consistent efforts to institutionalize the collection of environmental and environmental conflict resolution data.

Throughout this volume, contributors discuss how difficult it is to get data on a sufficiently high number of cases to allow quantitative statistical analyses that use significance testing. Regardless of the methods used to collect the data, whether surveys, interviews, observations, expert judgements, or archival records,[28] researchers tend to come up against the uniqueness of each ECR case. Gail Bingham's groundbreaking 1986 effort to examine patterns across a collection of individual case studies demonstrates the benefit of aggregating cases, but this meta-analytic effort was hampered by the wide variety of variables and data collection methods used by researchers.[29] These problems point to the need for the ECR field to work toward the institution-alization of data collection. In other words, we need to create the equivalent of a Bureau of Labor Statistics for ECR. As ECR programs themselves become institutionalized and their funding sources develop higher expecta-tions of demonstrated performance, the field will need to build an infrastruc-ture to support the next generation of research and evaluation.

The response to calls for data collection is often, "Yes, of course, but we cannot collect it all." The premise of the Tumamoc Hill meeting, discussed by Kirk Emerson and Christine Carlson in Chapter 10, was that it is possible to collect data that can be used both for program management and for program evaluation and research. In an effort to rein in researchers whose preference would be to collect a great deal of potentially useful data on every case, agency representatives and conflict resolution professionals at the meeting asked why they should expend resources collecting any of it and how the information might be used.[30] To date, each program has made its own choice on what to collect and how. Over time, a consensus may emerge on what information is most useful to an ECR agency, fostering an array of variables or indicators as part of a comprehensive case tracking system. The system will follow each ECR case longitudinally. It might include many of the vari-ables and indicators that researchers and evaluators suggest in their contribu-tions to this book.

Even if the field could agree to collect only a subset of these variables and indi-cators, if data collection were comprehensive, electronic, routine, decentralized, and longitudinal, the research and management potential would be profound.[31] The following discussion looks at each of these data collection elements.

Comprehensive

Data collection is getting easier. Albeit in fits and spurts, courts and agencies are moving toward consistent, uniform systems that capture every court or agency location and every case. These systems have as a goal the collection of population data and therefore have the potential to assist the field in meeting the need for comprehensive data. Despite this advance, the typical dataset produced by an ECR evaluator or researcher contains sample information for a given time period. This is true of most of the contributions in this volume. They are important, but these time-limited studies are less useful for a court or agency than a continuous information management system would be. At best, they may help answer specific questions at a single point in time. In contrast, a comprehensive information management system would be an ongoing management tool—a way for the court or agency to know where it has been, where it stands, and where it is headed. Such a system could become a vital instrument for administrators who must make decisions about resources. For the comprehensive system to be vital, the court or agency must accurately and consistently maintain the information. Such a system would be a researcher's treasure trove.

Electronic

When data are maintained in electronic format, they are easier to manipulate for empirical analysis. They are easier to transport, and it is easier to alter and organize the data's format, isolate specific variables, and include additional dimensions for comparison. Whereas most programs have developed paper tools for data collection (i.e., surveys and tracking forms), it is already possible to use hand-held digital devices for ECR data collection. Facilitators used these new technologies at a recent mass public participation event in New York City in connection with future land use of the site of the World Trade Center. More than 5,000 people, assisted by 500 volunteer mediators and facilitators, engaged in deliberations about this critical public policy issue. Similarly, for ECR and other forms of dispute resolution, the originators of the information (disputants, mediators, facilitators, stakeholders, and public officials) can personally input data, which the agency could download directly into the database. This would eliminate labor-intensive data entry. Until such technological advances are implemented, it is important that information be maintained in viable electronic database formats.

Routine

Data collection should be a daily, fully integrated part of every ECR program, not something that happens every three or four years, when a funding source commissions or supports a study. Spurred by the Maxwell School of Syracuse University's Government Performance Project, the National Performance Review, the Reinventing Government Movement, and similar efforts, government agencies at the national, state, and local levels have made serious efforts to evaluate their performance with various management and benchmarking techniques. New federal laws such as the Government Performance and Results Act require federal agencies to set goals and timetables with which they can judge their own performance. Similar mandates are popping up at the state and local levels. These mandates provide a substantial incentive for agencies to integrate information systems into their programs because they can use the systems to document their performance and identify areas for additional focus and improvement. Adopting routine data-gathering instruments and procedures provides a unique opportunity to integrate new ways of measuring dispute resolution efforts.

In the federal sector, the Paperwork Reduction Act places some limits on data collection from the public. The Office of Management and Budget supervises and enforces these limits. However, the limits do not apply to internal reporting requirements (data the agency's staff collects or enters). In contrast, most administrators of third-party dispute resolution organizations already require neutrals to participate in some form of reporting as a condition of membership on the roster or as a condition of payment for their services. This is true of the Federal Mediation and Conciliation Service (FMCS); the American Arbitration Association (AAA); and the U.S. Postal Service (USPS) mediation program, REDRESS. Both the FMCS and the AAA mandate the completion of data collection forms at the end of an arbitration case, so that these organizations can summarize and analyze their activities periodically. The USPS mediation program requires mediators to characterize the nature of the dispute and to report how many disputants and representatives were at the table, how long the process took, and whether there was a settlement.

Another common method of routine data collection is the distribution of a survey at the conclusion of a training session or dispute resolution process. The response rate typically is much higher when participants complete the survey immediately upon distribution as opposed to completing it later. There is often a moment of euphoria immediately after settlement that may inflate results, but these methods can be combined with other sources of information and longitudinal checks. If an agency establishes its survey distribution and collection process so that every participant receives one and completes it immediately after the session, the agency can begin to build up a useful body of information. However, this data collection process works only if efforts are routine and seen as part of everyone's job rather than as something sporadic, optional, and mandated by people from outside the organization.

Decentralized

As the number of information gatherers increases, accuracy becomes harder to control. However, if the research goal is to create a picture of an entire system, it is better to have more data that are less rich than to have fewer data that are richer. More data can be collected if more people are collecting it. Providing constrained data entry tools can address some accuracy issues. Properly structured databases can control how data are recorded, and properly maintained databases with decentralized stewardship of information can reduce error rates and provide feedback to researchers, program managers, neutrals, and disputants.

A major issue that will need to be addressed is the dissemination of data, especially to those researchers outside of the program or organization. Some visionaries in the field have imagined the construction of a central repository of ECR data, accessible by researchers, evaluators, and the general public. This "meta-database" could take a form similar to that employed by the

Bureau of Labor Statistics, the Department of Justice Juvenile Crime Data, or even the database housed at the University of Michigan that contains hundreds of datasets from researchers across the spectrum of the social sciences.

Longitudinal

Funding constraints impose time limits on many evaluation projects.[32] A time-limited evaluation may provide an accurate description of what is happening in a court or program over the course of one year, but it does not account for the fact that all courts and dispute resolution programs are moving targets. Point-in-time data have value, but that value is limited. This type of static data collection and analysis is an intensive and time-consuming process, and a lag always exists between the time period studied and the publication of a report. On the other hand, longitudinal data collection allows researchers to analyze over time how changes in the court or program affect outcomes. With longitudinal data collection, important changes that affect the agency or the process no longer threaten the efficacy of collected data, but merely become additional variables. These types of time-series analyses will become vital and valuable tools for ECR researchers and program directors.

Moreover, the initial and long-term effects of a program are likely to differ significantly. The ECR literature offers some information about programs' initial or short-term effects, but little or no information is available about long-term effects. As a result, it is difficult to respond to charges that a program's influence is simply a honeymoon effect that will diminish or disappear over time. If the interest is systemic-level data, and the reality is that systems are big, unwieldy, and slow to change direction, then a much longer time frame will be needed than we traditionally have used in our ADR research.[33]

How Do We Get There?

Imagine how much richer analyses of environmental and public policy conflict resolution processes and programs could be with comprehensive, electronic, routine, decentralized, longitudinal data collection.[34] Moreover, this quality of data would allow researchers and evaluators to control systematically for context and system design.[35] It is common, for example, to compare mediated cases with litigated ones. This comparison is flawed, because approximately 90% of all cases filed in court settle without a trial. How does one know that the case that settled in mediation would have gone to trial without mediation? Moreover, although mediation is voluntary, as Cary Coglianese points out in Chapter 4, we cannot assume that people are more satisfied with it simply because they chose it.

If we step back and look at the system that ECR is designed to affect as a whole, we can ask whether the system is functioning better on some dimension after it affords people the choice to mediate. Is it disposing of more cases in a given length of time? Are there fewer trials? A number of studies by the Federal Judicial Center and the Rand Institute of Civil Justice have used multivariate methods to examine these questions.[36]

Half the battle is envisioning the goal. Rosemary O'Leary and Susan Summers Raines concluded in Chapter 13 that for ECR to succeed in the EPA, it must become part of the dominant culture. The same can be said about ECR data collection. The "default setting" for courts and other public organizations that use ECR needs to be ECR data collection. Researchers and evaluators need to meld data collection with an information system that gives managers and practitioners what they need and something they routinely use on their own. Intense collaboration among court or agency top managers, records administrators, computer systems technicians, and researchers is key to attaining a well-designed, integrated, and implemented system. For this, mindsets will need to shift at all levels of organizations: court or agency personnel will need to take some responsibility for gathering research data, and researchers will need to take some responsibility for organizing and maintaining administrative data. Researchers, evaluators, program administrators, mediators and other neutrals, stakeholders, and the public all have a job to do; they will only succeed if they collaborate to get it done.

Notes

1. This volume, however, has two limitations. The first is that it does not address the use of binding or nonbinding arbitration as a form of ECR. Second, it primarily has a domestic, national orientation, not an international one. Each of these limitations poses important challenges for future researchers.

2. For recent examples of quality empirical work on ECR see Neil G. Sipe, "An Empirical Analysis of Environmental Mediation," *Journal of the American Planning Association*, vol. 64, no. 3 (Summer 1998), pp. 275–285; Ann E. Tenbrunsel and David M. Messick, "Sanctioning Systems, Decision Frames, and Cooperation," *Administrative Science Quarterly*, vol. 44, no. 4 (1999), pp. 684–707.

3. Rosemary O'Leary and Maja Husar, "What Environmental and Natural Resource Attorneys Really Think about ADR: A National Survey," *Natural Resources and Environment*, vol. 16, no. 4 (Spring 2002), pp. 262–264.

4. See, for example, Robin Gregory, Tim McDaniels, and Daryl Fields, "Decision Aiding, Not Dispute Resolution: Creating Insights through Structured Environmental Decisions," *Journal of Policy Analysis and Management*, vol. 20, no. 3 (2001), pp. 415–432.

5. See, for example, Thomas C. Beierle and David M. Konisky, "Values, Conflict, and Trust in Participatory Environmental Planning," *Journal of Policy Analysis and Management*, vol. 19, no. 4 (2000), pp. 587–602.

6. For an examination of what happens to agency roles when ECR is applied, see Care M. Ryan, "Leadership as Collaborative Policy-Making: An Analysis of Agency Roles in Regulatory Negotiations," *Policy Sciences,* vol. 34 (2001), pp. 221–245.

7. Clearly, these points raise questions about the objectivity of the evaluator. Asking that we accept the inherent value of ECR as largely a matter of faith and focus our efforts on determining when, where, and how it works best presumes a conclusion that many people would want to establish empirically. Engaging evaluators more directly as "advisors" (perhaps even as "participant-observers") in ECR processes removes some of the detachment we would normally expect from an evaluator. Ongoing feedback by critical observers who know the context and the issues often is of more practical value than feedback coming years later from independent evaluators whose understanding of the situation may be weak, even if their methods are analytically superior.

8. Morton Deutsch and Peter T. Coleman, eds., *The Handbook of Conflict Resolution* (San Francisco: Jossey-Bass, 2000).

9. Michael Alan Sacks, Karaleah S. Reichart, and W. Trexler Proffitt Jr., "Broadening the Evaluation of Dispute Resolution: Context and Relationships over Time," *Negotiation Journal,* vol. 15, no. 4 (1999), pp. 339–345.

10. For example, Tenbrunsel and Messick, "Sanctioning Systems, Decision Frames, and Cooperation."

11. Robert Axelrod, *The Evolution of Cooperation* (New York: Basic Books, 1984).

12. See, e.g., Catherine Honeyman, *Gacaca Jurisdictions: Interim Report of Observations* (August 20, 2002), draft report of Rwanda observation project supervised by Jens Meierhenrich, Lecturer in Departments of Government and Social Studies, Harvard University, Cambridge, MA (on file with editors).

13. For an example of the systems approach applied to natural resources management initiatives, see Jennifer A. Bellamy, Daniel H. Walker, Geoffrey T. McDonald, and Geoffry J. Syme, "A Systems Approach to the Evaluation of Natural Resource Management Initiatives," *Journal of Environmental Management,* vol. 63 (2000), pp. 407–423.

14. Judith E. Innes and David E. Booher, "Consensus Building and Complex Adaptive Systems: A Framework for Evaluating Collaborative Planning," *Journal of the American Planning Association,* vol. 65, no. 4 (1999), pp. 412–423; Judith E. Innes, "Evaluating Consensus Building," in Lawrence Susskind, Sarah McKearnan, and Jennifer Thomas-Larmer, eds., *The Consensus Building Handbook: A Comprehensive Guide to Reaching Agreement* (Thousand Oaks, CA: Sage Publications, 1999).

15. Adrienne E. Eaton and Jeffrey H. Keefe, eds., *Employment Dispute Resolution and Worker Rights in the Changing Workplace* (Champaign, IL: Industrial Relations Research Association, 1999).

16. Lisa B. Bingham, Gregory Chesmore, Yuseok Moon, and Lisa Marie Napoli, "Mediating Employment Disputes at the United States Postal Service: A Comparison of In-House and Outside Neutral Mediators," *Review of Public Personnel Administration,* vol. 20, no. 1 (2000), pp. 5–19.

17. Lisa B. Bingham and Mikaela Cristina Novac, "Mediation's Impact on Formal Discrimination Complaint Filing: Before and After the REDRESS Program at the United States Postal Service," *Review of Public Personnel Administration,* vol. 21, no. 4 (2001), pp. 308–331.

18. Leo F. Smyth, "Identity-Based Conflicts: A Systemic Approach," *Negotiation Journal,* vol. 18, no. 1 (2002), pp. 147–161.

19. Deborah R. Hensler, "Revisiting the Monster: New Myths and Realities of Class Action and Other Large Scale Litigation," *Duke Journal of Comparative and International Law,* vol. 11 (2001), pp. 179–213.

20. Christopher Honeyman, "The Wrong Mental Image of Settlement," *Negotiation Journal,* vol. 17, no. 1 (2001), pp. 25–32.

21. Peter T. Coleman, Ying Ying, and Joanne Lim, "Research Report: A Systematic Approach to Evaluating the Effects of Collaborative Negotiation Training on Individuals and Groups," *Negotiation Journal,* vol. 17, no. 4 (2001), pp. 363–392.

22. Innes and Booher, "Consensus Building and Complex Adaptive Systems."

23. Udaya Wagle, "The Policy Science of Democracy: The Issues of Methodology and Citizen Participation," *Policy Sciences,* vol. 33 (2000), pp. 207–223.

24. William Ascher, "Resolving the Hidden Differences among Perspectives on Sustainable Development," *Policy Sciences,* vol. 32 (1999), pp. 351–377.

25. For examples of this analysis and a review of some of the literature as to employment arbitration, see Lisa B. Bingham, "On Repeat Players, Adhesive Contracts, and the Use of Statistics in Judicial Review of Arbitration Awards," *McGeorge Law Review,* vol. 29, no. 2 (1998), pp. 223–260.

26. For an interesting discussion of the difficulties in merging the policy sciences with ecological sciences, see Ascher, "Resolving the Hidden Differences among Perspectives on Sustainable Development."

27. Peter S. Adler, "Science, Politics, and Problem Solving: Principles and Practices for the Resolution of Environmental Disputes in the Midst of Advancing Technology, Uncertain or Changing Science, and Volatile Public Perceptions," *Penn State Environmental Law Review,* vol. 10 (Summer 2002), pp. 323–341; Peter S. Adler, Robert C. Barrett, Martha C. Bean, Juliana E. Birkhoff, Connie P. Ozawa, Emily B. Rudin, "Managing Scientific and Technical Information in Environmental Cases: Principles and Practices for Mediators and Facilitators" (Tucson, AZ: RESOLVE, U.S. Institute for Environmental Conflict Resolution, and Western Justice Center Foundation, 2002, www.ecr.gov (accessed January 17, 2003).

28. Joseph S. Wholey, Harry P. Hatry, and Kathryn E. Newcomer, eds., *Handbook of Practical Program Evaluation* (San Francisco: Jossey-Bass, 1994).

29. Gail Bingham, *Resolving Environmental Disputes: A Decade of Experience* (Washington, DC: Conservation Foundation, 1986).

30. Christopher Honeyman, Bobbi McAdoo, and Nancy Welsh, "Here There Be Monsters: At the Edge of the Map of Conflict Resolution," *The Conflict Resolution Practitioner,* Office of Dispute Resolution, Georgia Supreme Court, 2001, www.convenor.com/madison/monsters.htm (accessed January 17, 2003).

31. Lisa B. Bingham, "Why Suppose? Let's Find Out: A Public Policy Research Program on Dispute Resolution," *Journal of Dispute Resolution,* vol. 2002, no. 1 (2002), pp. 101–126; Lisa B. Bingham, "The Next Step: Research on How Dispute System Design Affects Function," *Negotiation Journal,* vol. 18, no. 4 (2002), pp. 375–379.

32. Deborah R. Hensler, "ADR Research at the Crossroads," *Journal of Dispute Resolution,* vol. 2000, no. 1 (2000), pp. 71–78.

33. Jeffrey M. Senger, "Turning the Ship of State," *Journal of Dispute Resolution,* vol. 2000, no. 1 (2000), pp. 79–95.

34. The recent formation of a new research section at the Association for Conflict Resolution is a positive development for collaboration between researchers and practitioners; see www.acresolution.org (accessed January 17, 2003).

35. Craig A. McEwen, "Toward a Program Based Research Agenda," *Negotiation Journal*, vol. 15, no. 4 (1999), pp. 325–338.

36. For searchable databases and annotated bibliographies of ADR evaluation literature, see the following websites: Center for Analysis of Alternative Dispute Resolution Systems, www.caadrs.org (accessed January 17, 2003); Indiana Conflict Resolution Institute, www.spea.indiana.edu/icri (accessed January 17, 2003); and the Conflict Resolution Information Source, www.crinfo.org (accessed January 17, 2003).

Index

About the Editors

ROSEMARY O'LEARY and LISA B. BINGHAM are cofounders of the Indiana Conflict Resolution Institute at the Indiana University School of Public and Environmental Affairs. Supported by a grant from the William and Flora Hewlett Foundation, the Institute conducts applied research and program evaluation on mediation, arbitration, and other forms of dispute resolution.

Formerly a codirector of the Institute, *Rosemary O'Leary* is currently professor of public administration and political science and coordinator of the Ph.D. program in public administration at the Maxwell School of Citizenship and Public Affairs at Syracuse University. An elected member of the U.S. National Academy of Public Administration, she was recently a senior Fulbright scholar conducting research on environmental policy in Malaysia. She has worked as an environmental attorney, a consultant to state and federal agencies, and as director of policy and planning for the Kansas Department of Health and the Environment.

O'Leary's book, *Managing for the Environment: Understanding the Legal, Organizational, and Policy Challenges*, written with Robert F. Durant, Daniel J. Fiorino, and Paul S. Weiland, was the winner of the Academy of Management's Best Book in Public and Nonprofit Management for 2000 and the American Society for Public Administration's Best Book in Environmental Management and Policy for 1999.

Lisa B. Bingham is the Keller-Runden Chair in Public Service and director of the Indiana Conflict Resolution Institute at the Indiana University School of Public and Environmental Affairs. After practicing labor and employment law and becoming a partner in the law firm of Shipman and Goodwin, Bingham joined the law faculty of Indiana University in 1989. In 1992, she joined the faculty of the School of Public and Environmental Affairs.

Bingham is the director of the National REDRESS Evaluation Project for the United States Postal Service, a research project on transformative mediation of employment discrimination disputes. In 2002, she received the Association for Conflict Resolution's Willoughby Abner Award for excellence in research on dispute resolution.